MASONRY & CONCRETE CONSTRUCTION

Revised Edition

By Ken Nolan

Craftsman Book Company
6058 Corte del Cedro/ P.O. Box 6500 / Carlsbad, CA 92018

Acknowledgments

APA - The Engineered Wood Association, PO Box 11700, Tacoma, WA 98411

Dayton Superior Corporation, 721 Richard St., Miamisburg, OH 45342

DUR-O-WAL, Inc., 3115 N. Wilke Road, Suite A, Arlington Heights, IL 60004

Indiana Limestone Institute of America, Inc., Stone City Bank Building, Bedford, IN 47421

Symons Corporation, PO Box 5018, Des Plaines, IL 60017

York Manufacturing, Inc., PO Box 1009, Sanford, ME 04073

This book is dedicated to all those of us who were told, back when we were young, that we'd never amount to anything.

Well . . . What do you say now?

Library of Congress Cataloging-in-Publication Data

Nolan, Kenneth J., 1943-
Masonry & concrete construction / by Ken Nolan. -- Rev.
p. cm.
Includes index.
ISBN 1-57218-044-7
1. Masonry--Handbooks, manuals, etc. 2. Concrete construction--Handbooks, manuals, etc. I. Title.
TH5311.N64 1997
693í.1--dc21 97-43739
CIP

Second printing 2000

Contents

1

Planning, Site Work and Surveying

This book is a guide to methods, materials and techniques used by professional masonry and concrete contractors. Masonry is an ancient profession that in some ways has changed very little for centuries. But it's also a modern profession, with techniques and materials improving as science makes new discoveries. The emphasis throughout this new edition is on modern practices. But that doesn't mean the tried and true methods that really are still the best are neglected.

I've written this manual to cover all phases of masonry and concrete construction. You'll find everything from on-site preplanning, through footings, foundations and walls. From fireplaces and chimneys to seismic reinforcement. From brick and limestone veneer construction to techniques for stain removal. There are in-depth discussions of the materials you'll use — the properties and characteristics of ingredients found in a batch of cement, for example.

Throughout this revised edition, I've either expanded or updated every section. Often I've done both. And I've emphasized safety. There's an entire chapter devoted to the Occupational Safety & Health Act. In this section you'll find all the applicable OSHA sections pulled together into a handy condensed form. And finally, there's information that will help you make bid-winning estimates for all kinds of masonry and concrete construction.

Of course, this book can't cover everything you might ever want to know about masonry and concrete. No single volume could. I wrote it for the busy, working masonry and concrete contractor who needs a handy reference to keep in the cab of his truck. When you're out on the job site and need to know a formula or how to deal with some problem, you've got the help you need close at hand. And back at the office, you're sure to use it as you're going over blueprints and writing estimates.

Planning the Job

In this chapter, we'll cover what you need to know and do *before* you start working at the job site. Since you'll be involved at the very beginning of a building project, with the foundation, you have to make sure it's in the right place. The framers will simply build on the footing you laid. If you've built it partly on the neighbor's lot, or a few inches over the setback line, and it's not discovered until after the roof is finished, you're not going to like their choice of who to blame.

Let's begin at the beginning — with the building code.

Check the Building Code

All residential and commercial construction must comply with local building codes and requirements. So your first step, before starting any project, is always to check the code. The local building inspector or city engineer should have copies of the current code available.

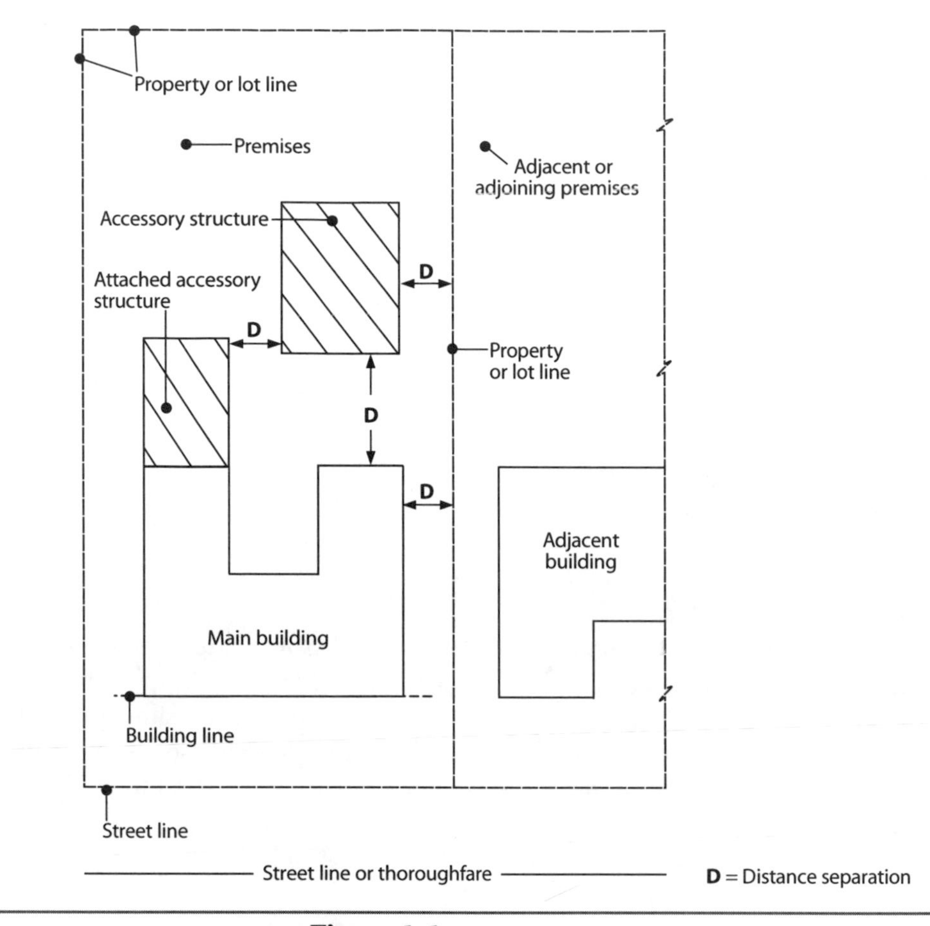

Figure 1-1
Terms used in the building code

The first sections of the building code cover the general provisions and define the terms used in the code. The rest of the code covers space and structural requirements, fire safety and many other conditions and limitations. Building codes often spell out each and every detail. That makes it easy to work to the code once you know where to look.

Building codes vary in their details, but most use the same standard terms. Let's look at some of these definitions. Many of them are illustrated in Figure 1-1.

Building line — A legally-determined boundary that no part of the building can cross. Exceptions are common, but the details vary widely. Never assume that what's allowed in your town is also OK in the next county. Always check the code before you start work.

Distance separation — This describes the amount of open space required between buildings. Open space helps keep fire from spreading from one structure to its neighbors.

Lot line — A surveyed and recorded boundary that separates one piece of property from another. The same phrase also describes the legally determined boundary that separates a piece of private property from a public street or other public property.

Premise — A term used to describe collectively a piece of property as well as any buildings or structures on it.

Property line — The legal boundaries marking a lot or parcel of property.

Setback — The open space required between a building line and the street centerline.

Street line—A boundary separating a lot or parcel of land from the street. The street line and building line are the same if there's no setback required.

Basement—A space in a building that meets both of the following requirements: First, it's partly below grade. Second, more than one-half of its height, measured floor to ceiling, is above the average outside grade. Most codes allow habitable space in a basement if the basement floor isn't more than 4 feet below the average outside grade. Most codes also treat a basement as a story if the floor directly above it is at least 7 feet above the finished grade. See Figure 1-2.

Cellar—A space in a building that's similar to a basement, except for the following differences. First, the floor level is more than 4 feet below the level of the average outside grade. Second, less than one-half a cellar's floor-to-ceiling height is above the average outside grade. Most codes don't allow habitable space in cellars, although a recreation room is usually allowed. Finally, cellars are rarely counted as a story. See Figure 1-3.

Check the Site Deed

There are several more tasks I recommend completing at this early stage. First, check the site deed. Watch out for covenants and/or easements. If you need to file a plot plan with a local official or agency, do so now, before you start work. Hire a licensed surveyor to check the site at this time. You *must* build within the established property lines and on the correct lot. Having a current survey of the lot provides a margin of protection for you as a contractor and for the property owner. Lawsuits are expensive and become even more so when you include your lost productivity.

Consider the Building Permit

All contractors, including those in masonry and concrete, need to be aware of the potential problems with building permits. In most areas, no work may begin without a permit in hand. Failure to comply with permit-issuing procedures or the terms of the permit itself can lead to heavy fines. Worse yet, these fines are often retroactive to the date that construction began.

Once you have a building permit, you'll notice it lists specific required inspections. These happen throughout construction as specific tasks are completed. It's your responsibility as a contractor or subcontractor to notify the inspector's office whenever your work nears the point of needing an inspection. Always give the inspector's office as much advance notice as possible.

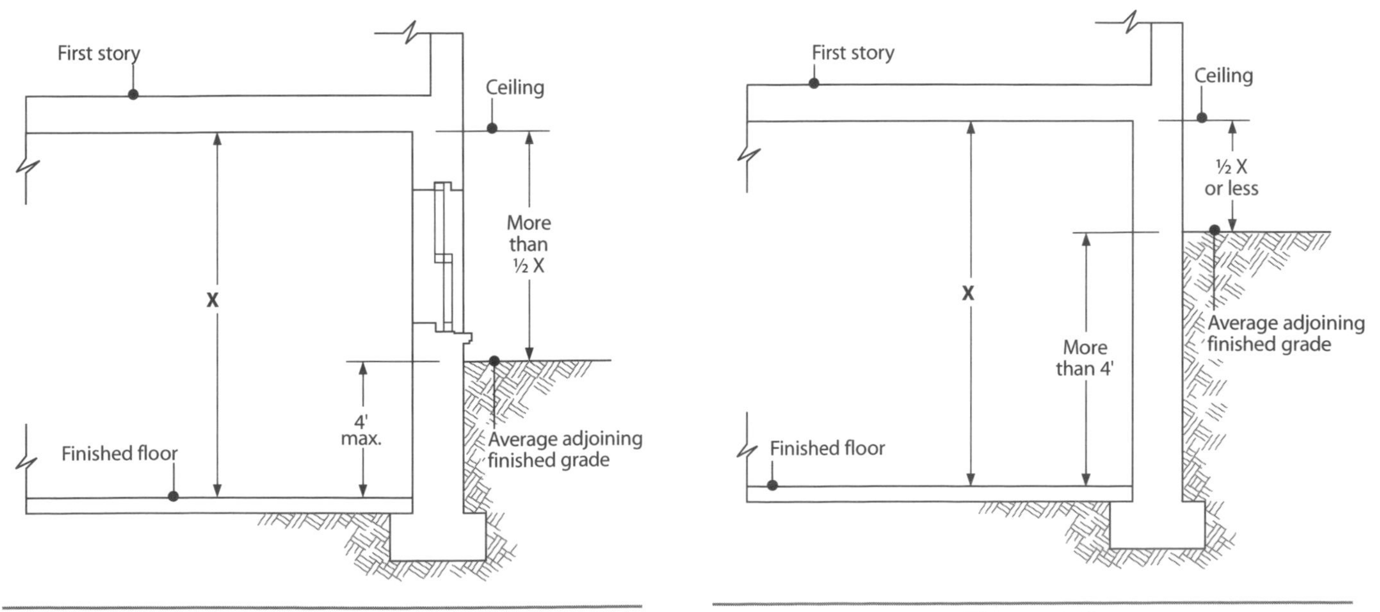

Figure 1-2
Basement location with respect to grade level

Figure 1-3
Cellar location with respect to grade level

That way, work won't be held up while you wait for an inspection.

Issuing a certificate of occupancy is the final step in the inspection/permit process. This certificate shows that the construction covered by your building permit is complete. It also certifies performance and passing of the final inspection.

Meeting the Standards

If you and your company undertake large projects, you'll probably be required to apply standard references. A standard reference, or a standard, is a specification, code, guide or procedure recognized and accepted throughout the industry. Some organizations that issue standards that apply to concrete and masonry construction are the American Society for Testing and Materials (ASTM), American National Standards Institute (ANSI), American Concrete Institute, National Concrete Masonry Institute and the Brick Institute of America.

Doing a Site Survey

We've already touched on the reasons for having a survey made by a licensed surveyor before any work is started. At this point I want to mention several good reasons for making your own survey as well.

Double-checking the setback against what the code requires is a smart move. Why? If there's an error, a building inspector is sure to spot it and slap a stop work order on the whole job site. Then you'll find yourself embroiled in legal action with the city, the general contractor, the land owner or all three. Double-checking the separation distance against what the code calls for is another good idea. Here again, an error in measurements is likely to lead to a dispute that ends up in court.

Doing your own surveying gives you familiarity with the site that you just can't get any other way. Plus, this early survey gives you a head start on the survey work for the foundation.

Both of the measurements just mentioned are easy to check. Do it! At this point any error you find can be fixed on paper. All it takes is a minute or two and your eraser and pencil. Fixing an error like this after you've started work costs far more. In fact, it could cost you your business.

While you're checking the site plan, give some thought to where you'll store materials on this job site. Remember, they're heavy. You can't pile them just anywhere. Take the time to check the site plans for underground tanks. At the job site, check any paved areas. Can they withstand the materials' weight? Also be sure you don't overload your vehicles — that's likely to cause an accident. Figure 1-4 shows the weights of the most common materials you'll use as a concrete or masonry contractor. Some of these figures will vary with the material's moisture content or texture of the material.

Soil Surveys and Analysis

Before you plan, let alone build, footings, foundations or walls, you need to know if the soil can support the structure. This is called the *loadbearing capacity* of a soil, and it varies with the kind of soil. How do you find out what kind of soil you're working with on a job site?

The U.S. Soil Conservation Service's soil survey for the area is your best bet. Their surveys cover all the information you need, and more. The Soil Conservation Service collects soil, climate and geographic data worldwide. Their maps plot this data over the top of an aerial photograph. They also publish written reports to match the mapped areas, which have even more detailed data. But the maps alone usually have all the information you'll need. They show:

- soil types
- soil pH
- soil's grain size
- depth to bedrock
- soil permeability
- boundaries between soil types
- seasonal high water table levels

The maps and reports aren't very expensive and you can order copies by calling (202) 205-0026 or writing to:

Superintendent of Documents
United States Government Printing Office
Washington, DC 20402

Material	Weight
Common brick, $1\frac{1}{2}$" x 4" x $8\frac{1}{4}$"	5.4 lb each, 2.7 tons per 1,000
Fire brick, 9" x $4\frac{1}{2}$" x $2\frac{1}{2}$"	7 lb each, 3.5 tons per 1,000
Face brick, $2\frac{1}{4}$" x $4\frac{1}{4}$" x $8\frac{1}{2}$"	6.48 lb each, 3.24 tons per 1,000
Paving brick, $2\frac{1}{4}$" x 4" x $8\frac{1}{4}$"	6.75 lb each, 3.37 tons per 1,000
Portland cement	94 lb/bag, 100 lb/CF, 2,700 lb/CY
Concrete	100-160 lb/CF, 3,800-4,100 lb/CY
Crushed stone	100 lb/CF, 2,700 lb/CY
Gravel	95-120 lb/CF, 2,565-3,240 lb/CY
Hydrated lime	50 lb/bag
Masonry cement	70 lb/bag
Sand (dry)	97-112 lb/CF, 2,600-3,000 lb/CY
Sand (moist)	112-127 lb/CF, 3,000-3,400 lb/CY
Sand (wet)	127-140 lb/CF, 3,400-3,800 lb/CY
Sand (shovel full, dry)	15 lb

Figure 1-4
Weights of common construction materials

If there's an agricultural extension bureau in the area, you can visit their office. The staff there can often answer your questions about local soils. They may also have maps that cover the information you need.

The best foundation-bed soil is one that:

- ❖ supports the building's weight
- ❖ doesn't swell when wet
- ❖ doesn't shrink as it dries
- ❖ isn't affected by frost heave

You're probably not going to find that ideal foundation bed soil. What you hope to find is the next best thing: a dry, well-compacted, sandy clay soil. Figure 1-5 lists some common soils and their loadbearing capacities.

Let's look at different kinds of soils now. We'll see what sorts of problems there are and how you can deal with them.

Rock

Rock isn't always bedrock, although it's easy to mistake a thin layer of rock for bedrock. Under the layer of rock is a bed of soft clay or sand and that's what the building really rests on. But can a bed of soft clay or sand support the building? Take another look at Figure 1-5. The loadbearing capacity of soft bedrock is 16,000 lb/SF. But soft clay, at best, has a loadbearing capacity of only 2,000 lb/SF.

Here's another pitfall that catches beginners: mistaking a large, buried boulder for solid bedrock. This isn't a safe bed for footings because the boulder may break loose when the weight of a building is added.

Sand

Sand swells or flows when wet. Then, as it dries, it shrinks and settles. All of these (settling, flowing, swelling and shrinking) are bad news. Footings can be ruined by any movement. The only time sand is safe to build on is when the moisture level is stable. If that's not the case, you can bet on the sand moving sooner or later.

Clay

This soaks up moisture like a sponge. And, like a sponge, clay soils expand as they take in more and more moisture. Footings and foundations can be lifted right

Type of soil	Loadbearing capacity (lb/SF)
Hard bedrock, such as granite	30,000
Soft bedrock, such as shale	16,000
Well-compacted gravel or gravel/coarse sand mix	12,000
Dry, hard clay or well-compacted coarse sand	8,000
Moderately dry clay or coarse sand/clay mix	4,000 to 6,000
Ordinary clay/sand mix	3,000 to 4,000
Silt, sandy loam or soft clay	1,000 to 2,000

Figure 1-5
Loadbearing capacities of common soils

up by this swelling action. And clay is slippery and unstable when it's wet. Add some weight to a footing on a bed like this and the soil squeezes right out from under it. That's not good. The foundation will either fail or become so unstable that the building won't be safe. But you can raise a clay soil's loadbearing capacity by improving the soil's drainage. You just add a layer of gravel to the top of the soil and then compact both soil and gravel.

Peaty or spongy soils

Peaty or spongy soils need specially-designed foundations. When it comes to planning foundations or structures for a site with soil like this, you're out of your depth. It's a job for a structural engineer, not a mason.

Fill

Avoid fill if possible. If it's very well-settled there's a chance you won't have too many problems now or later. But differences in the depth and makeup of fill make it settle unevenly. Fill made from lots of different materials may have as many different loadbearing values as it has materials.

Acid or Alkali? What pH Testing Tells You

I mentioned that soil pH (a measurement of relative acidity or alkalinity) is one of the pieces of data given on soil maps. As a mason you need to know the pH of three things on the job site:

- soil
- ground water
- water used for mixing

If the water or soil's pH is less than 7, it's acidic. The lower the pH, the more acidic the soil or water. For example, a soil with a pH of 6.5 isn't very acidic. However, a soil with a pH of 4.5 is very acidic and may need special handling.

At the other end of the scale are pH values greater than 7. Soil or water in this range is alkaline. The higher the pH, the more alkaline the water or soil. What do you care if the soil or water is acidic or alkaline? Ground water or soil with a pH of 9 will quickly break down concrete or mortar made with Type I Normal portland cement. That's why you care. You'll have to use Type II, or better yet, Type V portland cement in the concrete and mortar. You'll see in the next chapter that these types of portland cement are sulfate resistant.

A pH of 7 is neutral. Something with a pH of 7 isn't acid or alkaline, but it's not a likely pH for soil or ground water.

Drainage

Most state health departments require safe sanitation practices for drinking and waste water. If you're building in an area without sewers, make sure you know all the regulations involved. If you can, take a look at your job site in spring or during wet weather when the water table is at or near its peak. This is the best and easiest time to spot any problems with drainage, such as areas where water collects or places where seepage might be a problem later on.

To test the drainage, you can make a percolation test. This test will tell you how well waste water will disperse into the soil. Here's how:

1) At the job site, dig a hole. If you can't do this test in wet weather when the water table is high, saturate the hole with water before you do the test. Where I live the hole must be 2 feet deep and 18 inches in diameter. Check the regulations in your area.
2) Fill the hole with water. Then time how long it takes the water to drop 1 inch, 2 inches and 3 inches. Use a yardstick to measure the water.
3) Do steps 1 and 2 again and average the results. I look for about 5 to 7 minutes per inch.

If the water level drops quickly, it may mean that waste water could flow into drinking water at some distant location. If it drops slowly you may have poor drainage. Be sure these problems are solved at the beginning of the job. They can be very expensive later on.

Frost Heave

Frost heave describes the way soil is lifted up and disturbed when the water in it freezes. Water expands (by about 9 percent of its volume) when it freezes and pushes everything up. The soil in an area usually freezes to a certain depth and rarely below that. This depth is called the *frost line.* Below the frost line the soil isn't affected by the freeze-thaw cycle, so that's where you want to put the footings.

Surveying for Footings and Foundations

After all inspection work is complete, the next step is excavating. A good excavator will use a transit or some other sophisticated leveling device to position and locate the tops of footings, foundation walls and retaining walls. Don't let any excavator eyeball your job. No man alive can get it 100 percent correct, especially if the job site is uneven.

Even after a good excavator finishes his work, you probably should use your own instrument to make sure all the forms are level. Out-of-level forms or forms not set to the correct starting height will make it hard for any mason. For example, all mortar joints should be no more than 3/8 inch thick for structural strength. It's a bad practice for a mason to use thick joints to make up for depressions. And it uses up a lot of mortar. On the other hand, if a mason has to cut blocks to fit humps or correct the elevation on the first course, he'll use extra time and produce a lot of broken pieces.

Surveying Equipment

The surveying work you'll want to do calls for two basic tools: a *transit level* and a *graduated leveling rod.* Let's discuss both tools a bit more before we start telling you how to use them.

Transit Level

There are many different types of this precision measuring instrument. At the top of the scale, in price as well as precision, are electronic levels with such features as automatic leveling, laser-guided targeting and digital readout/input. However, most masonry and concrete contractors don't need this much precision. A transit is probably adequate for your surveying needs. A transit level has three main parts: the telescope, the leveling vial, and the circle.

Telescope (or scope)

This is a precision sighting optical device. It makes the images you see through it bigger. You take a sighting on a point simply by centering it in the vertical and horizontal *cross hairs* of the scope.

Leveling vial

This is a bubble-type level that works just like the bubble in an ordinary carpenter's level. However, it comes in different sensitivities. If you need precise readings, you'll need a sensitive leveling vial on the transit.

Circle

The horizontal circle is part of the plate that the scope rotates on. The circle, vertical or horizontal, is basically a scale that measures angles in degrees, marked by the divisions on the circle. There are 360 degrees in a circle. More precise transit levels have a second scale, called a *vernier*. A vernier lets you measure angles more precisely because it divides degrees into minutes. There are 60 minutes in a degree. The best and most precise transits have a second vernier that divides minutes into seconds. There are 60 seconds in a minute.

The first step in any surveying operation is to center and level the transit. Follow the instructions in the manufacturer's user's manual. Generally these manuals are quite complete, clearly written and well-illustrated. Your manual is the best resource for information that's specific to your instrument. Read it, use it and take care of it.

If you make readings using an out-of-level transit, the readings won't be true. Surveying with an out-of-level transit is a waste of time. Carelessness here can cost you everything, especially if it results in a stop work order for the whole project.

Graduated Leveling Rod (or Rod)

This is the second of your basic tools for surveying work. In a pinch, an ordinary 6-foot rule might work. But a rod is better because it's longer, by 4 to 9 feet, and it's easier to read accurately from a distance. The background color of a rod is white. Divisions for feet are in large red print while the other divisions are in black print. There are two types of rods which vary in the type of divisions used. Figure 1-6 shows an architect's rod. The divisions marked on this type of rod are feet (in red), and inches and eighths of an inch (in black). The engineer's rod is shown in Figure 1-7. The divisions marked on this rod are feet (in red), tenths of a foot and hundredths of a foot (in black).

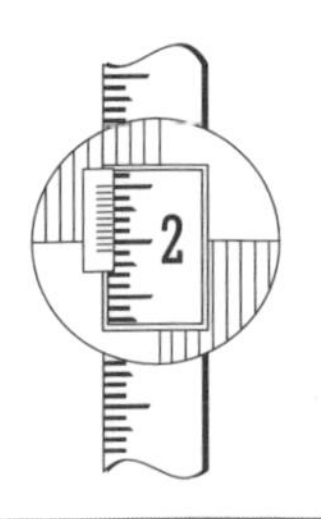

Figure 1-6
Architect's rod with target

Figure 1-7
Engineer's rod

Figure 1-8
Correct way to hold a leveling rod for surveying work

Target

This usually comes with the rod but once again you've two choices: the oval vernier or the snap-on target. Both kinds of targets have cross hairs and both work by sliding up and down the rod. Figure 1-6 shows a target on an architect's rod. Use the target's cross hairs to pinpoint elevation readings on the rod's scale.

I've known masons who use both sorts of rods as well as both types of targets in any combination. Choose the equipment you prefer or are most comfortable using.

Survey Teams

It usually takes at least two people to take a reading with a transit and rod. You'll need someone to hold the pole and move it around as necessary. Have an assistant (or rod holder) do these tasks, following your directions. Make sure the assistant holds the rod as shown in Figure 1-8 with the fingertips, taking care not to cover the scale.

Sometimes you and the assistant will be so far apart you can't communicate with each other easily. If you're not equipped with electronic transceivers, you'll have to use hand signals. Figure 1-9 shows the most common hand signals for surveying work. There aren't many of them and most are pretty obvious, so they're easy to learn. Remember, both members of a surveying team must use the same signals for hand signals to work.

It's possible for one person to do surveying alone. You can take the sightings and make the rod readings with what's called a self-reading rod. The one-man system

Figure 1-9
Standard surveyors' hand signals

sometimes is faster and it will save you an assistant's wages. Look at the survey needs of each job before you choose between a team, or soloing on the surveying work. Estimate the time you'll spend trotting back and forth moving the rod. Balance what your time is worth against the wages for an assistant.

The Question of Units

The measurements you'll find on site plans for heights and linear distances are usually in units of whole and decimal parts of a foot. The dimensions you'll find on building plans and blueprints are usually feet, inches and fractional parts of inches for units. Here are a few tips on how to convert between these two systems:

- ❖ 8 one-hundredths of a foot (written as 0.08) is about 1 inch
- ❖ 1/8 inch is about 1 one-hundredth of a foot (written as 0.01)

Figure 1-10 lists inches and the most common fractions as decimal parts of a foot. Use it to find decimal equivalents in feet alone for measurements that are in inches and fractions. On the next page are a few examples.

Whole inches	Fractional parts of an inch							
	0	1/8	1/4	3/8	1/2	5/8	3/4	7/8
0	0.00	0.01	0.02	0.03	0.04	0.05	0.06	0.07
1	0.08	0.09	0.10	0.11	0.12	0.14	0.15	0.16
2	0.17	0.18	0.19	0.20	0.21	0.22	0.23	0.24
3	0.25	0.26	0.27	0.28	0.29	0.30	0.31	0.32
4	0.33	0.34	0.35	0.36	0.38	0.39	0.40	0.41
5	0.42	0.43	0.44	0.45	0.46	0.47	0.48	0.49
6	0.50	0.51	0.52	0.53	0.54	0.55	0.56	0.57
7	0.58	0.59	0.60	0.61	0.62	0.64	0.65	0.66
8	0.67	0.68	0.69	0.70	0.71	0.72	0.73	0.74
9	0.75	0.76	0.77	0.78	0.79	0.80	0.81	0.82
10	0.83	0.84	0.85	0.86	0.88	0.89	0.90	0.91
11	0.92	0.93	0.94	0.95	0.96	0.97	0.98	0.99

Figure 1-10
Converting inches and fractions of inches to decimal equivalents in feet

Example 1

Find the decimal equivalent, in feet alone, for 2 feet 7 1/8 inches.

1) Find 7 in the column under the heading *Whole inches.*
2) Read across the 7 row to the column labeled 1/8 under the heading *Fractional parts of an inch.*
3) The value listed there (0.59) is the decimal equivalent, in feet alone, for 7 1/8 inches.
4) Now add the 2 feet to get the decimal equivalent in feet alone, 2.59 feet.

Example 2

Let's take one more example. What is 8 feet 4 1/2 inches in feet alone? Remember not to consider the 8 feet until the end.

1) Find 4 in the column labeled *Whole inches.*
2) Read across the 4 inch row to 1/2 in the column under the heading *Fractional parts of an inch.*
3) The value, in feet alone, for 4 1/2 inches is 0.38.
4) Add the 8 feet to get the decimal equivalent in feet alone, 8.38 feet.

Benchmarks and Elevations

Most buildings you work on as a masonry contractor have all the elevations specified. The elevations are based on a known elevation, called a *benchmark.* Usually one benchmark is enough, but on large jobs it's helpful to have several. The best benchmark is one that's easy to spot and difficult to move, like a bolt on a fire hydrant, the corner of a stone monument or a metal spike driven into a tree root. There's one final feature that's important in choosing a benchmark. Be certain it's located a good distance away from any of the construction action.

Keep accurate and up-to-date records of your survey work for each job. Good record keeping is a hallmark of a good businessman. It's also the best form of insurance you could possibly have. Finding the data later will be easier and you'll know it's current.

Finding Elevation Differences

You'll often have to find the difference in the elevations of two points. Let's work our way through a few examples to see how to do it.

Example 1

For this first example let's assume you can see both points from one location. Set up, center and level your transit there. Then take readings of both points. The difference between the two readings is the difference in their elevations. We'll use Figure 1-11 to demonstrate this basic technique. Remember there are two questions here. First, which is higher, point A or point B? Second, how much higher is the higher point? It's obvious, from looking at Figure 1-11, that B is higher than A. But it often won't be so obvious on a job site. Here's how it works:

1) The reading for point A is 69"
2) Expressed in feet only, that's 5.75'
3) The reading for point B is 40" or 3.33'
4) To find the difference between these elevations, subtract 3.33 from 5.75 to get 2.42

So point B is higher than point A by 2.42 feet, or 29 inches.

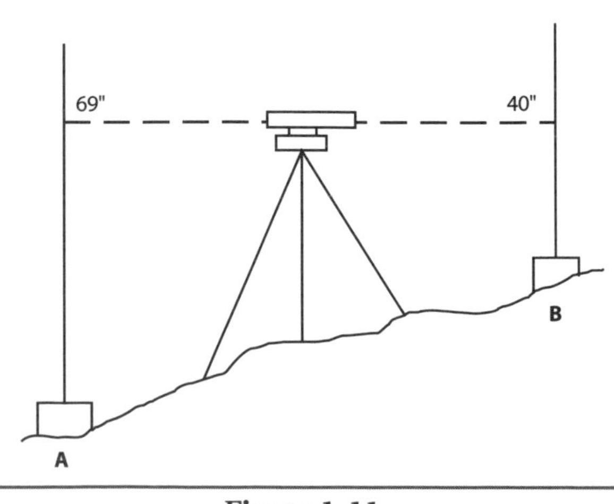

Figure 1-11
Finding the difference in elevation of two points

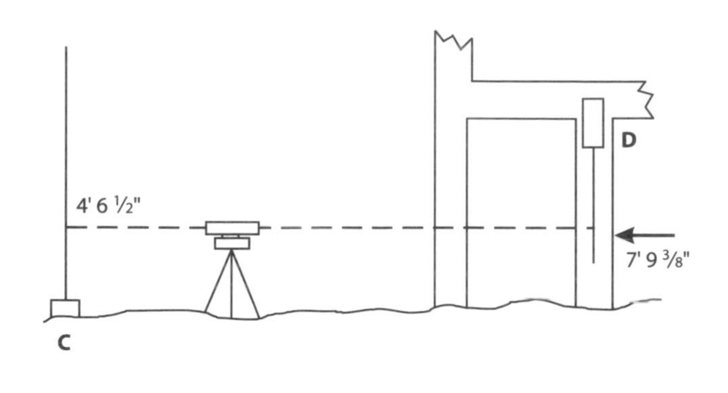

Figure 1-12
Finding the difference when one point is above the line of sight

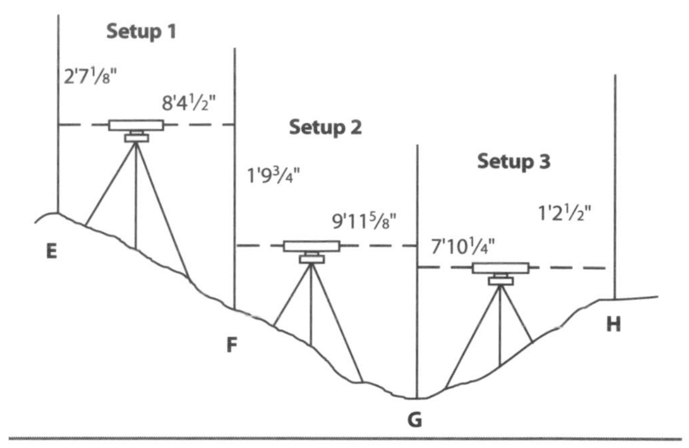

Figure 1-13
Finding the difference when two points can't be observed from one setup

Example 2

Let's look at Figure 1-12 for a problem that's a little more complicated. Taking a reading on point C is no problem. Point D, however, is located on the underside of the floor joist, above your line of sight. Let's see how you find its elevation, and then find the difference between the two elevations.

1) The reading for point C is 4'6½"
2) Expressed in feet only, that's 4.54'
3) To take an elevation reading for point D, place the foot of the rod against point D on the bottom side of the floor joist. That's right, hold the rod upside down, then take your reading.
4) The reading for point D is 7'9⅜" (above line of sight)
5) Expressed in feet only, that's 7.78'
6) To get the difference in elevation, add the two elevations (4.54' and 7.78') to get 12.32'

So point D is higher than point C by 12.32 feet.

Example 3

Let's look at one more example. This time we'll assume that the points are so different in elevation that it's impossible to make sightings on both from one transit setup. We'll use Figure 1-13 for this example, and we'll find the elevation difference between point E and point H. This example also uses two new terms: *plus sight* and *minus sight.* Plus sights are readings taken from a point to the line of sight. Minus sights are readings taken from the line of sight to a point.

In this example, you'll use three transit setup locations to find the difference in elevation. At each of these locations, take two readings — one plus sight and one minus sight. Then add the plus sights from the three locations together. Then add the minus sights together. If the sum of the plus sights is larger, point H is higher than point E. If the sum of the minus sights is larger, point E is higher. To find the elevation difference, follow these easy steps:

1) Convert all measurements to feet only
2) Add all the plus sights together:

   2.59' + 1.81' + 7.85' = 12.25'
3) Add all the minus sights together:

   8.38' + 9.97' + 1.21' = 19.56'
4) Subtract the total minus sights from the total plus sights to find the difference in elevation point E to point H:

   12.25' – 19.56' = –7.31'

The minus sign tells you point H is 7.31 feet lower than point E.

Note: When you convert feet and inches to feet only, the resulting value is an approximation. If you work out this problem using feet and inches, there's a ¼" difference. The precise answer would be –7'3½".

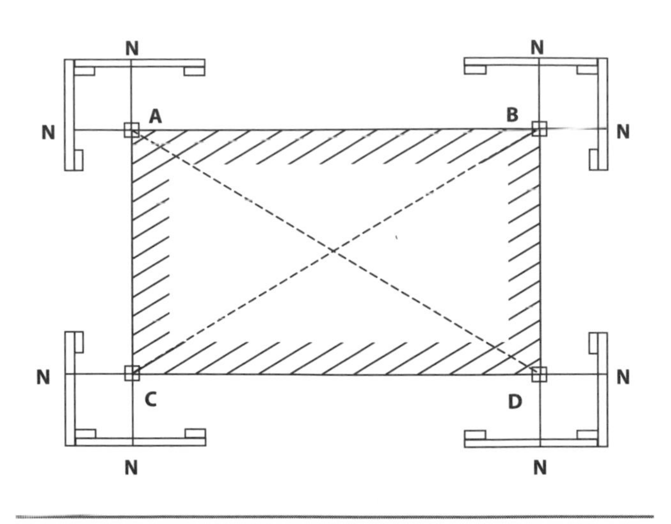

Figure 1-14
Building layout and batterboard setup

Staking Out a Building

This process is a good deal more than just pounding four stakes into the ground. It's also an important part of your early on-site survey work. We'll go through the process in stages, and take the procedure step by step. We'll refer to Figure 1-14 often, so let's begin by identifying its main features.

- ❖ Points A, B, C and D mark the corners where you place the stakes.
- ❖ The shaded area shows the future location of the foundation.
- ❖ Lines AB, BD, DC and CA are the building lines.
- ❖ The diagonal dashed lines, AD and BC, mark measurements you use to check for squareness.
- ❖ The right-angle shapes, shown outside the building lines at the corners, are the batterboards.
- ❖ Each N marks the location of a nail driven into the top of the batterboard.
- ❖ The lines that extend the building lines out to each nail location show part of the path followed by the string lines. The rest of a string line path matches that of the building line.

Let's also assume we know:

- ❖ line AB is the building's frontage
- ❖ the location of point A
- ❖ point B's direction, relative to point A
- ❖ all angles are 90 degrees
- ❖ the length of all four building lines

To stake out the building:

1) Level and center your transit on point A.
2) From A, sight on point B. Do that by turning the transit 90 degrees to the left.
3) Set point B at the known distance from A. Mark point B with a corner stake.
4) Leaving the transit at point A, sight on point C. That means you turn the transit 90 degrees to the right.
5) Set point C at the known distance from A. Mark point C with a corner stake.
6) Move the transit to point B. Center and level the transit on point B.
7) From B, sight on point D by turning the transit 90 degrees to the left.
8) Set point D at the known distance from B. Mark point D with a corner stake.

Before you move on to setting up the batterboards, stop and check the work you've done so far. There are two ways to double-check this part of your work. The first way is to compare the lengths of a set of parallel sides. Measure the lengths, for example, of line CD and line AB. If they're equal, then your work's accurate.

The second way to double-check the layout is to compare the lengths of the diagonals. In Figure 1-14, these are the dashed lines AD and BC. I recommend using the diagonals method because it also checks the layout for squareness. Equal diagonals mean you've set the corners accurately and square. If any of the four angles isn't 90 degrees, the diagonals won't be equal. If your diagonals aren't equal, go back and check the angles with your transit. What does it mean if all four still read exactly 90 degrees? The problem is probably an out-of-level transit setup. Start over with step 1 and this time be more careful!

Placing Batterboards

Now that you've set and staked your four corners, let's move on to setting up batterboards. Masons use batterboards to define building lines. Your carefully marked corner points A, B, C and D all disappear when the foundation is excavated. But the string lines that you run from the batterboards extend the building lines and cross each other exactly over the corner points. They make the job of re-establishing the corner points a piece of cake. The whole reason batterboards and string lines work is that they're set up outside the building lines. Excavation or other site work won't happen near them, so their positions aren't disturbed. A comfortable separation, the distance between the batterboards and the building lines, runs about 4 feet.

Set up the batterboards shown in Figure 1-14 as follows:

1) Level and center the transit in a convenient spot near the building's center.
2) Set three posts, the batterboard uprights, about 4 feet out from the building lines at each corner.
3) Take sightings at each corner on the foundation's top.
4) From this reading subtract the clearance — the distance between the foundation top and string lines.
5) If the result isn't a whole number of feet, add as needed to make a whole-foot number and set the rod target for this elevation. Adjusting this elevation to read as a whole number of feet makes it much easier to sight with the transit in the following steps.
6) Holding the preset rod at each of the batterboard uprights (posts), raise or lower the rod and center the target reading in the transit's cross hairs.
7) Mark each post with the location of the rod's foot. This marks the correct position for the top edges of the batterboard crosspieces.
8) Attach the crosspieces to the uprights following the markings.

The next step is setting the nails in the batterboard to attach the string lines:

1) Level and center the transit over each corner position.
2) Take sightings in turn on the two adjacent corner points.

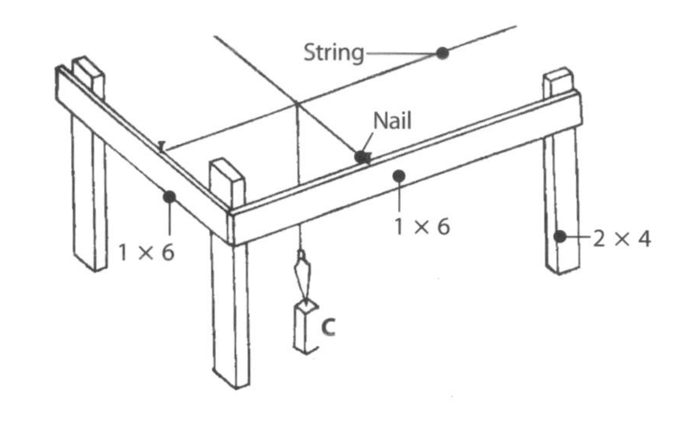

Figure 1-15
Checking batterboard and string line placement

3) Mark the points on the top edge of the batterboard.
4) Drive a nail into the batterboard at each mark.

Now, for the final step, simply run the string for the string lines from nail to nail. The string lines extend the building lines out to the batterboards and crisscross exactly above the corner points. It's easy to check the positions of the string lines. Take a look at Figure 1-15, then just follow these steps:

- ❖ Tie a plumb bob to a short length of line.
- ❖ Attach the line to one of the string line crossing points.
- ❖ Lower the plumb bob until its tip touches the top of the corner stake.

Your string line and batterboard setup is correct if the plumb bob tip touches the corner stake at its center.

Measuring and Laying Out Horizontal Angles

All of the surveying measurements we've discussed so far have been for vertical angles. The steps for measuring a horizontal angle are somewhat different. As an example we'll use the angle shown in Figure 1-16. This is angle EFG and point F is the pivot point. Here goes:

1) Level and center the transit over point F.
2) Attached to the transit's circle you'll find the horizontal clamp screw. Loosen this screw.

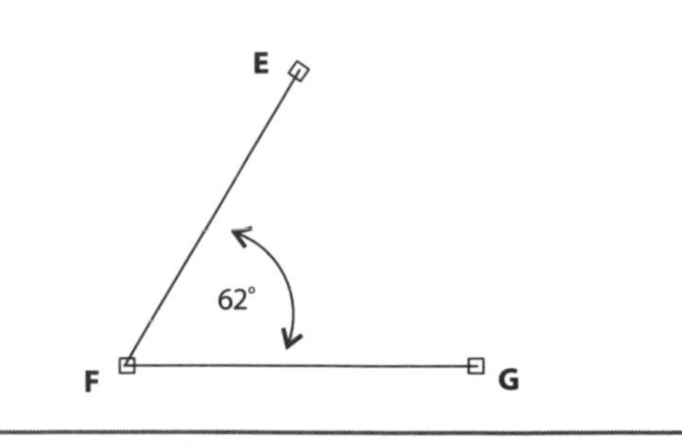

Figure 1-16
Horizontal angle of 62°

3) Turn the scope and sight on point E.
4) Align point E with the scope's horizontal cross hair.
5) Tighten the horizontal clamp screw.
6) Turn the tangent screw to align point E with the scope's vertical cross hair.
7) Reset the circle's scale, by hand, to zero. Either the circle's scale or the index (pointer) moves. This varies from transit to transit.
8) Loosen the horizontal clamp screw again. (Be careful not to move the circle in the process.)
9) Turn the scope and sight on point G.
10) Align point G with the horizontal cross hair.

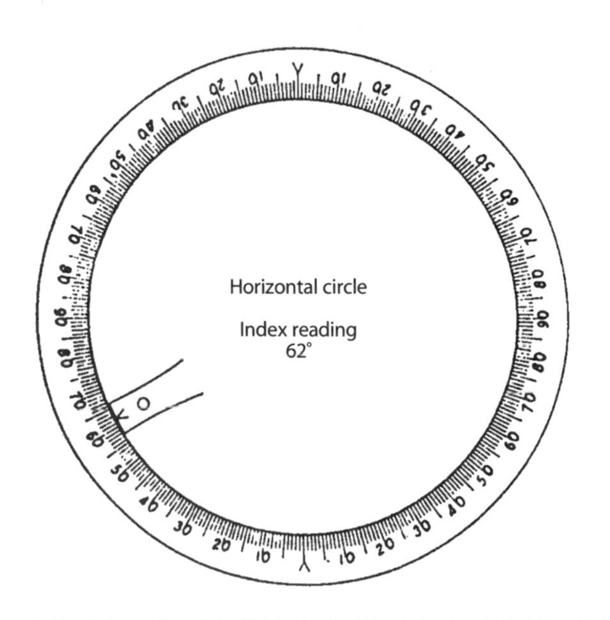

Figure 1-17
Horizontal circle and index showing 62° angle turned to the right

11) Tighten the horizontal clamp screw. (Remember not to disturb the circle.)
12) Turn the tangent screw to align point G with the scope's vertical cross hair.
13) Read the value of the horizontal angle turned on the circle's scale. See Figure 1-17.

Out on job sites, you'll find most angles are 90 degrees. With that in mind, here's a quick run-through to show you how to set this angle. We'll use angle HIJ in Figure 1-18. Let's also take a shortcut by assuming that the locations and elevations of points H and I are known. Point I is the pivot point. We need to find point J. Here's how to set this point:

1) Center and level the transit over the pivot point I. Turn the scope and sight on point H.
2) Align point H with the scope's horizontal and then the vertical cross hairs.
3) Reset the circle's scale, by hand, to zero.
4) Turn the scope and sight on point J.
5) Align point J with the scope's horizontal and then the vertical cross hairs.
6) Read the horizontal angle that was turned from the circle's scale.

Sometimes you'll want more accurate readings than these. Why? An angular error equaling 1 degree over a distance of 100 feet causes a 1 3/4 foot error. But a transit with a vernier divided into minutes gives much more accurate readings. How much more accurate? Sixty times as accurate, since 1 degree is made up of 60 minutes. An angular error of 1 minute over 100 feet causes an error of only 3/8 inch. It's unlikely that you'll ever need

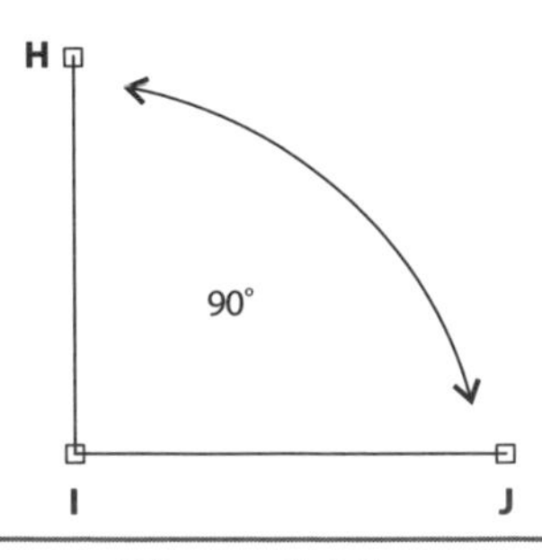

Figure 1-18
Horizontal angle set for 90°

the level of accuracy that's possible using a vernier divided into seconds. An angular error of 1 second over 100 feet results in a total error of only $^{1}/_{200}$ inch.

Reading a Vernier

Reading a vernier takes some practice. Start by looking at Figure 1-19. The vernier (minutes) and the circle (degrees) scales have been set to zero by aligning their indexes. The index is the zero on each scale.

On the vernier there's an R to the left of the 60-minute mark and an L to the right of a second 60-minute mark. What's going on? The index at the zero point divides the vernier in half and makes two scales. Read from the R side of the scale for angles turned to the right or clockwise. Read from the L side of the scale for angles turned to the left or counterclockwise. Don't worry if it seems backward at first to have the L side of the scale on your right and the R side to your left. This is correct. Your vernier isn't on backwards. The strangeness wears off with practice.

Now that you know the parts of a vernier, let's talk about how to read one. The vernier's index is also a pointer, or the marker you use to read degrees from the circle scale. The next question is: What marker do I use to read minutes from the vernier scale? The answer brings us, at long last, to the secret of the vernier!

The division marks on these scales (vernier and circle) are very finely calibrated so that no matter what angle you turn with the transit, there will always be one, and only one, pair of division marks in precise alignment. The first step in reading minutes for an angle is to find this unique point. Then you simply read the minutes value that's marked on the vernier's scale. Sounds pretty easy, doesn't it? All you need now is a bit of practice at reading the fine divisions on the scale.

That brings us to the second point. A vernier scale with a separate mark for every minute would be quite difficult to read. For that reason, in the following examples, we'll use two different vernier scales. The first is a 5-minute vernier, shown in Figure 1-20. This means that each division represents 5 minutes of a degree. The second example uses a 15-minute vernier, shown in Figure 1-21. Each mark on its scale represents 15 minutes of a degree.

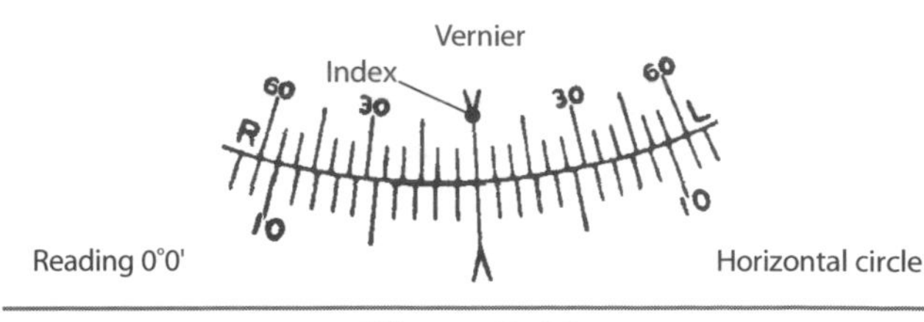

Figure 1-19
Initial setting of horizontal circle and vernier

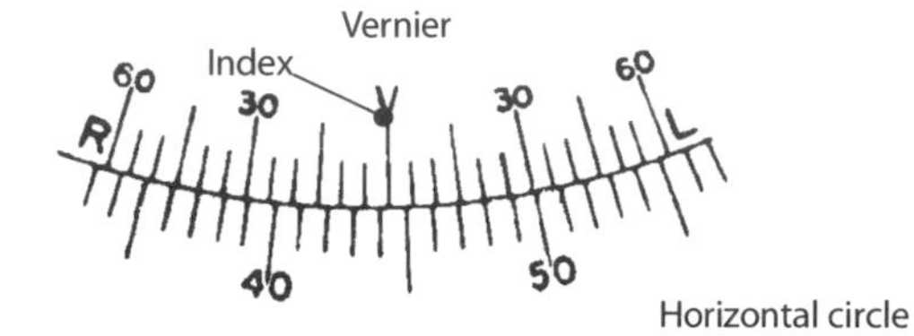

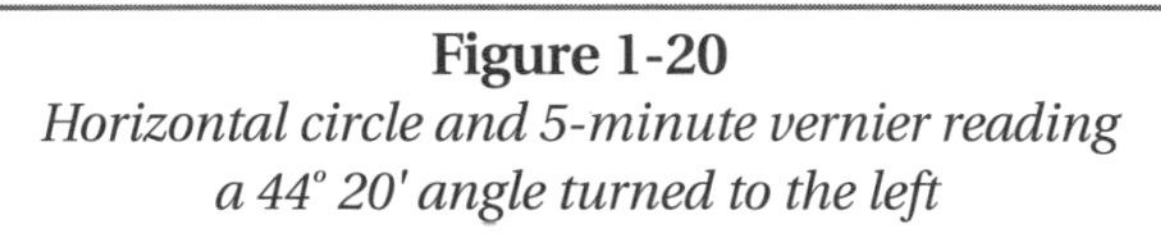

Figure 1-20
Horizontal circle and 5-minute vernier reading a 44° 20' angle turned to the left

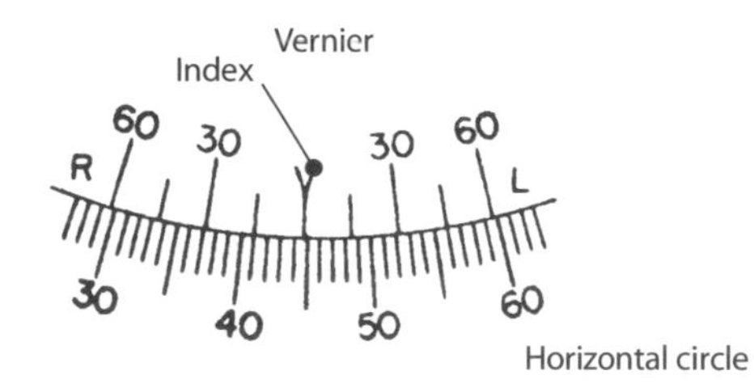

Figure 1-21
Horizontal circle and 15-minute vernier reading a 44° 45' angle turned to the left

Example 1

For this example we'll use the 5-minute vernier in Figure 1-20. The transit is turned for an angle to the left. Here are the steps:

1) Read degrees from the circle's scale as marked by the vernier's index V. The answer is 44 degrees.
2) The angle was turned to the left. So, on the L side of the vernier scale find the pair of exactly aligned division marks, and read the minutes.

3) The alignment is at the vernier's fourth mark to the left from 0 (the index).

4) Since this is a 5-minute vernier, the fourth division from 0 equals 4 × 5. That's 20 minutes.

5) Combine the degree reading with the minutes reading. The answer is an angle of 44 degrees 20 minutes turned to the left.

Example 2

For our second example, the angle is also turned to the left, but this time we'll use the vernier in Figure 1-21, with a 15-minute scale. What's the angle in degrees and minutes?

1) Read degrees from the circle's scale as marked by the vernier's index V. The answer is 44 degrees.

2) On the L side of the vernier scale, find the pair of exactly aligned division marks and read the minutes.

3) The alignment is at the vernier's third mark to the left from 0 (the index). The answer is 45 minutes.

4) Combine the degree reading with the minutes reading. The answer is an angle of 44 degrees 45 minutes turned to the left.

The Vertical Vernier

You use a vertical vernier to read the minutes (and seconds if the transit has two vertical verniers) of an angle that's turned up or down from the zero point. They're read in the same way as horizontal verniers. The vertical circle (and the vertical vernier) are usually located off to the side of the scope. On a vertical vernier, the scale to the right of the index is marked up and the scale to the left of the index is marked down. You read minutes for angles of elevation on the up side of the scale. You use the down side of the scale to read minutes for angles of declination.

All this information on surveying may seem a bit complicated and not particularly important to your work. But checking out these details could keep you out of trouble with the building inspectors.

Now that you've gotten your feet thoroughly wet, let's dive into the real substance of this book — concrete and other kinds of masonry work.

2

Concrete Characteristics and Properties

Properly-cured concrete is a strong, economical, low-maintenance building material. It's durable and resists abrasion, fire and weather. And you can cast concrete into almost any shape.

Concrete has three main disadvantages: low tensile strength, high thermal conductivity and permeability to water vapor. But thanks to advances in science and materials engineering, these have been largely overcome. Steel reinforcement reduces the tensile strength problem. Modern insulation materials solve the problem of high thermal conductivity. And there's a waterproofing material or surface treatment on the market for every water vapor problem.

To make concrete, you combine portland cement, fine aggregate (usually sand), coarse aggregate (usually gravel) and water. Sometimes you'll also need special materials called *admixtures*. They help you make good concrete even when weather or site conditions are difficult. The conditions and requirements of each job site help you decide how much of each material you'll use in that particular mix. If the weather changes or if the job specs change, you may need to adjust your mix so it still matches the job.

The more you know about the materials and the ways they work together, the easier it is for you to choose the best concrete mix for the job. Now let's take a closer look at each material.

Portland Cement

Portland cement is usually sold in bulk and then delivered to the job site in ready-mix trucks. However, it also comes in 94-pound bags that hold 1 cubic foot (CF) each. You'll probably only use it this way for small jobs and for special projects. Don't store bags of portland cement for any length of time. It gets lumpy and you can't use lumpy cement.

There are five types of portland cement used in construction.

Type I Normal

This is a general-purpose cement. It's usually the only type you'll need. It comes in three varieties; regular, air-entrained and white. We'll discuss air-entrained cement later in this chapter. Regular portland cement is gray in color and costs less than the white variety.

Type II Modified

This type of portland cement generates less heat than Type I as it cures, and it makes concrete that's moderately resistant to sulfate damage.

Type III High Early Strength

Use this when you want concrete to set and cure as soon as possible. Here's an example to explain what I mean. Let's say you're on a job near Chicago. It's early October and the job's running behind schedule. Then a cold front rolls in. You can't stop work and wait out the weather. But you can mix concrete using Type III portland cement. This concrete will cure in just three days instead of the seven needed to cure concrete made with Type I cement.

Type IV Low Heat

This generates even less heat than Type II as it cures, but it's expensive. That's why it's only used on jobs such as dams that call for very large masses of concrete. If you use ordinary Type I portland cement here, it may crack open from the heat generated as it cures. If it gets hot enough it can even explode!

Type V Sulfate-Resistant

This portland cement has a higher resistance to sulfate damage than Type II. Use it, instead of Type II, when job site conditions are more extreme.

Aggregate

About three-fourths of a concrete mix is aggregates. They add strength to a mix and help it resist damage from freezing. They're also important to the quality of the concrete's finish. And they help you cover the biggest possible area with the mix.

It's best to use round or cube-shaped aggregates. They're easier to mix thoroughly. No matter what the aggregate is, it must be clean. Dirty aggregate makes concrete that's weak or unusable. This happens because the cement paste will bind with the dirt on the aggregate instead of the aggregate.

Concrete uses two sizes of aggregate, fine and coarse. Let's look at each one.

Fine Aggregate

This is usually sand, but small pebbles, or particles of crushed stone may also be used. Sand or other materials must be clean, hard and well graded. Clean aggregate is free of silt, clay or organic material. Well-graded aggregate is made up of grains or pieces that span a particular size range. The top of the size range for fine aggregate is ¼ inch diameter. The minimum depends on the fine aggregate material. As a rule of thumb, dust is too small to work properly as aggregate, so that's used as a cutoff point. For grading, aggregates are sifted through a series of progressively finer-mesh screens.

Coarse Aggregate

This is usually gravel, but crushed stone or slag is also used. The material must be hard, clean and sound (the pieces don't have any flaws, fissures or cracks). Coarse aggregate pieces range in diameter from ¼ inch to 2 inches.

Clean coarse aggregate is free of any silt, clay, organic material or sand. Bank or creek gravel, for example, usually can't be used straight from the pit because there's too much sand in it. Here's the rule of thumb for coarse aggregate size: Never use coarse aggregate that's larger than one-quarter the thickness of the work. For example, say you're pouring a 4-inch patio slab. Don't use aggregates with a diameter above 1 inch.

A product that's handy on a small rush job is mixed aggregate. It's a packaged combination of sand and gravel in the right proportions for making concrete. It's sold at supply yards.

Coarse Aggregates for Lightweight Concrete

Concrete normally weighs about 150 pounds per cubic foot. But you can make concrete that weighs about 100 pounds per cubic foot by using lightweight aggregate such as cinders, expanded shale, or slag. You can even make concrete that weighs as little as 50 pounds per cubic foot by using pumice or expanded mica (also called vermiculite) as the coarse aggregate. Lightweight concrete has many uses. For example, use it to:

- fill between floor sleepers
- make precast blocks and roof slabs
- fireproof other surfaces

Don't use lightweight concrete where you need waterproof, abrasion-resistant or high-strength concrete. Lightweight concrete lacks these qualities.

If you're using cinders as aggregate, be sure they're hard, vitreous clinkers. They must be free of soot and sulfide and contain no unburned coal or ash. They must also be clean. You can check how clean the cinders are by washing them. First, soak the cinders in water for 24 hours. Then pick up a small handful and rub them briskly between your palms. If your hands stay clean, the cinders are OK to use. If your hands are blackened, don't use them. They're too dirty.

Don't use ashes for lightweight aggregate. They're much too alkaline. Contact with alkaline compounds damages concrete and makes it disintegrate.

Lava rock, in general, is another aggregate material to avoid. This is especially true if you're not certain exactly what it is or where it came from. Lava rock from Oregon or Washington usually is satisfactory. Rhyolite (a light-colored volcanic rock) and other darker basaltic lavas are good aggregates. But other lava-type rocks are too light or contain too many impurities to give good results in concrete.

Water

Water is a very important part of concrete. Be sure you measure water accurately. It affects the quality of the concrete during both the mixing and the curing stages. Use too much water when you're mixing concrete and you'll get weak concrete. Concrete made with 6 gallons of water per bag of cement is 40 percent stronger than concrete made with 8 gallons of water per bag of cement. Use too little water and the concrete will be difficult to work with. Only the right amount of water at both stages will give you strong, water-resistant concrete.

Use clean water only. That's water without any oil, organic matter, strong acids or alkaline compounds. Don't use sea water or brackish water. There are alkaline salts in both. Be very careful here. Even a very small amount of alkaline salt in the water will ruin the concrete. As a general rule, if you wouldn't drink the water, don't use it for making concrete.

Material Proportions

Different concrete mixes have different uses. Some very important information about a prepackaged concrete mix is included in the three-digit number printed on the bag. These numbers give the amounts of cement, sand and gravel in the mix. A 1:2:3 mix contains one part cement, two parts sand and three parts gravel.

Figure 2-1 is a table of common concrete mixes that tells you how much water to add per bag of cement. You need this information when you make your concrete mixes. You also use it in choosing a packaged concrete mix. It's OK to make minor adjustments to these amounts so that your mix is more workable. However,

Kind of work	Proportions (to 1 sack of cement)		Water per sack of cement, when sand is:		
	Sand (CF)	Gravel (CF)	Wet (Gal)	Moist (Gal)	Dry (Gal)
Thin work, 2" to 4" thick	2	2	3½	3¾	4½
Waterproof and wear-resistant, 4" to 8" thick	2	3	3¾	4½	5½
General purpose, reinforced and waterproof, 8" to 12" thick	2½	3½	4½	5	6½
Mass concrete with moderate strength, not waterproof	3	5	5	6	7

Figure 2-1
Trial concrete mixture for various kinds of work

Proportions of materials			Quantities of materials		
Cement	Sand	Gravel	Cement (sacks)	Sand (moist) (CY)	Gravel (CY)
1	1.5	3.0	7.6	0.42	0.85
1	2.0	2.0	8.2	0.60	0.60
1	2.0	3.0	7.0	0.52	0.78
1	2.0	4.0	6.0	0.44	0.89
1	2.5	3.5	5.9	0.55	0.77
1	2.5	4.0	5.6	0.52	0.83
1	2.5	5.0	5.6	0.46	0.92
1	3.0	5.0	4.6	0.51	0.85
1	3.0	6.0	4.2	0.47	0.94

Figure 2-2
Approximate material amounts needed to make 1 cubic yard of concrete

avoid making major changes. Remember that the water-to-cement ratio affects the concrete's quality.

As you study Figure 2-1, you'll see that you need to know how wet the sand is. You need to include the water in the sand as part of the total in the water-to-cement ratio. Here's a handy way to estimate the moisture level of sand. Pick up a handful of the sand, squeeze it, then open your hand. If the sand forms a firm ball, it's wet. If it crumbles after briefly forming a ball, it's moist. If it spills freely from your opened hand and forms no ball at all, it's dry.

Depending on the aggregate used, the quantities of materials required may vary by as much as 10 percent. Figure 2-2 lists approximately how much cement, sand and gravel you'll need to make one CY of concrete.

Let's look at average concrete to see how to use this information. You'll get to know this type of concrete very well because it's the type to use for foundation slabs and similar work. Here's the formula:

- 1 part portland cement (use at least 5 sacks of cement per CY in average concrete)
- 6 parts (maximum combined volumes) of fine and coarse aggregate
- 6½ gallons (maximum) water per 94 lb bag of cement

Let's try this out in an example. Say you're going to pour a 1 CY slab for a storage building in your yard using average concrete. From Figure 2-1 you see that the proportion of cement to sand and gravel is 1:2½:3½. Now here's a list of what you'll need to make 1 CY:

- 5 sacks or 5 CF portland cement
- 5 × 2½ or 12½ CF of dry sand (fine aggregate)
- 5 × 3½ or 17½ CF of gravel (coarse aggregate)
- 5 × 6½ or 32½ gal water

To convert the sand and gravel quantities from CF to CY, divide each by 27.

Preparing Small Batches of Concrete

It's not hard to prepare and mix small batches of concrete. Here are some hints to help you get organized and some guidelines for mixing and measuring. Use them to get predictable and dependable results.

Planning ahead is important. Be sure you have on hand all the tools and equipment you'll need. And be sure they're all in good working order. Then, before you head out to pick up materials you need for a small job, stop to consider how much those materials weigh. Take another look at Figure 1-4, the table of construction material weights. Now, let's pause for a moment and add up the weight of materials you'll use for a CY slab.

470 pounds portland cement (5 sacks)
1,213 pounds dry sand (12.5 CF)
1,662 pounds gravel (17.5 CF)
3,345 pounds total weight

Maybe you better have the supply yard deliver the materials. It's often cheaper to buy ready-mix when you factor in your labor cost.

Measuring Materials

You can measure materials by weight or by volume. Measuring by weight is better because it's more accurate, it's easier and you won't need any special equipment. You can get by with a standard household scale and some recycled containers. Ideally, all the containers hold the same amount. I like to use old drywall compound pails. They meet all the requirements and they're also lightweight and rustproof.

Begin by weighing the empty pail. Then, while the pail is still on the scale, adjust the scale so that it reads zero. This is called *zeroing*. The measurements you make, after zeroing, show only the weight of the pail's contents. I recommend using a separate pail for each different material.

After you've weighed out each ingredient, it's a good idea to mark the correct levels for the material on the pail. Marking the pails cuts down on how much time you spend measuring materials for the next batch.

If there's no scale handy, you'll need to measure your materials using the volume method. When you're measuring by volume, all of your containers must hold the same volume. Here's a quick example to show the way measurement by volume works. Let's say you want to prepare a 1:2½:3½ mix.

1) Measure 1 pail of portland cement.
2) Measure 2½ pails of dry sand.
3) Measure 3½ pails of coarse aggregate.
4) Start with about ½ pail of water.

Figure 2-3
Concrete mixer for use on small jobs

Mixing Small Batches

The best way to mix a small batch of concrete is to use a concrete mixer, like the one shown in Figure 2-3. Mixers come in many sizes. The smaller ones hold as little as ½ cubic foot of concrete. Bigger mixers, on the other hand, can handle up to 6 CF per batch. If your batch size falls somewhere between two mixer sizes, get the bigger mixer. Overloading is hard on a mixer.

You can rent a concrete mixer and most of the other tools you need for concrete work from an equipment rental yard. Mixers are powered by gas or electricity. A gas-powered mixer is a good choice because you can run it almost anywhere. An electric mixer will work fine if you're certain that the power's hooked up at the job site.

Concrete must be thoroughly mixed. To be sure, let the mixer run for at least three minutes after you've added the last ingredient. Here are the steps you follow to mix a batch of concrete:

1) Turn the mixer on.
2) Put about 10 percent of the water into the mixer drum to keep the dry materials from sticking.
3) Add the dry materials one at a time, and follow up by adding some water each time.
4) Now, let the mixer run for at least another three minutes.

Clean the mixer and your tools regularly. Dirty, concrete-encrusted tools or mixers are not the signs of a good mason.

Common Flaws in Concrete

Here are the two most common flaws you find in concrete work:

Segregation — This usually happens when you overwork the concrete so it settles and separates after you pour it. The fine materials, including most of the cement paste, rise to the top. Meanwhile, most of the coarse aggregates settle to the bottom. The layered concrete you get in the end is weak and unsatisfactory. The easiest way to avoid this problem is to use air-entrained concrete. As you'll see in the next section, air-entrained concrete is much less likely to segregate than other types of concrete.

Bleeding — This happens when you let concrete sit too long between mixing and placing, and when you overwork it. This causes free water mixed with dissolved portland cement and the finest particles of aggregate to come to the surface. The layer that forms on the concrete surface, called *laitance*, is very weak.

Bleeding is usually to blame when a concrete surface becomes crazed (covered with widespread, very fine cracks). Bleeding also causes concrete to scale or peel away, flake by flake. And bleeding also leads to dusting, when the surface turns to powder. However, bleeding isn't a problem when you use air-entrained concrete. This leads us to the next topic: concrete admixtures.

Admixtures

Admixtures are chemicals you add to concrete to:

- entrain air
- speed up or slow down setting
- make it easier to work with
- cut down on segregation and bleeding
- change its color

It's very important to use admixtures correctly. If you don't, you'll get weak concrete. It's a good idea to avoid combining admixtures before adding them to the mix. I recommend that you add only one admixture at a time. To save yourself time and money, you should start out by making a small batch first. Test the small batch to be sure it has the qualities you want. Make as many small batches as you need for practice. Work out any kinks, figure out in what order to add the admixtures, and how much of each one to use. Work it all out, then when you're sure of the mix, go ahead and make a full-size batch.

But before we get into the details of admixtures, let's look briefly at a subject that will come up repeatedly: *hydration*. That's the chemical reaction that happens when portland cement and water combine to make a paste or gel. The cement paste coats each particle of aggregate and then binds with it. The paste also fills any voids between pieces of aggregate, making a single, solid mass. That's concrete.

The quality of the paste depends on the:

- ratio of cement to water
- time allowed for hydration
- conditions hydration takes place under

Unfortunately, the last two are the most important and most difficult to control. Hydration is a very difficult reaction to manage. But this is where admixtures can really help you out. The air-entraining agents, accelerators, retarders, water reducers and superplasticizers are all aids for managing hydration.

Air-Entrained Concrete

Air-entrained concrete is more durable and more weather-resistant than ordinary concrete. An air-entrained mix has millions of tiny air bubbles in it. These bubbles act like flexible ball bearings. They lubricate and plasticize the mix. Concrete work that's exposed to repeated cycles of freezing and thawing performs best and lasts longest if you use air-entrained concrete. To make air-entrained concrete, either add an air-entraining admixture to the mix, or use Type I air-entrained portland cement to make the mix.

Manufactured, slag or lightweight aggregates used in air-entrained concrete yield good results. Used in other types of concrete, these kinds of aggregates produce coarse concrete. You'll probably also use a smaller amount of fine aggregate in an air-entrained mix.

Some more advantages to using air-entrained concrete:

- It resists deicing salts, sulfates, sea water, and alkaline water. Contact with any of these causes serious damage to other kinds of concrete.

- ❖ It goes further, so you can use less cement. That's good, because cement is the most expensive ingredient in concrete.
- ❖ It bleeds less than other concrete.
- ❖ It shrinks less than other types of concrete.
- ❖ It's less likely to segregate.

Usually you use between ¾ and 3 fluid ounces of air-entrainment admixture per 100 pounds of cement. The exact amount you use will depend on:

- ❖ the mix
- ❖ the air temperature at the job site
- ❖ the type of cement
- ❖ the grade of the sand
- ❖ how much extra-fine material, such as fly ash, you use as aggregate

Accelerators

Accelerators are admixtures that speed up hydration. Use them when you're doing concrete work in cold weather. When you speed up hydration, you shorten the setting time. That way you don't have to protect the fresh concrete from freezing for as long. Remember, however, that accelerators aren't antifreeze for concrete. You'll still have to protect the concrete from freezing. But you won't have to do it for nearly as long.

Most accelerators are liquids and their active ingredient is calcium chloride. You can't use these admixtures in any concrete where the presence of chloride compounds is unacceptable.

Retarders

Retarders are the opposite of accelerators — they slow down hydration and extend the concrete's setting time. Use retarders when you're doing concrete work in hot weather.

When concrete dries too quickly, it forms a stiff, hard skin on the outside, while the inside is still liquid. During hydration, the liquid concrete expands and shrinks. But if a hard surface skin hinders this, a lot of stress builds up in the concrete. Eventually this stress cracks open the hard skin on the surface. You can't do decent finish work on cracked concrete.

Retarders delay the initial set of your concrete and buy you some time. Enough time so that all the concrete has finished expanding and shrinking before any of it sets.

The other difficulty you'll face doing concrete work in hot weather is that too much water tends to evaporate from the concrete. Water evaporation is caused by:

- ❖ high air temperature
- ❖ heat generated by the concrete as it sets
- ❖ low humidity
- ❖ wind

Here's a worst-case scenario with the highest evaporation rates. Imagine a dry day, air temperature over 100 degrees F, and a strong steady warm wind blowing. You're on a job site that's running behind schedule, so you have to pour today and you have to use a high temperature mix. These conditions are all too familiar to anyone who's worked in the southwestern U.S.

Unless you control evaporation, your concrete will be weak. The water that's evaporating is needed for hydration. Concrete can't develop its full design strength if it isn't completely hydrated.

Retarders are liquids and you usually use about 3 fluid ounces of retarder per 100 pounds of portland cement. At 70 degrees F, a retarder adds about two to three hours to concrete's initial set time.

Both accelerators and retarders reduce the amount of water you'll need in the mix. But you shouldn't use them for that purpose alone. There are other admixtures developed just for cutting down on the amount of water your mix needs. They do a better job and don't have any negative side effects. We'll look at them next.

Water-Reducing Admixtures

Water-reducing admixtures are chemicals that help control and optimize the degree and rate of hydration. When you use a water-reducer, you can cut the water in the mix by about 8 to 10 percent. The water-reducer allows you to do this by keeping the mix workable. If you cut the water in a mix that much without using a water-reducer, the resulting mix will be hard to pour, work and finish. Worst of all, the concrete would be weak. However, when you make concrete with a water-reducing admixture, it's stronger and more plastic.

Water-reducing admixtures also extend concrete's setting time. This makes the mix easier to work, place, pump and finish. And most water-reducing admixtures don't contain calcium chloride. Most water-reducing admixtures won't have any problems working together with air-entraining.

My own favorite water-reducing admixture is: *WRDA with Hycol* from W. R. Grace & Company. If you have problems finding a local supplier, try writing to the company. Their address is:

W. R. Grace & Co.
7721 West Parkland Court
Milwaukee, WI 55223

Many suppliers of ready-mix concrete add a water-reducing admixture to all their batches, often at no extra charge.

Superplasticizers

Superplasticizers are a fairly new group of admixtures. You add them to a mix to keep it from forming clumps. Clumps, made up of dry materials that didn't get mixed in, make concrete weak. In a mix with no clumps, all the materials combine better, making very strong concrete. Using a superplasticizer in the mix will also cut down on friction in the mix, making it easier to pour, work and finish. Superplasticizers also act as water-reducers.

Coloring Concrete

You add coloring agents to concrete the same way you would any other admixture. Add all the dry materials to the mixer (cement, aggregate and coloring), then let it mix for three or four minutes. By then, the coloring agent is evenly spread through the whole batch of cement. Then just add water, and mix for four more minutes.

Coloring concrete is an art that has many practical uses. However, it's important that you be careful selecting your coloring materials. These materials must meet all of the following requirements:

- ❖ alkali resistant
- ❖ lightfast
- ❖ chemically inert
- ❖ low calcium sulfate content
- ❖ insoluble in water
- ❖ fineness
- ❖ uniformity

We'll discuss each of these requirements in detail.

Alkali Resistant

Some organic pigments in coloring materials react chemically with alkaline compounds and as a result lose their color. Here's how to test pigments for this alkaline reaction:

1) Mix 20 parts cement with one part of coloring.
2) Use enough water in the mix to form a buttery paste.
3) Divide the mix into several samples and set them aside.
4) Keep the samples moist while they cure. This takes several days.

Keep an eye on the samples as they cure. You especially want to keep tabs on any apparent change in their color. If there's considerable fading, that means your coloring material isn't alkali resistant. It also means the material isn't suitable for coloring concrete. A word of warning about this test: The samples may develop *efflorescence*, a powdery white coating. If efflorescence develops, brush it off. Only then can you judge how colorfast the pigment is.

Lightfast

The coloring materials used in concrete must also withstand the bleaching effects of exposure to direct sunlight. Unfortunately, there's no quick or simple way to test how lightfast your coloring materials are. There are high-tech, high-priced testing instruments that measure how lightfast colors are. But not many concrete contractors have access to either a fadeometer or a weatherometer. Unless you do have access to either of these instruments, the best advice I can give you is to trust the published technical data and depend on the experience of the manufacturers.

Chemically Inert

Your coloring materials can't be chemically reactive with either the cement or any other admixtures in a way that causes the concrete to lose strength.

Calcium Sulfate Content

You shouldn't use any coloring material that contains more than 15 percent by weight of calcium sulfate. Why? Calcium sulfate, you may recall, is the main ingredient in many accelerators. If the coloring agent has too much calcium sulfate, it changes the effect the accelerator has on setting times.

Insoluble in Water

Your coloring materials must not dissolve in water. If they do, they'll wash out of the finished concrete.

Fineness

You can test coloring material for fineness by passing it through a number 325 mesh screen. If 98 percent of the coloring material passes through the screen, it's OK.

Uniformity

Only use coloring agents made by manufacturers with good quality-control standards. That's the only way you can count on getting the same product and results each time you use the product.

Here's one last piece of advice on mixing colored concrete. To get the same color in every batch, always measure the coloring agents by weight, not by volume.

Using Iron-Oxide Coloring Agents

Many, but not all, natural and synthetic iron-oxide coloring agents meet all of these requirements. You can use them to color concrete many shades of red, yellow, brown, and black. Whether you use a natural iron oxide, a synthetic iron oxide, or a combination of the two will depend on what color you want. Synthetic iron-oxide colorings end up costing a little less because they go farther.

Testing over the years has shown that using iron-oxide coloring agents in concrete won't decrease the concrete's compressive strength. In some cases, adding an iron-oxide coloring agent actually increases compressive strength.

When you mix a small amount of coloring agent with white portland cement, the resulting color is a tint. The tinting powers of different coloring agents vary widely. For example, adding 3 pounds of natural yellow ochre to a bag of portland cement barely changes its color. However, adding the same amount of synthetic yellow oxide to a bag of cement produces concrete that's definitely yellow. The color of the portland cement also affects a tint's relative depth and clarity. The lighter the cement color, the brighter and clearer the tint.

The color of the aggregate also affects the color of the concrete. You've probably seen colored concrete that's colored by the aggregate instead of with a coloring agent. Even when you're using a coloring agent, the color of the aggregate also affects the color of the concrete. The lighter the aggregate color, the clearer the concrete color. Gray cement and dark aggregates combined with a coloring agent produce grayed colors.

How Much Do You Use?

How do you figure out how much coloring to use to get just the right color? This always takes some experimenting. As a rule of thumb, use these amounts. By following them you'll at least start out inside the ballpark.

- ❖ For light, pastel colors: Use ½ to 1 pound of coloring agent per 90 pound bag of white portland cement combined with light-colored aggregates.
- ❖ For deeper, stronger colors: Use 3 to 7 pounds of coloring agent per bag of gray portland cement and dark-colored aggregates.

There are over 150 shades of red, yellow, and brown oxide. And to make it more confusing, each manufacturer is free to name their colors as they please, because there's no industry standard. That means that although two companies may make a coloring agent that both call tan, the color that you get may not be at all the same. It also means that light brown, buff, ecru, beige and Taos tan could all be the very same color.

Color	Type of coloring agent (synthetic or natural)	Amount of coloring agent (% by weight) lighter colors	darker colors
Yellow	synthetic yellow oxide	2	5-6
	ferrite yellow oxide		
Red	synthetic red oxide	½ - 2 (white cement)	3 - 6
	natural red oxide	3 (white cement)	
Brown	synthetic brown oxide	4 - 6	6 - 10
	natural burnt umber		
	natural raw umber	3 - 5	5 - 10
	natural burnt sienna		
Gray	synthetic black iron oxide	3	5
Black	synthetic black iron oxide		10
	lamp black		2
Note: This table assumes the use of gray portland cement except where noted.			

Figure 2-4
Proportions of coloring agents in concrete

The best starting point when you're selecting a color is either the manufacturer's color cards, or samples. Most manufacturers are helpful and a good source of information. Describe the results you want and they'll tell you which of their products to use, and in what proportions. However, if you're naturally curious, or just want to check their advice, do a little experimenting. Just follow this simple procedure.

1) Start out by mixing up a small batch of your regular mix. Keep the batch small enough that you do all the mixing in a single steel pan or pail.

2) Measure your coloring agent, allowing either 5 percent by weight for gray portland cement, or 1 percent by weight for white portland cement.

3) Add the coloring agent and thoroughly mix it in.

4) Add the right amount of water and mix thoroughly. Thorough mixing is important in steps 3 and 4 to avoid streaky color.

5) Pour the mix into a steel pan or mold and allow it to dry.

6) Make three or four mixes, but increase or decrease the amount of coloring you use by about 25 percent in each batch.

It takes about two days for a trial batch of concrete this size to cure enough for you to accurately tell its color.

With so many different colors and the lack of standard names, it's impossible for me to give specific amounts of coloring agents to use to produce a certain color. The best guidance I can give you appears in Figure 2-4. The percentages given are for mixes made with gray portland cement. If you use white portland cement, 1 or 2 percent of pigment is usually enough.

Color Permanence

Iron-oxide colorings for concrete are known for their longevity. What seems to be color loss is usually a breaking down of the concrete surface, not fading of the coloring. As colored cement paste weathers, more and more of the aggregate is exposed to view. If the aggregate color contrasts with the color of the paste, it makes the color of the concrete appear to have faded when it hasn't.

Efflorescence is easy to mistake for color fading, because the powdery white deposit collects on the concrete's surface. It's caused by leaching of hydroxides from the concrete. When these compounds reach the surface, they react with the carbon dioxide in the air and change to calcium carbonate. White calcium carbonate crystals show up more on colored concrete than on natural gray concrete.

Colors Other Than Iron Oxides

You can also color concrete green and blue. Before we discuss these coloring agents, however, I want to emphasize that they are entirely different from the iron-oxide agents we've discussed until now.

Green — DuPont makes a phthalocyanine green that's available in one color only. Contact their customer relations people for help regarding the amount to use to get a specific shade of green.

Blues — Use blue coloring agents with caution. One blue coloring agent that looks promising is phthalocyanine blue. However, at this point the data isn't in on how well it holds up to exposure. Tests of an ultramarine blue gave erratic results. There are cases of color failure without any apparent reason. And the ultramarine blue seems to react with the lime in cement during curing.

The Curing Process

Before we discuss surface treatments, let's define what we mean by curing. Curing begins at the same time as hydration. It continues as long as hydration does, or until hydration is complete. Although this may seem like two names for one process, it isn't. Here's why. Hydration describes the chemical and physical reactions that occur between the ingredients in concrete. Curing describes the way the concrete hardens and develops its strength. Curing is a result of hydration.

The first 12 hours after the pour are the most critical to the hydration and curing processes. This is the time to be extra careful. You need to make sure the conditions are right for hydration and curing to happen as they should. Any lack of care here shows in the concrete's quality later.

Both curing and hydration continue long after the first 12 hours. Concrete actually develops most of its strength in the first seven days. However, it won't reach its full, planned, 28-day strength unless hydration continues for that entire time. Everyone in the business knows it's important for fresh concrete to retain water. What you may not know is how important water is throughout hydration. That's between 7 and 28 days. Everyone in the business also knows that leaving the forms on the concrete helps it retain water. But it's often either impractical or impossible to leave forms on for 7 days, let alone the entire 28 days. That's the reason there are so many curing methods and so many cure promoter compounds for sale. In this section we'll look at the older, popular curing methods. Then we'll cover the newer chemical approaches for controlling curing.

Curing Methods

Many traditional curing methods are still widely used. But they all have their drawbacks. Let's take a critical look at four time-tested methods.

1) Wrap or cover fresh concrete with pieces of burlap. Keep the burlap wet until the concrete is fully cured. There's a problem here that we'll find again and again. This method only works if the burlap doesn't dry out. If someone forgets or waits too long to wet the burlap, it dries out.

2) Sprinkle water directly onto new concrete. Repeat frequently so the concrete stays damp until it's completely cured. This has the same drawback — drying out — as using burlap, but even more so, as there's no burlap to hold the moisture in.

3) Cover fresh concrete with moist earth. Wet the soil often without washing the soil away. The disadvantage with this method is that it's messy and can leave ugly stains. And again, if too much time passes between wettings, the concrete gets too dry.

4) Cover fresh concrete with waterproof paper or plastic film. This works, but the materials are expensive and may take more than one person to handle.

Curing Compounds

The modern chemical solutions for curing concrete have many advantages. Here are some good reasons to use one of them:

- ❖ low material cost
- ❖ easy application
- ❖ no wetting or other followup
- ❖ reliable results
- ❖ low labor cost

There are many different curing solutions. However, when you look at the main ingredient, they all fall into one of four groups. Here's a list of the groups, including the disadvantages of each:

Resin curing compounds — These solutions tend to discolor concrete. Don't use them on concrete surfaces when appearance is important.

Sodium silicate curing compounds — These don't usually meet ASTM specifications for moisture retention, so you can't use them if the building code or plans require all materials to meet ASTM specifications.

Chlorinated rubber curing compounds —These compounds turn yellow as time passes, making the concrete appear discolored. Don't use them where appearance is important.

Acrylic curing compounds — These are the best curing solutions to use. They offer all the advantages without any disadvantages.

Surface Treatments for Concrete

There are many compounds you can spray onto a concrete surface to:

- ❖ make it harder
- ❖ make it smoother and more uniform
- ❖ make it stronger so it lasts longer
- ❖ seal out stains and dirt

We'll discuss the uses, varieties and any difficulties you might have with each type of surface treatment.

Hardening Compounds

Add a hardening compound to concrete in high-traffic areas such as entrance-level floors of multistory buildings, and in heavy-use areas such as warehouse floors and loading docks.

You'll also want to use a hardener on any concrete surface that gets heavy use while construction is going on. By applying a hardener to curing concrete, you'll save time and money on maintenance and repairs that you'd have to make otherwise.

Sealants

Because the surface of concrete is porous, it's easily stained and soiled. A quality sealant quickly pays for itself through savings in cleanup and maintenance costs.

Before you apply a sealer, make sure the concrete is clean. No matter how fresh or perfectly cured the concrete is, there's always a surface layer of dust. Clean this dust off first or you'll be sealing the dust instead of the concrete.

Multipurpose Compounds

These are combination treatments for concrete that are really easy to use (Figure 2-5). My favorite is *Cure & Hard* from Symons Corporation. This multipurpose, spray-on compound dries clear and stays that way. But the best thing is how much you get for so little effort. Spray it onto clean concrete and it does all of this in just one treatment:

- ❖ aids curing
- ❖ hardens
- ❖ dustproofs
- ❖ seals

Testing Concrete

There are several important tests to do on a batch of concrete. In this section we'll look at each test and how you do it.

Figure 2-5
Applying a spray-on surface treatment

Figure 2-6
Slump test using a cone-shape mold

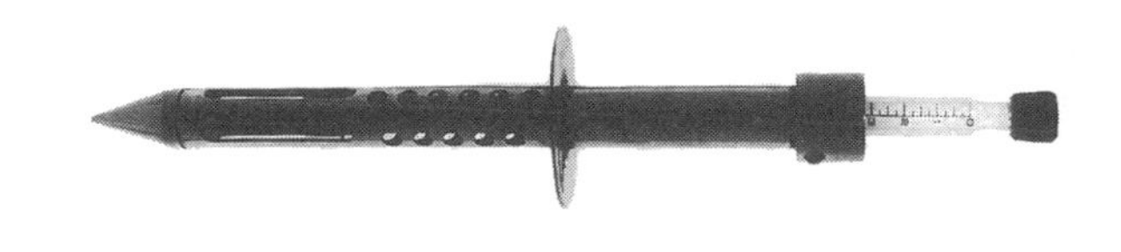

Figure 2-7
Slump tester

Slump Tests

A slump test tells you many things you need to know about a batch of concrete. The greater the slump, the weaker the concrete. When you order ready-mix, they'll ask what slump you want. You can use the slump test to:

- ❖ judge the concrete's consistency
- ❖ tell how wet or dry the concrete was when poured
- ❖ compare batches of concrete for the grade and amount of aggregates
- ❖ compare water content between batches of concrete

Here's how to make a good slump test:

1) Fill a cone-shaped mold, like the one shown in Figure 2-6, one-third full with fresh concrete.
2) Rod the concrete 25 times using a $^5/_8$-inch diameter rod.
3) Add more concrete to fill the cone mold to two-thirds full.
4) Rod it 25 times more.
5) Finish filling the cone to the top with concrete.
6) Rod it again.
7) Screed the concrete so it's level with the top of the cone.
8) Turn the cone over and carefully remove the cone from the concrete, and set it next to the concrete.
9) Lay the rod across the top of the two cones, as shown in Figure 2-6.
10) The distance between the rod and the top of the concrete is the slump, measured in inches.

There's also another way to do a slump test, and that's with a slump tester like the one shown in Figure 2-7. The Humboldt Manufacturing Company of Norridge, Illinois makes this slump tester. To use it you simply insert the probe into a concrete mass and then, after 60 seconds, take readings of slump and workability. The only limitation to using a slump tester such as this is the concrete mass has to be at least 6 inches thick. You can use this lightweight, convenient tool in a wheelbarrow, a ready-mix chute, or even in a slab.

If either type of slump test shows that your batches aren't the same, here's an important point to remember. If you need to adjust the water content of your mix, do it by changing the amount of aggregates. Never try to correct a water content difference by changing the amount of water you use per bag of cement in the mix.

Concrete Test Cylinders

Figure 2-8 shows standard concrete test cylinders. Keep them upright and protected from the elements. You use these 6- by 12-inch cylinders to test concrete's strength. Here's a simple way to make a test cylinder:

1) Fill the cylinder one-third full.
2) Rod the concrete 25 times using a 5/8-inch-diameter rod.
3) Fill the cylinder two-thirds full.
4) Rod the concrete 25 times.
5) Finish filling the cylinder with concrete.
6) Rod it 25 times.
7) Screed the concrete even with the top of the cylinder.
8) Cover the top of filled molds with plastic to prevent evaporation from happening too fast.

Figure 2-8
Concrete test cylinders curing

Here are few more hints on making test cylinders:

- Take your test samples in mid-pour, not at the beginning or at the end.
- Always store test cylinders upright.
- Don't disturb test cylinders for at least 24 hours.
- Store test cylinders in an area protected from freezing, high temperature and rain.
- After 24 hours, send your test cylinders to the testing laboratory.

Estimating Concrete for Small Jobs

You measure, price and estimate concrete work in units of cubic yards or cubic feet. All measurements must be in whole or decimal parts of one of these units. It's easy to change cubic feet to cubic yards. All you do is divide by 27 because there are 27 cubic feet in 1 cubic yard. It's just as easy to go the other way, and change cubic yards to cubic feet. Simply multiply by 27.

Now let's do some estimating. The first step is to find the area of a job. Area is simply length × width × depth (or thickness). Let's try this out using some numbers. We'll do an estimate for a patio slab for Mr. Hammer. He wants a patio that's 16 feet long, 6 feet wide and 5 inches thick.

Your first step here is making all the units the same — either feet or yards. Get them mixed up and you'll either lose the job or wish that you had. So we'll change 5 inches into feet. Of course, 5 inches is equal to 5/12 foot. But the arithmetic we'll do next is lots easier when that measurement's a decimal instead of a fraction. Figure 1-10 in Chapter 1 lists the decimal equivalent, in units of feet, for whole inches and the most common fractional parts of an inch.

Let's go back to Mr. Hammer's patio. From Figure 1-10, you know that the decimal equivalent of 5 inches is 0.42 foot. To find the volume of the slab, you need to multiply its length × width × thickness. Here's the arithmetic:

$$16' \times 6' \times 0.42' = 40.32 \text{ CF}$$

Then convert 40.32 CF to CY by dividing by 27. You get 1.493 CY. Round that up to 1.5 CY. Rounding up is the easiest way to include an allowance for waste in your estimates.

The next step is to find out what kind of mix you use for a patio slab and how much material you'll need to make 1.5 CY of that mix. To do this, we'll go back to Figures 2-1 and 2-2.

Figure 2-1 tells you the right mix for the job and how much mixing water it takes:

1) In the column labeled *Kind of work*, the best description of the job is the second one down: Waterproof and wear-resistant, 4" to 8" thick.
2) Read across this row and you'll find a 1:2:3 mix recommended for this job.
3) You'll also find that (assuming the sand is moist), you'll use 4½ gallons of water per sack of cement to mix the concrete.

Figure 2-2 gives the amounts of cement, sand and gravel you need to make 1 CY of 1:2:3 mix:

- ❖ 7 sacks of Type I portland cement
- ❖ 0.52 CY of moist sand
- ❖ 0.78 CY of gravel

For Mr Hammer's patio with a 1.5 CY volume, multiply each quantity by 1.5 and your estimate's done:

7 sacks × 1.5 = 10.5 total sacks cement
0.52 CY × 1.5 = 0.78 CY total moist sand
0.78 CY × 1.5 = 1.17 CY total gravel

For smaller projects, using less than 1 CF of concrete, there's another option. Use dry premixed concrete that comes in bags. Building supply stores carry several different mixes in two sizes of bags, 45 pound and 90 pound. One 90-pound bag holds everything you need, except the water, to make ⅔ CF of concrete.

3

Mortar

Mortar makes up 7 to 15 percent of a wall's total area. Although that's a small percentage, the role played by mortar isn't minor. Everyone knows that mortar holds the unit masonry together so it forms a stable structure. But it also does a lot more. Mortar protects masonry from water damage and weathering, so it lasts longer. Mortar gives stone, brick or block work a clean, finished, uniform look.

The key properties you need in good mortar are workability, water retention, bond strength and durability. Since each of these depends on the others, you can't change just one. If, for example, you add more water to improve workability, you'll decrease the mortar's strength. Changing one changes the others, often in ways you can't predict. Be extra careful if you plan on customizing a mortar mix.

When you choose mortar, you have to consider the brick or block you'll use it with. For example, you need to know how much water the masonry will absorb. Mortar that works well with glass block, which doesn't absorb water, doesn't work with concrete block, which absorbs a lot of water from the mix. Concrete block is so absorbent that if you don't take this into consideration, there won't be enough water in the mix for it to finish hydration and curing.

Here are some more things to keep in mind when you choose mortar:

- site conditions
- structure type or use
- level of exposure
- types of loading
- types and amounts of stress
- reinforced or unreinforced construction

To choose the best mortar for each job, you'll need to know mortar inside and out. That's exactly what this chapter's all about.

Mortar Components

Mortar is made of four basic ingredients: portland cement, hydrated lime, sand and water. We'll cover each one in this section. We'll also take a good look at masonry cement. This is a manufactured, premixed, packaged combination of portland cement, hydrated lime and optional admixtures. Masonry cement is so much easier to use that you don't often find anyone mixing mortar from scratch at a job site anymore.

Portland Cement

The binding agent in mortar is portland cement. As you already know from Chapter 2, there are five types of portland cement. Usually you make mortar with Type I normal portland cement. Once in a while you'll use Type III portland cement to make mortar that sets up faster than normal. In near freezing weather, for example, you have to use fast-setting mortar.

Hydrated Lime

Hydrated lime is the dry, powder form of *slaked* or *quick lime*. It's added to masonry cement, usually when it's manufactured, to help it hold water. When you make mortar from scratch at the job site, your first step is to mix hydrated lime with portland cement. This means that the mortar:

- ❖ sets more slowly, cracks less easily and forms stronger bonds
- ❖ is easier to work with so the whole job runs smoother and faster
- ❖ doesn't dry out so quickly so there's less remixing needed
- ❖ sticks better to the brick or block, making stronger, more waterproof bonds
- ❖ fills in small voids so the work is stronger and lasts longer
- ❖ heals hairline cracks, making bonds stronger and more waterproof (if you keep it moist)

That last function is really a big help anytime you work with fast-setting mortar (made with Type III cement). It tends to get lots of hairline cracks, right at the bond point. Where there are cracks, sooner or later there's water damage too. That's why it's a good idea to add some extra hydrated lime to mortar you make with Type III cement. Don't get carried away though; this is a trade-off situation. It's true that mortar with hydrated lime seals joints better. But the more hydrated lime you have in a mortar mix, the weaker your mortar. It takes judgment to make the right mix.

ASTM specification number C207-49 defines two types of hydrated lime. Type N is normal hydrated lime. Type S is special hydrated lime. The difference is that the standards for Type S are stricter than for Type N.

Type S must have no more than 8 percent unhydrated oxides. There are no requirements for Type N. Unhydrated oxides in mortar are a problem. Sooner or later they hydrate. When that happens they get larger and make the mortar expand. But there's no space for the mortar to expand. So stresses build up that may crack the masonry.

Type S hydrated lime, when tested 30 minutes after mixing with water, has a plasticity figure of at least 200.

Another advantage of Type S lime is how long you have to soak it in water before using it. Type N hydrated lime has to soak for several hours or overnight to become as plastic as Type S lime with only 30 minutes of soaking. Type S hydrated lime is clearly the better choice.

Masonry Cement

This is a factory-prepared mix of portland cement, hydrated lime and, sometimes, one or more admixtures. Masonry cements come in 70-pound bags. You mix one bag of masonry cement with sand and water to make about 1 CF of mortar. There are lots of advantages to using masonry cement:

- ❖ quality-controlled proportions and materials
- ❖ good appearance
- ❖ reliably good results
- ❖ cost effectiveness
- ❖ less waste

You don't add hydrated lime to masonry cement. It's already included. Adding more would only weaken the mortar. But to make Type M or S mortar, you'll have to add portland cement to masonry cement. We'll cover that later.

Admixtures for Masonry Cement Mortars

There are many brands and types of admixtures to add to masonry cement mortar. But don't be taken in by the ads. You don't need most of them to make good mortar. All you really need, besides the masonry cement, is well-graded, clean mason's sand and clean water. Mix it all together and you'll end up with a batch of top-quality mortar.

The only admixture for mortar that's worth the money is air-entrainment. Air-entrained mortar, like air-entrained concrete, is full of millions of tiny air bubbles that work like ball bearings. This makes the fresh mortar easier to work with and the cured mortar last longer.

Wetting agents are admixtures that claim to have all the good effects of air-entraining. What's not mentioned is that some of them also weaken bonds. Bonds made with wetting agents have to be tested to find out if they're strong enough to be safe. That takes 28 days and several pieces of the masonry. If you think wetting agents sound like more trouble than they're worth, you're probably right.

Some latex-based compounds claim that they make bonds stronger. But some are moisture sensitive, which isn't a good property for mortar.

Sand for Mortar

Sand is cheap and so common in all masonry work that it's easy to take for granted. But don't make the mistake of thinking that quality doesn't count when you're buying sand. And don't let anyone tell you that all sand's the same. You can't make good mortar with just any old sand. Use only clean, well-graded sand.

Sand is graded by sifting it through finer and finer mesh sieves. As the sieve number or size goes up, the grains that pass through get smaller. See Figure 3-1. Well-graded sand is made of grains that cover all sizes between coarse and fine.

Poorly-graded sand may cost less, but it's no bargain. You can't make workable mortar with it unless you add extra cement. But cement's the most expensive ingredient in mortar. Save a few pennies on sand — spend several dollars more on cement. Find a supplier who carries mason's sand that meets ASTM standards (Specification C144, Aggregates for Masonry Mortar).

Building supply stores carry washed sand for masonry work. How well washed the sand is changes with the supplier. I'd advise you to find a good supplier and stick with them.

There are two tests to help you judge how clean sand is: the silt-and-clay-content test and the colorimetric test. We'll look at both tests to see what they tell you about the sand. We'll also cover, step-by-step, how to do each one.

You use the silt-and-clay-content test to find out how much silt and clay is in the sand. Besides a sample of the sand and some water, you need a 1-quart clear glass jar with a lid (a clean mayonnaise jar is ideal) and a ruler.

To do the test:

1) Put about 2 inches of sand in the bottom of the jar.
2) Fill the jar about three-quarters full with clear water.

Sieve number (size)	Percentage of sand passing through	
	Average	ASTM limits
4	100	100
8	100	95 - 100
16	80	60 - 100
30	50	35 - 70
50	5	15 - 35
100	7	0 - 15

Figure 3-1
Grading standards for mason's sand

3) Screw the lid on and shake the jar hard.
4) Let the jar sit, undisturbed, for about 12 hours. During this time, the sample separates into layers. Each layer is a different material with a different weight. Grains of sand are heaviest and so they fall to the bottom first. Silt and clay are lighter. They take longer to settle out, and make a separate layer on top of the sand.
5) Use the ruler to measure how thick the silt and clay layer is.

Note: Be careful not to move the jar. The particles of silt and clay on the top are the smallest, lightest ones of all. Just a little bump and they'll float loose, spoiling your measurement.

If the silt and clay layer is less than 1/8 inch thick, the sand's clean enough to use in mortar. But if the silt and clay layer is more than 1/8 inch, the sand's not clean enough to use in mortar.

You use the colorimetric test to find out how much organic matter is in the sand. Don't be put off by the test's technical-sounding name. It just means you'll use color as the yardstick to decide whether or not to use the sand. Be very careful as you measure materials for this test. A measuring error throws everything off.

For this test you'll need a sample of the sand, water, two clear glass jars (one quart and one pint), and some lye. Use either household lye or sodium hydroxide from the drugstore. The important thing is the concentration — at least 94 percent sodium hydroxide.

Be very careful handling lye or sodium hydroxide. They're dangerous, caustic materials. Direct contact with skin or mucous membranes causes serious burns.

Always keep these materials as far away from your eyes as possible. If you have safety glasses, wear them! If you don't own a pair, buy some. They're not expensive and they could save your sight.

Here's the step-by-step procedure:

1) Fill the pint jar half full with 8 ounces of clear water.
2) Add a heaping teaspoon of lye to the water and stir to dissolve.
3) Put a half pint (8 ounce) sample of the sand in the bottom of the quart jar.
4) Add the water and lye solution to the sand in the quart jar.
5) Screw the lid on tight and shake the quart jar hard for two minutes.
6) Set the jar in a safe place where it won't be disturbed for several hours.
7) Check the color of the liquid. It tells you how much organic matter is still in the sand. The darker the color of the liquid, the dirtier the sand.

If the liquid is clear, the sand is clean and completely safe to use in mortar. If the liquid is a pale straw-like color, there's some organic material in the sand but not enough to cause problems. If the liquid is darker than that, there's a lot of organic matter in the sand and it's not clean enough to use in mortar. Wash it and retest it until you get it clean enough.

Sand's cheap, but that's no reason to waste it. Here's an easy tip that saves sand. At the job site, use tarps under the sand as shown in Figure 3-2. Storing sand this way has several benefits:

- ❖ Cleanup is as easy as folding up a tarp, as shown in Figure 3-3.
- ❖ Leftover sand is easy to save and use on the next job.
- ❖ Small rocks aren't accidentally scooped up with sand. This is a common problem when sand's stored on bare earth.
- ❖ You can reuse the tarps.

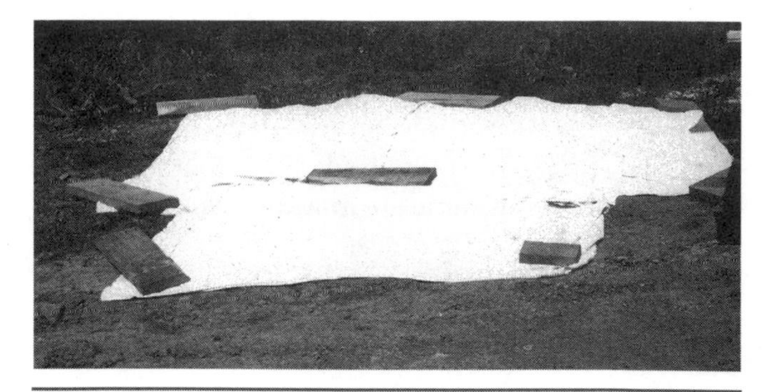

Figure 3-2
Place tarps on the ground before dumping sand

Figure 3-3
Fold tarp around remaining sand at end of job

Properties of Mortar

Masons define good mortar by its properties. Some are there right from the start and the finished mortar keeps them for life. Color's a good example. If the mortar's gray when you mix it, it stays gray.

Other properties take time to develop. These properties are often linked with one of mortar's chemical reactions. If the reaction keeps going, the property goes on developing, sometimes for weeks. Strength is a good example of a developed property. Most mortars take 28 days to reach full strength.

Let's take a good look now at each one of mortar's properties.

Plasticity and Body

Fresh mortar's ideal feel under the trowel is very smooth and buttery. Mortar like this is very easy to work with. It goes onto masonry with so little effort that it almost seems to spread itself.

Fresh mortar's ideal texture is not too stiff and not too thin. Good mortar clings to the surface of the brick or block on contact and at the same time clings to itself. But mortar that doesn't have enough body will ooze out of the joints. That's a big problem. It's unsightly and can cause stains. But worst of all, it causes the work to settle unevenly and can turn a good, level wall into a demolition job.

Water Retention

Mortar has to hold the water it needs for full hydration and a good cure. Mortar that falls short here is weak and hard to work with. Good mortar resists water loss to the unit masonry. Some kinds of brick and block are naturally very absorbent. Concrete block and brick, for example, are almost sponges.

Bond

Bond is the surface-to-surface link between hardened mortar and unit masonry. It's what makes masonry construction work. The more complete the bond, the stronger the work. Full, complete bonds make masonry work waterproof, strong and long-lasting.

Strength

This is one of masonry construction's biggest assets. Mortar's strength develops over time, taking as long as 28 days to complete. Its strength makes masonry work able to resist many kinds of force, including pulling, twisting and crushing.

Water Resistance

Mortar's ability to keep water out of masonry work depends on the quality of the bond. Well-bonded masonry is waterproof masonry. Leaks, dampness and stains on or near the work are signs that water has found a way inside the masonry. When this happens, any of the following may be the reason:

- poor flashing
- poor design
- poor bond
- hairline cracks in mortar

Color

This property is the only one that has nothing to do with how well mortar does its job. Mortar's color has absolutely no effect on a structure's safety or strength. But that doesn't mean you can afford to ignore it or treat it lightly. How something looks is always important to people. It's a big part of client satisfaction. Most people don't know or care about any of mortar's other properties. Like it or not, your work's judged mostly on how it looks.

You can color mortar with precolored masonry cement or coloring admixtures. Colored masonry cements are better because you can easily get them in a wide choice of colors. The other big advantage is the color stays the same, batch after batch, no matter what size the batch is or how much sand you use in the mix.

The other option, coloring mortar with admixtures, isn't nearly as easy to do well or consistently. The amount of admixture you use has to be refigured each time anything in the mix changes. Even so, the color won't quite match the other batches. But the biggest problem with admixtures is the risk you take of overusing them. The result is weak mortar that never reaches full strength.

A popular, modern look in masonry work uses black mortar. The best masonry cement for this, and the darkest I've found, is *Dark* from the Marquette Cement Company. Their special processing makes a masonry cement that makes a waterproof mortar that's a very deep and even color. Marquette makes two other colored masonry cements, *Buff* and *Dark Brown*. All three of these are Type N masonry cements. That means you

Mortar type	Parts by volume of:			Aggregates (measured in a damp loose condition)
	Portland cement	Masonry cement	Hydrated lime	
M	1	1	—	Not less than 2¼ and not more than 3 times the sum of the volumes of the cements and lime used.
	1	—	¼	
S	½	1	—	
	1	—	Over ¼ to ½	
N	—	1	—	
	1	—	Over ½ to 1½	
O	—	1	—	
	1	—	Over 1¼ to 2½	

Figure 3-4
ASTM specifications for mortar for unit masonry

use them for general purpose masonry work. Type N masonry cements also have a 28-day compressive strength of 750 psi. If you can't locate a local dealer I recommend writing to the company directly:

Marquette Company, Gulf and Western Industries, Inc.
First American Center
Nashville, TN 37238

Mortar Types and Uses

Most of the time you'll work with mortars that meet ASTM standards (Specification C270 Mortar for Unit Masonry). The standard identifies four types of mortar: M, S, N and O, including their ingredients, proportions and designed 28-day strength.

Before we go on, let's take a closer look at the last item. To test the 28-day strength of mortar, it's mixed up and made into a 2-inch cube. After 28 days, a specific amount of crushing force is applied to the cube of mortar. To pass the test, the cube must not crack or crumble. The unit for measuring compressive force is pounds per square inch or *psi*.

Now back to the four types of mortar. We'll cover their properties and proportions. And we'll look at what kinds of work you use each one for, and why.

Type M

This mortar has the highest 28-day compressive strength rating, 2,500 psi. Because it's so strong, it's used in reinforced and unreinforced loadbearing walls. Architects or the building code often require using Type M mortar for any work that's subject to extreme loads. Here are some examples of extreme loads:

- ❖ strong or gusty winds (high-rises are often subject to this kind of extreme load)
- ❖ earthquakes
- ❖ heavy rainfall (this stresses the concrete catch basins of storm drain systems)

Study Figure 3-4 and you'll see that Type M mortar has the most cement and the least sand of all the types of mortar. It's made with equal parts of portland cement and masonry cement. That means that it's also the most expensive. This explains why you only use it when it's structurally required.

Type S

This mortar has a 28-day compressive strength of 1,800 psi. It's used for loadbearing walls in both reinforced and unreinforced masonry. What makes Type S mortar different from other mortars is the high tensile strength of its bonds. It's the mortar to use for any masonry work that's subject to bending forces. Here are a few examples:

- ❖ retaining walls
- ❖ foundation walls or slabs
- ❖ sewers
- ❖ manholes
- ❖ brick pavements
- ❖ sidewalks
- ❖ patios

Another case where you need Type S mortar's special bond strength is in veneer work, where mortar is the only tie between veneer and backing. Looking at Figure 3-4, you can compare Type S mortar to the others. It has ½ part portland cement to 1 part masonry cement. It's cheaper than Type M but costs more than Type N.

Type N

This is the mortar you'll use the most. It's plain masonry cement. It has a medium 28-day compressive strength of 750 psi. You can only use it in above-grade work such as:

- ❖ exterior walls above grade
- ❖ exterior exposed work (such as chimneys)
- ❖ non-loadbearing interior walls
- ❖ stone work
- ❖ glass block

Type O

This mortar's 28-day compressive strength is low, only 350 psi. Use it only in interior, non-loadbearing or partition-type walls. Don't use Type O mortar anywhere that it might freeze because it fails at low temperatures.

Mixing Mortar

Be sure to mix your mortar thoroughly or it'll be weak and hard to work with. You can mix mortar by hand or with a mechanical mixer. We'll look at both, and see when and where each works best.

Hand Mixing

Today's masons regularly use this age-old method. It's still the best way to mix batches for small or special projects.

First, mix the dry materials (masonry cement and sand) together. Or, if you're mixing mortar from scratch, portland cement and lime followed by the sand. You want a homogenous mix of masonry cement and sand and that's much easier to do with dry materials. If you don't mix mortar correctly at this stage, it'll have weak spots from pockets of too much sand and too little cement.

After the dry materials are thoroughly mixed, add the water. Measure water carefully. Too much water makes weak mortar. Mix the water in completely. Mortar that's not mixed enough isn't easy to work with and doesn't form full strength bonds.

The only tools you need to mix mortar by hand are a mortar box or a wheelbarrow, and a mason's hoe. Then just follow these steps:

1) In a wheelbarrow or mortar box, dry-mix the sand and masonry cement. A mason's hoe makes a great stirrer.
2) When all the dry materials are a uniform gray, they're well-mixed.
3) Begin adding water to the dry mixture, a little at a time.
4) Thoroughly stir the mixture with the hoe each time you add water. Continue until you've added all the water.
5) If the mix seems too thin you can thicken it. Just add more sand and cement in the same proportions you used at the beginning.

Mortar's been mixed this way for hundreds, if not thousands, of years. Not many building methods have survived so long with so little change. Still, it's easy to see that this isn't a practical way to prepare mortar on a large-scale project. When you've landed a big job, it's time to use a mechanical mixer. But don't toss out your old mortar box and mason's hoe! You'll still want them for the small stuff.

Machine Mixing

There are many good mechanical mixers that you can buy or rent. Using a mixer frees you from the dull work of mixing batch after batch of mortar, by hand, all day long. Mortar that's mixed by machine also tends to be more uniform from batch to batch. Here are the steps for mixing mortar with a mechanical mixer:

1) Turn the mixer on and add about half of the water to its barrel. This keeps the dry materials from sticking to its sides, making lumps.

Figure 3-5
Stand the mixer on cement blocks so it can dump into a mortar box

2) Put about half of the sand into the mixer to blend with the water.
3) After mixing the sand and water, add all of the masonry cement to the mixer's barrel. Mix well.
4) Add the rest of the sand to the mix. The contents of the mixer will stiffen and become harder to mix at this point.
5) Add the rest of the water, thinning down the mix.
6) Let the mixer run for at least another three minutes. Mortar turns out even better if you let this final mix cycle for a full five minutes. Those two extra minutes of mixing pay off in yield, workability, water retention and board life. If you're making mortar from scratch you'll need to make a few small changes:
   - Before step 1, dry mix the portland cement with the hydrated lime.
   - Add the combined portland cement and hydrated lime to the mixer at the same point, step 3, as the masonry cement it replaces.

Here are a couple of great time-saving tips:

- Place the mixer up on cement blocks as shown in Figure 3-5. Then it's high enough to dump directly into a mortar box — which gives your back a rest.
- Keep an extra mortar box and a damp cloth handy. They'll hold a spare batch of mixed and ready-to-use mortar. Just empty the mixer into the mortar box and cover it with the damp cloth. That spare batch will let you get ahead of the crews spreading mortar. And because you're ahead by a batch, there's time to clean the mixer without holding up progress.

Make a habit of always cleaning out your mixer at the end of the day. It only takes a few minutes to do and it adds years to the useful life of your mixer.

Retempering

Fast-paced construction calls for mortar that takes its first set quickly. That way there's no delay in the work. At the same time, mortar must take a set slowly enough for you to mix and use fairly large batches before it starts to set and stiffen. Mortar has to be wet and plastic or it won't bond well.

Masons deal with this by *retempering* the mortar. To retemper mortar you just mix in more water as you need it. It's OK to do this any time during the first 2½ hours after you mix the mortar. As long as you stay inside that time limit, the retempering won't weaken the mortar. But you can't retemper mortar after it sits more than 2½ hours. And don't even think about retempering mortar that sat overnight.

How long a mortar stays plastic without retempering depends on several things:

- how fast-setting a cement you use
- the temperature of the mortar
- the air temperature
- the temperature of the water in the mortar
- amount of moisture in the air (humidity)
- amount and speed of wind that blows on the mortar
- amount of direct sunlight that reaches the mortar

Mortar stays plastic longest on days that are cool, humid, overcast and windless. Mortar loses its plasticity fastest on hot, dry, sunny and windy days. If you use water that's warmer than 80 degrees F, the mortar also loses plasticity quickly. Cover mortar you're not using with wet burlap to cut down on water loss from evaporation.

Figure 3-6
Insufficient mortar in collar joint

Figure 3-7
Furrowed bed joint

How to Place Mortar

Good workmanship is important in masonry work because nearly everything's done by hand. Every single block, brick and trowel full of mortar is placed one at a time. This makes masonry different from most other modern methods of building. It's the least mechanized of all.

It's not easy to find clear signs of poor workmanship except during construction. They don't show in the finished work — until they cause a failure. The most common flaws are bed, head or collar joints that aren't complete, as shown in Figure 3-6. In the rest of this section we'll look at both the quick and dirty way, and the correct way, to make bed and head joints.

Bed and Head Joints

Furrowed bed joints are very common. Figure 3-7 shows what one looks like. It's a method many bricklayers use. Laying bricks is easier with a furrowed joint because mortar's spread outward to both sides of the wall. But furrowed bed joints aren't a good idea. It's too easy to end up with an invisible void right smack under the center of the brick. Nobody, including the mason, knows that the pocket's there. Sometimes the project architect will prohibit furrowed bed joints.

The correct and safe way to make bed joints is to spread them. Figure 3-8 shows this kind of bed joint. First spread the mortar flat, then make a shallow end-to-end slice with the tip of the trowel. This spreads the mortar from side-to-side without leaving deep hollows in the mortar the way furrowing does. There's no chance you'll leave a void behind.

Unfilled bed joints are likely to crack at the *bond point* (where the mortar and the masonry join). See Figure 3-9. Cracks here are an open invitation to water and the resulting water damage. If you're lucky, the damage will be limited to efflorescence. You might also hear this called *whiskering* or *saltpetering*. This is a serious problem for mortar so it's covered fully later in this chapter.

Figure 3-8
Slicing mortar in spread bed joint

Figure 3-9
Cracked bed joints in fluted blocks

Figure 3-10
A clip joint

Figure 3-11
Full end joint

Masonry that goes up in a big rush often has poor head joints. The masons on the job are under nonstop pressure to work faster, and the usual result is a lot of *clip head joints*. Figure 3-10 shows a clip head joint. You can see this joint doesn't stand much chance of ending up more than half filled. And that's being generous. You can't see this flaw from the front side. If you trowel mortar onto the end of the brick after you lay the brick into the wall, you'll get a clip head joint. Don't let anyone use this type of head joint on a job you're in charge of.

A clip head joint leaves a passage for water to get into a wall. This can lead to efflorescence which in turn can lead to *spalling*. Lots of small flakes, chips or scales (called spalls) form and they break off easily.

Pick and Dip Joints

Figure 3-11 shows a brick with mortar for an end joint applied the correct way. This method makes a fully-filled, solid, waterproof head joint. Masons call it the *pick and dip* and it's very simple. Here's how it works:

1) Spread mortar for the bed joint. *Buttering* is another name for this.
2) Pick up the brick and spread or dip the mortar for the head joint on its end.
3) Lay the brick into the wall.

Special Conditions

Masonry cements are made to work best in average conditions. That's fine if the weather cooperates twelve months a year. You don't work in that paradise? Then you'll need some of the tips and tricks we're going to cover next. You can use them to mix quality mortar that forms waterproof, full-strength bonds even when conditions are marginal.

Hot Weather

Mortar, like concrete, needs water to hydrate completely. The reasons for, and results of, low hydration are much the same for mortar and concrete. We covered both points in detail in Chapter 2. But the solutions and precautions you use for mortar are different. And that's what we'll focus on now.

First, you can use evaporation to slow down the rate that mortar loses water. Evaporation cools curing mortar just like evaporating sweat cools the human body. Cool mortar loses less water.

But don't run out and hose everything down just yet. You don't want to get the masonry cement or the masonry units wet. Wet brick and block cause mortar stains that you won't be able to completely clean off or cover up. Wet masonry units also spoil the bond because they don't react the same way with mortar. When they're dry, they soak up a little bit of water and some dissolved masonry cement from the mortar. This produces suction that makes the mortar and the masonry unit grip each other. This grip helps the mortar form strong, waterproof bonds.

But there are some things that you *do* want to get wet. Wet the equipment, such as wheelbarrows, mixing tubs, and mechanical mixers, mortar boxes and mortar boards. Use wood boxes and boards because they absorb water and the cooling effect lasts longer. Wet the whole wall after it's finished and has its first set. That cools the entire wall and keeps the mortar from taking its final set too soon.

Cold Weather

Cold weather slows down hydration in mortar just as it does in concrete. But that's not the big problem when you're working with mortar in cold weather. The big problem is the risk that the mortar, or the whole wall, will freeze. Mortar that freezes before it's taken a set has no strength at all. The work's wasted, so masons don't usually go on working when it gets that cold. Figure 3-12 provides some generally-accepted recommendations for mixing mortar and protecting newly-laid masonry construction in cold weather.

Mortar won't really start to freeze until about 28 degrees F. That's because the chemical reactions in mortar create some heat. An experienced mason can tell from the way mortar feels and spreads when it's about to freeze. Some of the signs are:

- ❖ mortar that's hard to spread
- ❖ bricks or blocks that feel like they're sliding in the mortar
- ❖ ice crystals that are big enough to see

Masons can keep on working when it's this cold. But it's not cheap to do and everyone must be careful and stay focused on the work. Just one mistake can make it all worthless.

Heating the Work Space and Materials

The best way to protect masonry work from freezing is to enclose the whole work area. Then heat the enclosed space to the right temperature and keep it there until the mortar has taken a set. Usually you'll also have to heat the water and the sand before you use them in mortar. Be very careful not to get the temperature too high. Chemical reactions happen faster at high temperatures and mortar's very likely to *flash-set* (set up

<table>
<tr><th>Work day temperature</th><th>Construction requirement</th><th>Protection requirement</th></tr>
<tr><td>Above 40° F</td><td>Normal masonry procedure</td><td>Cover walls with plastic or canvas at end of work day to prevent water entering masonry.</td></tr>
<tr><td>40 to 32° F</td><td>Heat mixing water to produce mortar temperatures between 40° to 120° F.</td><td>Cover walls and materials to prevent wetting and freezing. Covers should be plastic or canvas.</td></tr>
<tr><td>32 to 25° F</td><td>Heat mixing water and sand to produce mortar temperatures between 40° and 120° F.</td><td rowspan="2">With wind velocities over 15 mph, provide windbreaks during the work day and cover the walls and materials at the end of the work day to prevent wetting and freezing. Maintain masonry above freezing for 16 hours using auxiliary heat or insulated blankets.</td></tr>
<tr><td>25 to 20° F</td><td>Mortar on boards should be maintained above 40° F.</td></tr>
<tr><td>20 to 0° F and below</td><td>Heat mixing water and sand to produce mortar temperatures between 40° and 120° F.</td><td>Provide enclosures and supply sufficient heat to maintain masonry enclosure above 32° F for 24 hours.</td></tr>
</table>

Figure 3-12
Cold weather masonry construction and protection

suddenly). Retempering won't help at all so there's no point even trying it. Flash-set mortar is weak because it takes a set before it can hydrate and develop strength and a good bond.

Don't forget the temperature of the brick or block when you're warming materials. If you don't warm them also, you risk having the whole wall flash-freeze.

Using Additives

If you must work in cold weather, use mortar that's made with air-entrained Type III portland cement. It sets up faster. Another way that you can go on working in cold weather is by using an accelerator in the mortar. It works like an antifreeze. Accelerators shorten the set time of mortar. Using them isn't risk- or problem-free. So let's take a look at the biggest pitfalls you face.

You may recall from Chapter 2 that accelerators are mostly calcium chloride. This is a salt, and that's a problem. Salts make metal corrode. Don't use an accelerator in mortar for reinforced masonry work. The salt in the accelerator will make the rebar corrode and become weak. Salts can also cause efflorescence on masonry (if they're able to combine with water). Hairline cracks in mortar will pull water into the wall through the mortar. Salt in the mortar, from the accelerator, dissolves on contact with the water.

And there's another problem. These admixtures are expensive, and they aren't easy to use. Accelerators must be measured carefully. Too much accelerator makes mortar weak. Too little and the mortar will freeze anyway. Also, accelerators only work if they're proportioned correctly. Normally you add accelerator equal to 2 percent of the mix's total weight.

Preventing Efflorescence

This is the biggest problem that mortar causes in masonry work. Efflorescence shows up on masonry surfaces when two things are present. The first is salts that dissolve in water. There are at least ten different chemical combinations that cause efflorescence. The mortar is almost always the source of the salt. The second is water or some kind of moisture in the masonry work. Water usually gets in through cracks in mortar.

Here are some steps you can take that help prevent efflorescence:

- ❖ Use only clean water — no salt or brackish water.
- ❖ Use only washed sand.
- ❖ Store materials in a dry, covered place.
- ❖ Protect work during construction with a waterproof cover. Place it over the work at the end of each day and anytime it might get wet.

Estimating Mortar

The cost of mortar is one of the biggest expenses in masonry construction. So you need to know how many bricks or blocks you can lay with a certain amount of mortar.

It's easy to estimate mortar materials for hollow core unit masonry because you only use mortar on the outer edges of these blocks. This is called *face shell bedding*. I use the same method for estimating solid core masonry. But first you need to calculate how many bricks or blocks you'll use.

Rule-of-Thumb Estimating for Hollow Core Unit Masonry

Just follow these steps:

1) Divide the total number of blocks by 30 to get the number of bags of mortar cement you need. One 70-pound bag of mortar cement is enough to lay 30 blocks.
2) Divide the number of bags of mortar cement by 8 to get the tons of sand you need. One ton of sand is enough to mix eight 70-pound bags of mortar cement.
3) The only ingredient left to estimate is water. Multiply the total number of bags of mortar cement by 5. Five gallons of water will mix one 70-pound bag of mortar cement. Of course, you don't need to estimate water if it's readily available at no cost. We've included water to be complete.

Concrete or clay masonry unit size (inches)	Wall thickness (inches)	Mortar (CF)	Masonry cement (bags)	Sand (CF)
5 x 8 x 12	8	1.8	0.6	1.8
4 x 8 x 16	4	2.4	0.8	2.3
8 x 8 x 16	8	2.4	0.8	2.3
10 x 8 x 16	10	2.4	0.8	2.3
12 x 8 x 16	12	2.4	0.8	2.3

Figure 3-13

Materials for mortar for 100 masonry units (1:3 mix, 3/8-inch joint)

Now let's try this out with some numbers for a typical job using 1,500 blocks. To figure out how much mortar you need:

1) Divide 1,500 by 30 to get the number of 70-pound bags of mortar cement: 1,500 ÷ 30, or 50 bags.
2) Divide the 50 bags of mortar cement by 8 to get the tons of sand: 50 ÷ 8, or 6.25 (round off to 6.3) tons. Since 6.3 is greater than 3, modify the estimate by dividing 6.3 by 3 to get 2.1. Now subtract 1 from the 2.1 to get 1.1 and multiply by 0.5 to get 0.55. Round that off to 0.6 and add to the original number 6.3 to get 6.9. Round that off to 7 tons. This is the modified final estimate for sand, 7 tons.
3) Multiply the total number of bags of mortar cement by 5 to find out how many gallons of water you'll need for mixing: 50 × 5, or 250 gallons.

Modifiers for Rule-of-Thumb Mortar Estimates

Here are some special situations where you may want to change your estimate.

- ❖ If your mason is a real master, you can allow less for waste.
- ❖ If you're storing sand on tarps to conserve it, you'll need less.
- ❖ If you save the leftovers from opened bags of mortar cement, you'll need fewer bags. This only works if you keep it completely dry. Use resealable, waterproof buckets, like the ones drywalling compound comes in. If you don't store mortar cement properly, it'll get lumpy and then it's useless.
- ❖ If you're working with 6-inch block, overestimate all the materials so you'll have extra mortar. You're likely to need it since 6-inch blocks often have thinner faces.

Estimating Mortar Materials Per 100 Units

Figure 3-13 lists how much of each material you'll need to make enough mortar to lay 100 units of some common sizes of block.

Here's how you can use this information. Let's say the blocks are 10" × 8" × 16", mortar is a 1:3 mix and joints are 3/8 inch. You need a total of 500 blocks.

Using Figure 3-13 you know how much mortar you need to lay 100 of these blocks (2.4 CF). And how much masonry cement (0.8 bags) and sand (2.3 CF) makes that much mortar. Multiply these amounts by 5 to find out how much material you need for 500 blocks:

- ❖ total amount of mortar: 2.4 CF × 5 = 12 CF
- ❖ masonry cement: 0.8 bags × 5 = 4 bags
- ❖ sand: 2.3 CF × 5 = 11.5 CF sand

Concrete or clay masonry unit size (inches)	Wall thickness (inches)	Number of masonry units	Mortar (CF)	Masonry cement (bags)	Sand (CF)
5 x 8 x 12	8	220	3.7	1.2	3.7
4 x 8 x 16	4	110	2.6	0.9	2.6
8 x 8 x 16	8	110	2.6	0.9	2.6
10 x 8 x 16	10	110	2.6	0.9	2.6
12 x 8 x 16	12	110	2.6	0.9	2.6

Figure 3-14

Materials for mortar for 100 SF wall (1:3 mix, 3/8-inch joint)

If you're using a 1:2.5 mix instead of a 1:3 mix:

1) Multiply the amount of masonry cement (4 bags) by the factor 1.1553.
2) Multiply the result (4.6 bags) by 2.5 for the amount of sand.

Note: Be careful if you're using damp sand. More moisture in sand makes it take up more space because the water displaces sand. In other words, 1 CF of damp sand has less sand (and more water) than 1 CF of dry sand.

Estimating Mortar Materials per 100 Square Feet of Wall

Use Figure 3-14 if you want to estimate masonry by the 100 square feet of wall instead of by the number of masonry units. There's no allowance for waste included in either Figure 3-13 or Figure 3-14.

4

Sizing and Building Concrete Footings

❖❖

In Chapter 1 you studied the soil at your job site. You identified the soil type. You found the soil's loadbearing capacity and checked the site's drainage. You need all of that information to find the load on a footing. Finding the load takes some basic math and some data from a few tables. I think a bit of explaining helps too. Let's get started by looking at just what I mean by *load* — what it is, how many kinds there are, and how you estimate and measure it.

Footings and Load

Load is a physical force. Some kinds of load you may find easier to think of as weight. That's fine as long as you're a bit careful. Always remember that load isn't really just weight. It's a description of how a physical force, that may or may not have weight, is supported by the different parts of a building. This difference between weight and load is shown by the unit that load is measured in — pounds per square foot. From now on we'll abbreviate that as *lb/SF*.

Many parts of buildings carry or support loads of one sort or another. For now we're concerned with the following loadbearing members:

- ❖ footings
- ❖ foundations
- ❖ piers
- ❖ columns
- ❖ beams
- ❖ walls

There are many kinds of loads. For now, we'll only look at the three most basic ones. *Dead load* is the total amount of load that comes from the building itself. Think of it as the weight of the floors, walls, roof and foundation all added together. *Live load* is the total amount of load from whatever is on or inside the building. This includes furnishings, permanent machinery (such as HVAC equipment) and the people who use the building. *Design load* is the dead and live load added together. These three kinds of loads are universal — they're always there, in any structure.

Before you can find out how much load a footing carries, you need to know what kind of footing it is. That depends on the type of foundation. You use footings with pier foundations, continuous wall foundations or foundations that combine continuous walls and piers. You don't calculate load on a pier footing the same way as load on a continuous wall footing. This is true even when they're combined in one foundation.

We'll approach this as a four-step process:

1) Find the sources of load.

2) Estimate the amount of dead and live load.

3) Combine dead and live loads to find the design load for different kinds of footings.

4) Match the size of a footing to the load it carries.

Figure 4-1 shows a section view of the house we'll use as the sample house in the three following examples. The section view is of half the house, and we'll use the dimensions 28 feet by 20 feet (half of 28 feet by 40 feet)

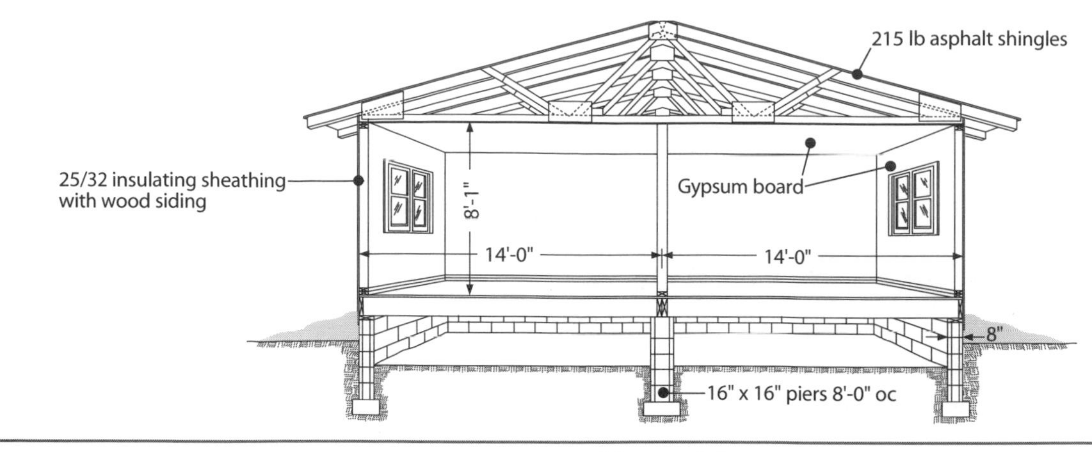

Figure 4-1
Sample house, one-story, truss roof

Source of load	Pounds live load/SF
First floor	40
Second floor	30
Attic floor (storage only, not habitable)	20
Roofs (general)	20 to 40

Figure 4-2
Live loads estimates for residential buildings

for our calculations. The sample house has a foundation that combines a continuous wall on the perimeter with piers down the center.

The first thing to do is to find out how much load of each kind we're working with here. Figures 4-2 and 4-3 list some average live and dead load values. We'll use these averages for our load estimates.

Calculating Loads on Pier Footings

This is the formula for finding the design load on a pier footing:

$$\textit{Design load} = \textit{Total dead load} + \frac{\textit{Total live load}}{2}$$

You reduce the live load by half because the sample building is a house. The whole live load in a house never ends up all on one pier. But if the footings are for a storage building, such as a warehouse, you use the full live load. In that case, one pier could end up carrying all or most of the load.

Follow these steps to find half of the live load:

1) Find how much floor area a pier carries (the area supported by one pier). Figure 4-1 shows that a pier supports a floor area that's 28 feet wide by 20 feet deep. The area is 20 × 28 = 560 square feet.

2) Now, using Figure 4-2, find all the sources of live load that apply to the sample house. In this one-story house, the only live load is from the first floor. The walls in a mixed foundation carry all the roof load (live and dead).

3) Total all the load values that fit the sample house. The answer is 40 lb/SF. (Yes, we could have skipped these last two steps — this time. But you won't always have it so easy.)

4) Now we're ready to find the live load carried by a pier. It's the area (560 SF) multiplied by the amount of estimated live load (40 lb/SF). The answer is 22,400 pounds (560 × 40 = 22,400).

5) To find half of the live load, divide the full load, 22,400 pounds, by 2. The answer is 11,200 pounds (22,400 ÷ 2 = 11,200).

6) Since there are two piers under each half of the house, we need to divide 11,200 by two to get the live load on each pier. Each pier will support 5,600 pounds (11,200 ÷ 2 = 5,600).

Source of load	Dead load (lb/SF)
Roofs: Gable, sheathed with ¾" boards, supported 2' on center, 15 pound felt, 215 pound asphalt shingles	7
Add to gable roof load for:	
Asphalt shingles	3
Built-up roof	5
Slate	7
Walls: Stud framing, plates and sills	
2 x 4s, 16" on center	2
Stud wall, plastered on one side	10
Stud wall, sheathed with wood siding	7
Stud wall, ⅜" drywall on both sides	6
Brick veneer	40
Concrete block, 8", lightweight aggregate	38
Concrete block, 8", normal aggregate	85
Floors: 2 x 10s, 16" on center	7
Miscellaneous: Concrete or stone (per inch thick)	12

Figure 4-3
Dead load estimates for residential buildings

Calculating Loads on Wall Footings

Because the sample house has a trussed roof, there's no dead load on the pier or floor. However, there is a dead load on the walls. Using the information in Figure 4-3, we can estimate the dead load on the walls as follows:

roof	7 lbs
shingles	3 lbs
studwall and wood siding	7 lbs
floor joists	7 lbs
foundation wall (8" concrete block)	85 lbs
	109 lbs/SF

We figured that our section view sample house has 560 square feet. We now need to multiply 560 by 109 pounds to come up with the total dead load on the walls. The answer is 61,040 pounds (560 × 109 = 61,040).

You can now calculate the design load using the formula:

$$Design\ load = Total\ dead\ load + \frac{Total\ live\ load}{2}$$

$$Design\ load = 61{,}040 + \frac{22{,}400}{2}$$

$$Design\ load = 61{,}040 + 11{,}200$$

$$Design\ load = 72{,}240$$

To calculate the load per lineal foot on a footing, you must first find all the sources of load on the footing and the dead load of each source. For our example, we'll use the dimensions given for the house section in Figure 4-1 and the load figures from Figure 4-3.

First we need to find the total lineal feet of the footing for the house section. The section shows two sides that are 20 feet deep and a width of 28 feet. That's 20 + 20 + 28 = 68 lineal feet of footing. Now we need to multiply 68 feet by the bearing area of the footing. The footing width for the sample house section is 16 inches. If you look at Figure 4-4, you'll see that the bearing area for a 16-inch footing is 1.33. Multiply the lineal feet by the bearing area to get the total square feet of bearing area (68 × 1.33 = 90.4).

We know from our calculations for the design load for a pier footing that the total dead load on the walls is 61,040 pounds. You divide the total dead load by the total square feet of bearing area to get the total dead load per lineal foot of wall footing (61,040 ÷ 90.4 = 675.2).

Now we need to calculate the live load per lineal foot. Again, we can use the same calculations that we used for the pier footing. The total live load is 22,400 pounds. Divide the live load by the total square feet of bearing area to get the total live load per lineal foot of wall footing (22,400 ÷ 90.4 = 247.8).

To calculate the total design load per foot of wall footing, add the total dead load per foot to the total live load per foot.

$$Design\ load = Total\ dead\ load + Total\ live\ load$$

$$Design\ load = 675.2 + 247.8$$

$$Design\ load = 923$$

Width of footing (inches)	Bearing area of footing (SF)	Soil's loadbearing capacity (lb/SF)					
		1,000	2,000	3,000	4,000	6,000	8,000
16	1.33	1,330	2,670	4,000	5,330	8,000	10,670
18	1.50	1,500	3,000	4,500	6,000	9,000	12,000
20	1.67	1,670	3,335	5,000	6,670	10,000	13,340
22	1.83	1,835	3,665	5,500	7,335	11,000	14,665
24	2.00	2,000	4,000	6,000	8,000	12,000	16,000

Figure 4-4
Safe loads for wall footings

Now you're set to move on to the next step — finding the right dimensions (width, length, and thickness) for these footings. The design loads you just found play an important part in this.

Sizing Wall and Pier Footings

Here's a rule of thumb for sizing footings. For small projects, footings should be twice as wide as the finished wall and as thick as the finished wall is wide. For large projects, don't rely on any rules. Have the footings designed by an engineer.

To plan footings you need to know:

- ❖ load bearing capacity of the soil
- ❖ wall footing design load
- ❖ pier footing live load

You're now ready to start running through the examples. In all three, the soil's loadbearing capacities are random choices taken from Figure 1-5 in Chapter 1.

Example 1

Here's what you already know:

- ❖ load bearing capacity of the soil is 1,000 lb/SF
- ❖ wall footing design load is 923 lb
- ❖ pier live load is 5,600 lb

Sizing the Wall Footings

First let's find the correct width for this wall footing. To do that, use Figure 4-4 and follow these easy steps.

1) Under the heading *Soil's loadbearing capacity (lb/SF)*, find the column labeled *1,000.*
2) Read down that column. Find the number that's closest to, but not less than, the wall footing's design load of 923 lb/LF. Since the footing's design load is less than the minimum safe load of 1,330 lb/SF, we can use 1,330 lb/SF.
3) Staying in that row, read across to the far left column under *Width of footing* to find the correct width of the wall footing to use. The answer is 16 inches. Most building codes will specify a minimum wall width of 8 inches and a minimum footing width of 16 inches.

What about the length of the wall footing? Until now, we've been working with the section view of the sample house, using only half the length of the house. Since the house is actually 28 feet by 40 feet, let's use the full dimensions for the calculations in the next examples. So, we have two long sides of 40 feet and two short sides of 28 feet.

1) Add the four measurements together:
40' + 40' + 28' + 28' = 136'
2) Convert 136 feet to inches by multiplying by 12:
136' × 12 = 1,632"

You now know two dimensions for these footings: their width and their length. The last step is finding the right thickness. There's a quick way to do this that also works for finding pier footing thickness. Just follow these rules of thumb:

- ❖ Make all unreinforced footings at least 8 inches thick.
- ❖ If the footing projects more than 4 inches beyond its wall or pier, increase the footing's thickness by at least 1.5 times the amount of extra projection.

Let's try this out on the sample house's wall footing:

1) You start by finding out how much wider the footing is than the wall. That's easy. Just subtract the wall width, 8 inches, from the footing width, 16 inches.
   16" – 8" = 8"

2) You find the amount of projection per side by dividing the answer, 8 inches, by 2.
   8" ÷ 2 = 4"

3) Since the projection isn't over 4 inches, we can use the minimum footing thickness of 8 inches.

Sizing the Pier Footings

Since pier footings are square, finding the right size for them is easy. All four sides are the same size. Use Figure 4-5 to find the right size. You already know the soil's loadbearing capacity is 1,000 lb/SF, and the pier footing live load is 5,600 lb/SF. So let's see how this works:

1) Under the heading *Soil's loadbearing capacity*, find the column labeled *1,000*. Read down that column. Find the number that's closest to, but not less than, the pier footing live load. The answer is 6,250 lb/SF.

2) Staying in that row, read across all the way back to the far left column, *Footing size*. Read the per-side measurement to use for the pier footings: 30 inches per side.

All you need to do now is find out how thick to make these pier footings. To do that, you use the same rules of thumb as you did to find the thickness for the wall footings. Here goes:

1) To find how far the footing sticks out beyond the pier, just subtract the pier width, 16 inches, from the footing width, 30 inches, to get 14 inches.

Footing size, per side (in.)	Bearing area of footing* (SF)	Soil's loadbearing capacity (lb/SF)					
		1,000	2,000	3,000	4,000	6,000	8,000
16	1.77	1,780	3,560	5,340	7,120	10,680	14,240
18	2.25	2,250	4,500	6,750	9,000	13,500	18,000
20	2.78	2,780	5,560	8,340	11,120	16,680	22,240
22	3.37	3,370	6,740	10,110	13,480	20,220	26,960
24	4.00	4,000	8,000	12,000	16,000	24,000	32,000
27	5.06	5,060	10,120	15,180	20,250	30,370	40,500
30	6.25	6,250	12,500	18,750	25,000	37,500	—
33	7.55	7,550	15,120	22,680	30,240	45,360	—
36	9.00	9,000	18,000	27,000	36,000	—	—
39	10.56	10,560	21,120	31,680	42,240	—	—
42	12.25	12,250	24,500	36,750	—	—	—

* To find the bearing area of a pier footing:
[width x length = area in square inches] ÷ 144 inches/SF = area in SF.

Figure 4-5
Safe loads for pier footings

2) To find the per-side projection, divide the total, 14 inches, by 2 to get 7 inches.

3) This answer, 7 inches, is more than the allowed maximum, 4 inches, that lets you use an 8-inch-thick footing. Now what do you do?

You apply part two of the rules of thumb for finding footing thickness. Part two said: If the footing projects more than 4 inches beyond its wall or pier, increase the footing's thickness by at least 1.5 times the amount of extra projection. Let's try it out:

1) Start by finding the amount of extra projection. To do that, just subtract the allowance, 4 inches, from your per-side projection, 7 inches, to get 3 inches.

2) To find how much to increase the footing thickness, multiply the extra projection amount, 3 inches, by the factor given in the rules of thumb, 1.5, to get 4.5 inches.

3) You finish off by adding the increase, 4.5 inches, to the minimum thickness, 8 inches, to get 12.5 inches, or 12½ inches.

Example 2

In this example the only change I'm making is in the type of soil at the job site. This time it's an ordinary soil of mixed clay and sand. Figure 1-5 in Chapter 1 gives the loadbearing capacity as 3,000 lb/SF.

There are a few shortcuts you can take in this example. You can recycle the design load data. Changing the soil's loadbearing capacity has no effect on these figures. Those loads are: 923 lb/LF for the wall footings, and 5,600 lb/SF for the pier footings. Another piece of data that's unchanged from Example 1 is the length of the wall footing. As we saw in Example 1, this is the same as the measurement of the sample house's perimeter, 1632 inches, or 136 feet.

Sizing the Wall Footings

You already know the length of the wall footing, so the first step is finding the correct width for the wall footing. Use Figure 4-4, following the same steps as in Example 1, to match the new loadbearing capacity of the soil with the design load.

1) Find the soil's loadbearing capacity: 3,000 lb/SF.

2) Read down that column, and find the number that's closest to, but not less than, the design load: 923 lb/LF. The answer is 4,000 lb/LF.

3) Read back across that row to the far left column to find the width for your wall footing. The answer is 16 inches. According to the rule of thumb, the footing has to be twice the width of the wall. 16 inches is twice the width of our 8-inch wall.

Balancing Footing Loads

Footings that don't have balanced loads aren't safe or stable. Balancing loads isn't hard to do. It's just common sense. Think of the load balance between the wall and the pier footings as being like a children's teeter-totter. Add or subtract something on one side and you upset the balance. All you do to restore the balance is make an equal change on the other side.

Example 3

This time let's make a small design change to the sample house in Figure 4-1. Your clients sent you a set of blueprints for a small building. It's almost the same as the sample house, but with one difference. They want a foundation that's entirely on piers. The contract you sign specifies that the foundation shall consist of exactly 22 piers as shown in Figure 4-6.

Sizing the Pier Footings

The most important thing to consider with this example is the total area of the perimeter footings' loadbearing surface. If the area of loadbearing surface is the same, there's no difference between the 18 perimeter pier footings and one continuous wall footing.

Start by finding the total area of loadbearing surface for a continuous wall footing. After you know that area, divide it evenly among the required number of pier footings. Let's take a little shortcut in finding the total area. We'll do that by recycling the soil loadbearing capacity from Example 1, which is 1,000 lb/SF. That means you already know width, 16 inches, and length, 136 feet, for this footing. Let's go on from there:

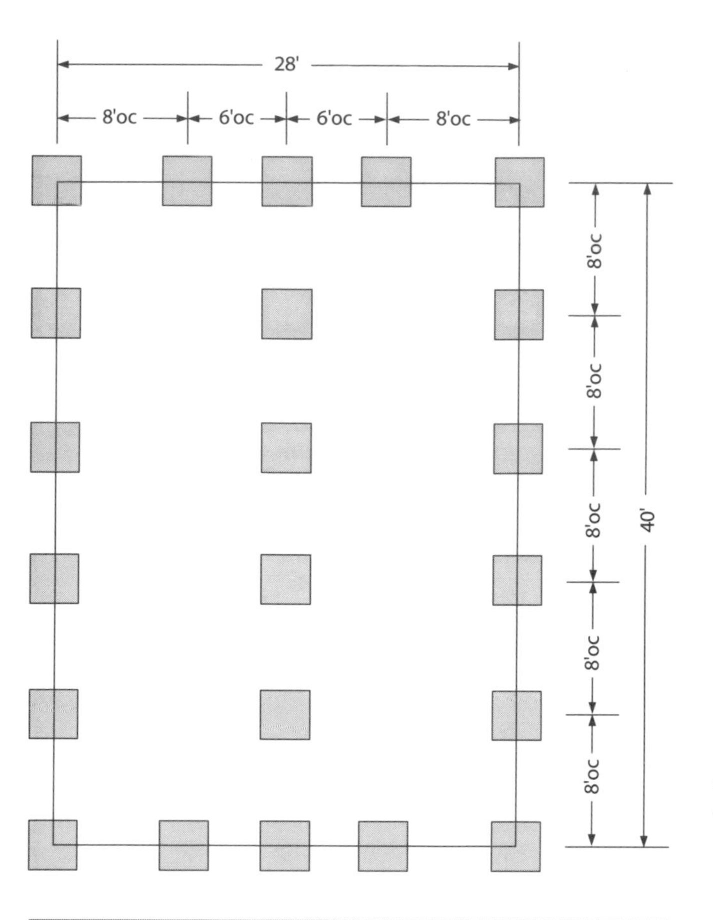

Figure 4-6
Location of piers for sample building

1) Change the width, 16 inches, into feet to get 1.33 feet.
2) Find the area of the loadbearing surface: 136 feet × 1.33 = 180.88 SF. Round that up to 181 SF.
3) Divide the total loadbearing area by the number of pier footings: 181 ÷ 18, or 10 SF per pier footing.
4) To turn that into a per-side measurement, find the square root of 10. The answer is 3.2 feet per side, or 38.4 inches on a side.

So each pier footing will be 38.4 or 38²/₅ inches on each side. But how thick will they be?

1) The pier footing width is 38.4 inches minus the pier width of 16 inches, or 22.4 inches.
2) Find the per-side projection by dividing the total, 22.4 inches, by 2 to get 11.2 inches.

An 11.2-inch projection is more than the rule of thumb allows for a footing of the minimum thickness, 8 inches. Your next step is finding out how much thicker to make the pier footings:

1) Subtract the 4 inches per-side allowance from the real amount, 11.2 inches per side, to get 7.2 inches per side.
2) Find how much thicker to make the footings by multiplying the extra amount of footing, 7.2 inches per side, by 1.5 to get 10.8 inches.
3) The total thickness for these footings is the sum of the minimum thickness 8 inches, and the extra thickness, 10.8 inches, for a total of 18.8 or 18⁴/₅ inches.

The dimensions of the perimeter pier footings are:

38.4" x 38.4" x 18.8"
or
38²/₅" x 38²/₅" x 18⁴/₅"

Now, what about the nonperimeter footings? A rule of thumb is to make all the footings on the nonperimeter piers the same as those on the perimeter.

Forms for Concrete Footings

Always oil or coat your forms before pouring concrete into them to make it easier to strip the forms from the hardened concrete. You can buy prepared coatings for forms that are also surface treatments for the concrete. They help cure, harden or seal concrete and make it easy to remove the forms. But you don't need dual-purpose coatings for footing forms. Just put a light coat of petroleum oil on footing forms. I prefer to use light-colored oils because they're cleaner so they're not likely to stain or discolor the concrete. Don't over-oil the forms — a light coating that sinks into the wood leaving the surface feeling slightly oily is just right. You don't want forms dripping with oil.

Here are some good financial reasons you should oil your forms:

❖ Removing unoiled forms from hardened concrete takes much longer. The forms won't easily break loose because they've bonded to the concrete. Your labor costs just went up.

Figure 4-7
Tying a fireplace footing to the wall footing with rebar

Rebar number	Diameter (in.)	Area (sq. in.)	Approx. weight of 100 ft. (lbs.)
2	1/4	0.05	17
3	3/8	0.11	38
4	1/2	0.20	67
5	5/8	0.31	104
6	3/4	0.44	150
7	7/8	0.60	204
8	1	0.79	267

Figure 4-8
Steel rebar data

- You'll usually want to remove forms after seven days, at the most. But concrete takes 28 days to reach full strength and hardness. Until concrete's at full strength and hardness, it's easily damaged by rough handling. Rough handling, unfortunately, is what it's going to take to get unoiled forms off the footing. You can count on having more than a few chips and deep scratches to repair by the time you've stripped the forms. Your labor and material costs just went up some more.
- In the process of getting the forms off the footing, you'll either destroy them completely or damage them badly. In either case it's unlikely that you'll be able to use the forms again. You'd counted on getting the normal two or three more uses out of those forms. Add the additional form material onto your job costs, and minus them from your profit.

Steel Reinforcement in Concrete Footings

Any concrete work is stronger when it's reinforced. Concrete reinforcing is either steel rod (rebar) or steel wire mesh. A reinforced concrete footing:

- is stronger
- isn't as likely to crack as it cures
- can be made thinner, so it weighs less
- lasts longer
- resists bending and flexing better

Also, footings for projecting features, such as fireplaces, can be tied into the main footing as shown in Figure 4-7. That makes them stronger and safer.

The project architect or structural engineer takes care of designing reinforced concrete footings. They'll give you plans for reinforced footings that have all the details clearly spelled out. Sometimes the plans include exact locations for the rebar within the footing. Be sure you follow the plans very carefully. Rebar placement usually comes right out of the building code. If you don't follow the plans right to the letter, your work won't comply with the code.

Placement of Rebar

Use two #4 bars for a wood-bearing partition in a one-story structure. Figure 4-8 matches rebar sizes (numbers) with measured diameters. It also gives the area and weight of 100 feet of rebar. For two stories, with either a wood-bearing partition or a nonbearing masonry partition, use two #5 bars. A separate foundation is required. See Figure 4-9.

Position rebar in the forms first. Then pour or cast concrete around the rebar. Be very sure that you have the rebar where you want it before pouring the con-

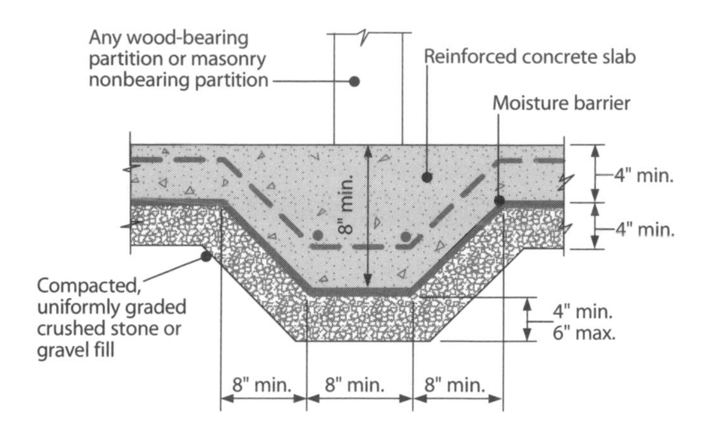

Figure 4-9
Masonry nonbearing partition on a separate footing

crete. It's almost impossible to move or remove rebar after concrete sets. You'll have to demolish the work and start over from scratch.

Figure 4-10 shows how to place rebar in a footing form. The boards that run across the forms and sit on top of them are called *spreaders*. The rebar is suspended from the spreaders by tie wires. The tie wires hold the rebar in position within the forms as the concrete's poured and as it cures and hardens. We'll cover tie wires in greater detail a little later.

Look again at Figure 4-10. Near the center and at the upper left you see two short pieces of rebar that run across the forms. They're tied to the other rebar, not suspended from the spreaders. These short pieces of rebar serve several important purposes. They keep the long rebar pieces separated from each other and keep them where they belong in the form. To comply with the building code, you need to be sure that the rebar stays where it's supposed to. But, during a pour, the moving mass of concrete picks up a lot of momentum. The cross-tied, short rebar pieces keep the flowing concrete from carrying or floating the rebar away or moving it out of place.

Whenever you need to connect two pieces of rebar, overlap the pieces by at least 12 inches. Use this overlap:

- to join two pieces and continue a straight run of rebar. See near the upper left in Figure 4-11.
- to turn a corner vertically and at the same time tie wall rebar together with the footing rebar. See Figure 4-11.
- to turn a corner horizontally, within the footing.
- to turn a horizontal corner as shown in Figure 4-12. The rebar from both sides of the footing overlap, forming the corner.

Tie Wires

Let's take a closer look at tie wires now. As you can see in Figure 4-10 or 4-11, it takes a lot of tie wires to hang rebar from spreaders. You can use a hand tying tool like the one shown in Figure 4-13. Use Figure 4-14 to find the correct lengths for tie wires.

Figure 4-10
Steel rebar suspended in footing form

Figure 4-11
Rebar splices showing the 12-inch overlap

Figure 4-12
Correct placement of rebar at footing corners

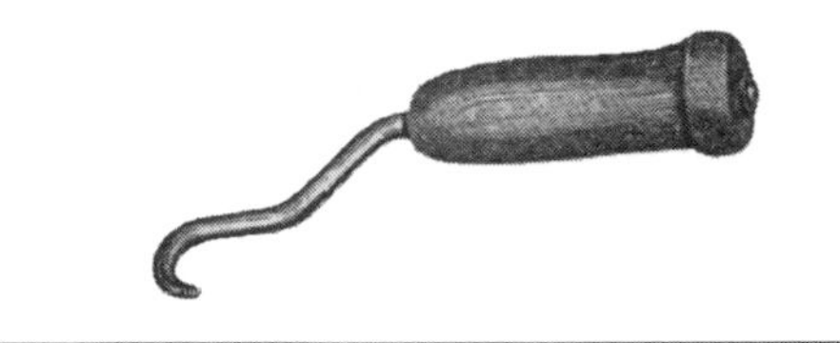

Figure 4-13
Hand tying tool

Using Wire Mesh

Sometimes you can use steel wire mesh reinforcing instead of rebar. This is a 6 x 6 gauge wire mesh that comes in rolls and in 5 x 10 foot panels, as shown in Figure 4-15. Because this form of reinforcing is easy to handle, I like to use the panels whenever I can. Be sure to check the building code before you order any wire mesh reinforcing. Some codes won't allow it. Also, be careful to follow the code's guidelines for overlap amounts at joints and corners.

Whether the reinforcing is wire mesh or rebar, store it in a clean dry place. Rebar or mesh coated with dust or rust won't form a strong bond with concrete.

Rebar in Stepped Footings

You use stepped footings and foundations on sloping job sites — a house with a partial basement that's built into a hillside, for example. Make the footings for stepped foundations level and plumb at the steps. Reinforce these footings with ½-inch (or larger) rebar. Place the rebar 8 inches on center for the whole width of the step. See Figure 4-16. We'll cover this sort of foundation in more depth in the next chapter.

Pouring Concrete Footings

Whenever you can, form and pour a footing at one time. That makes the footing a single unit. This also gives a footing some important advantages over footings that are poured in several parts. A single pour footing:

- is stronger
- is less likely to crack while curing and hardening
- saves you repeating the setup work
- saves you a second day's worth of charges for any rented equipment

Keyways for Continuous Wall Footings

The forms you use for continuous wall footings have a special feature that's shown in Figure 4-10. Running down the center of the form and attached to the underside of the spreaders is a *furring strip*. It forms an open U-shaped channel in the top of the concrete footing, called a *keyway*. The keyway, as we'll see in the next chapter, helps to interlock the footing with the wall built on top of it.

Rebar sizes	¼"	⅜"	½"	⅝"	¾"	⅞"	1"	1⅛"	1¼"
¼"	3½"	4"	4½"	5"	5½"	6½"	7"	7"	7½"
⅜"	4"	4½"	5"	5"	5½"	6½"	7"	7"	7½"
½"	4½"	5"	5"	5½"	6"	6½"	7½"	7½"	8"
⅝"	5"	5"	5½"	6"	6½"	7"	8"	8"	8½"
¾"	5½"	5½"	6"	6½"	6½"	7½"	8"	8½"	8½"
⅞"	6½"	6½"	6½"	7"	7½"	7½"	8½"	9"	9½"
1"	7"	7"	7½"	8"	8"	8½"	9"	9½"	10"
1⅛"	7"	7"	7½"	8"	8½"	9"	9½"	10"	10½"
1¼"	7½"	7½"	8"	8½"	8½"	9½"	10"	10½"	10½"

Figure 4-14
Length of tie wires for various rebar sizes

Figure 4-15
Panels of steel wire mesh reinforcing

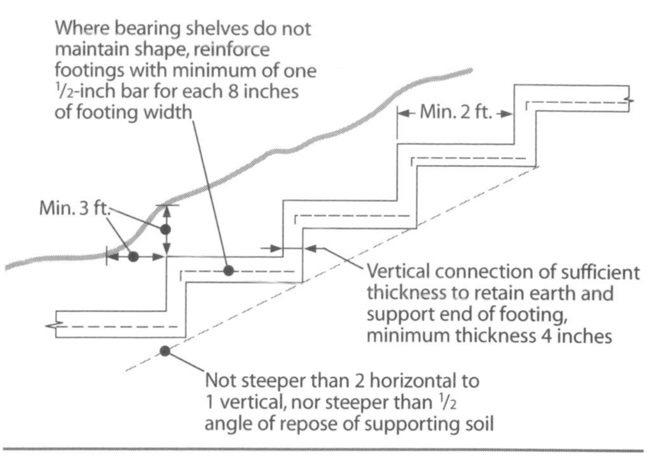

Figure 4-16
Reinforcement of stepped footings

After You Pour

After the pour and the finishing work are done, cover the fresh concrete footing to slow the curing process. Use a heavy, waterproof material, such as roofing paper, as shown in Figure 4-17. Roofing paper is sturdy enough to be reused. In the next chapter you'll see how to use it two more times.

Leave the forms on the footings as long as you can. They're an extra layer of protection and they help the concrete cure properly. The longer you delay stripping the forms, the better it is for the footings. Wait at least 24 hours before you remove the forms. Then wait at least three more days, or if possible seven days, before you put any load on the fresh-poured footings.

Figure 4-17
Fresh poured footings covered with roofing paper

5

Slab and Block Foundations and Retaining Walls

Foundations, like their footings, are designed to match the loads they must support and the soil that supports them. And while design is important, it's not everything. A poorly-built foundation, no matter how well-designed, is a worthless foundation. A foundation that's well-designed and well-built easily carries the load of the structure.

A sound foundation also resists other forces, including:

- soil movement caused by changes in moisture content
- frost heave (soil lifting up caused by the freeze/thaw/freeze cycle)
- overturn and uplift forces caused by the wind
- lateral force on below-grade parts caused by the surrounding soil

Foundations react to forces they can't resist. They crack, settle or shift. This is what's meant by foundation failure. A foundation that has failed is unstable, and the building it once supported is unsafe.

In this book you'll learn how to design and build strong, dependable foundations. Although there are many different kinds of foundations, only a small number are in general use. The others are custom work, which isn't typical. So we've focused on the kinds of foundations that masons build all the time. These foundation types are:

- continuous wall foundations (covered in Chapter 4)
- slab foundations
- stepped foundations
- basement wall
- grade beam and pier foundations (in the next chapter)

Before we start in with specifics for each different type of foundation, let's take a look at the materials used for foundations. They're not all the same. You'll see that some materials are used far more often than others. Common materials for building foundations are poured concrete and concrete block.

Most foundations in new residential construction are concrete block or poured concrete. Since the development of reinforced concrete construction, you rarely see brick foundations. This is even more true for fieldstone, which was once the preferred material for foundations. Stone foundations are so rare now that they could qualify as an endangered species.

Why have concrete block and poured concrete so completely taken over for foundations? The answer, in one word, is price. Concrete block and poured concrete are cheaper. Concrete (block or mix) costs less. That lowers your material cost. Building foundations with either concrete material is faster. Fewer labor hours needed per job lowers your labor cost. And masons working on concrete foundations don't need to be as

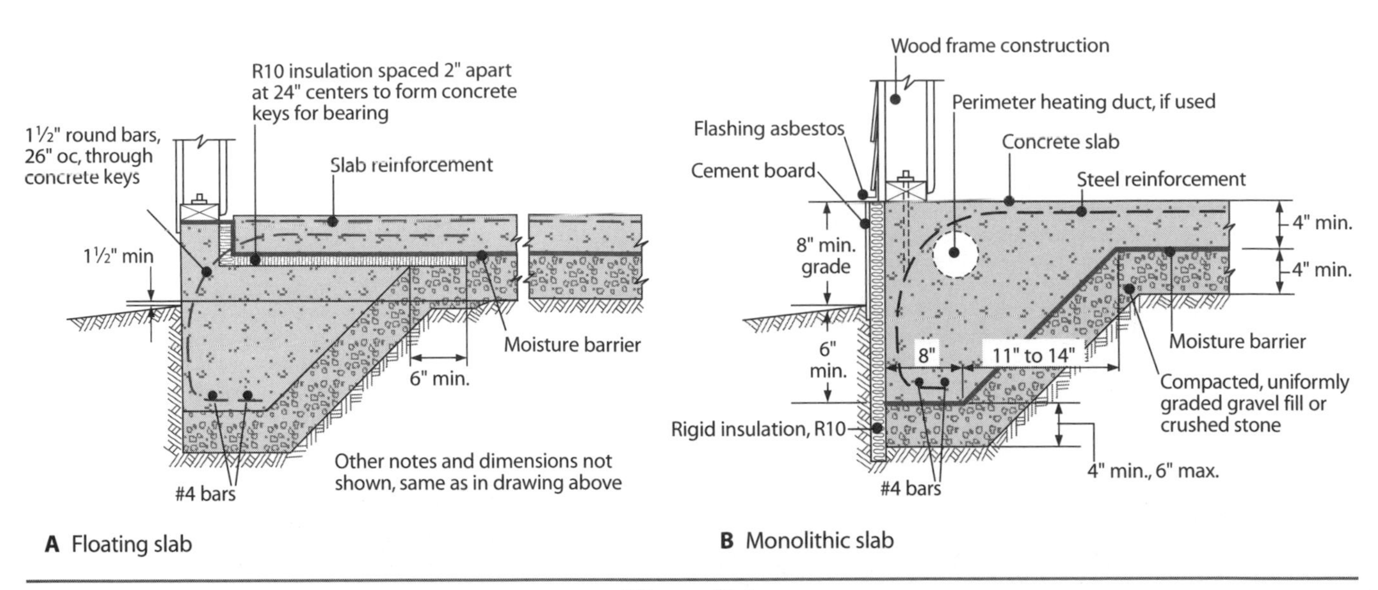

Figure 5-1
Slab foundations

experienced, as skilled, or as highly-paid, which also lowers your labor cost.

So let's start with the most common foundation in modern construction — the slab.

Slab Foundations

A slab foundation isn't the same thing as a slab floor. Here's a list of the most important differences.

- A slab foundation is cast as a single reinforced unit.
- A slab foundation is both foundation and floor.
- A slab floor is cast independently from the foundation.
- A slab floor is isolated from the foundation by a layer (or several layers) of rigid insulation.

You anchor a structure to a slab foundation using $^5/_8$-inch bolts that are set into the slab when it's cast. The weight of the slab itself is enough to anchor the building to the ground.

Slab foundation design, including which type to use and how to reinforce it, are complex decisions that involve many factors. The designer has to consider the weight of the structure, the type of soil beneath the slab, the drainage characteristics of the underlying soil and the climate. A structural engineer is trained to analyze all these variables, and more. To design a safe slab foundation you need this level of expertise and knowledge of the local soils. If you're not an expert, leave the design job to someone who is. This holds true even when the building is very lightweight. Do-it-yourself engineering doesn't belong in slab foundation design.

We'll be looking at two types of slab foundations in this section:

- *Floating slabs* are mostly used for single-story buildings. They have no separate footings, but do have insulation around the perimeter and under the slab. The floor to eave height of a building on this type of slab shouldn't be more than 12 feet. The floor to gable peak height of a structure on a floating slab foundation should be less than 20 feet. See Figure 5-1 A.
- *Monolithic slabs* are formed in a continuous pour with no construction joints, usually in large pans with thickened edges. They're only appropriate in warm climates. Even in mild climates, most building codes only allow you to use monolithic slabs for smaller structures such as garages. See Figure 5-1 B.

Preparing a Site for a Slab Foundation

Before you pour any slab foundation:

1) Compact the soil beneath the slab to prevent settling and cracking later.
2) If you add any fill, allow time for it to settle thoroughly and become as compacted as the undisturbed soil below.
3) Provide the slab with a well-drained subgrade by placing a 4- to 5-inch layer of gravel on top of the soil.
4) To protect the slab from frost damage, slope adjoining grades away from the slab in all directions. You'll further minimize the chance of frost heave damage if the subsoils are mostly sand or gravel.
5) Avoid silty sand or clay soils when possible.

If the water table is too high, you can either install footing drains or a sump and sump pump.

Concrete for Slab Foundations

For slab foundations, use concrete with a 28-day compressive strength of 2,000 psi. If there are further special requirements they'll be cited in your building code.

Reinforcing Slab Foundations

How much reinforcing you use, the way you distribute it in the foundation, and how you lap it all depend on the type of reinforcing material. Here are some things to remember when you use welded steel fabric reinforcing. They're based on ASTM's Standard Specifications for welded steel wire fabric for concrete reinforcement.

- ❖ Use structural grade or better materials.
- ❖ Evenly distribute the reinforcing material in both directions.
- ❖ Lap all pieces at least 6 inches at all edges.
- ❖ Use at least 20 pounds of reinforcing per 100 SF of concrete.

If you're working with plain or deformed bar or rod reinforcing:

- ❖ Use structural grade or better materials.
- ❖ Use bars or rods that are at least 1/2 inch in diameter, size No. 4.
- ❖ Place rebar at least 18 inches on center.
- ❖ Distribute reinforcing evenly from the slab's center to the thicker outside edges. See Figure 5-1.
- ❖ In general, plan to use at least 30 pounds of reinforcing per 100 square feet of concrete. For more specific requirements, check Section 1815.7 and Figure 18-III-1 in the UBC.

Waterproofing Slab Foundations

There are two types of widely-used moisture barrier materials. Both separate the concrete foundation slab from surrounding soil. Use at least 35-pound asphalt roofing felt or one of the many membrane-type barrier products. For either material, lap the edges at least 6 inches.

Insulating Slab Foundations

You should put a layer of insulation between the vapor barrier and the slab. The material generally used is an extruded polystyrene material called Formular Brand. Use the chart in Figure 5-2 to choose the right grade and thickness. Figure 5-3 shows the insulation in place.

Grade	Where used	Thickness available	R Value per inch
#1	Against foundation walls	1" 1½" 2"	R5
#2	Beneath slabs	1" 2"	R5

Figure 5-2
Insulation board for foundations

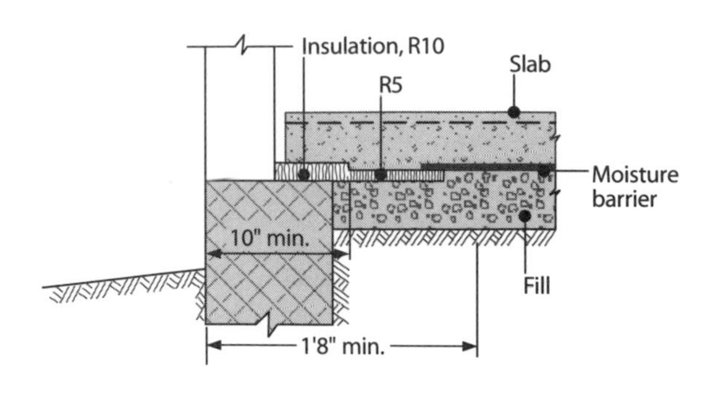

Figure 5-3
Slab foundation insulation

Stepped Foundations

Stepped foundations are used for sloping sites and buildings with partial basements. The steps keep the foundation and its footing on solid, stable ground. A stepped foundation also helps keep uphill parts of the foundation from being undermined during construction.

The stepped foundation shown in Figure 5-4 is on a sloping rock surface. Here, it's possible to cut plumb, level steps directly into the rock. The foundation can't possibly slip.

You'll often be able to chip out the steps on sites with gentle slopes. If the rock is solid, you can drill holes with a hammer drill then epoxy steel dowels into the holes to anchor the steps. Loose, brittle or crumbling rock (for example, decomposed granite) isn't a stable base for a stepped foundation. Remove this material, then cut the steps in the underlying solid, stable rock.

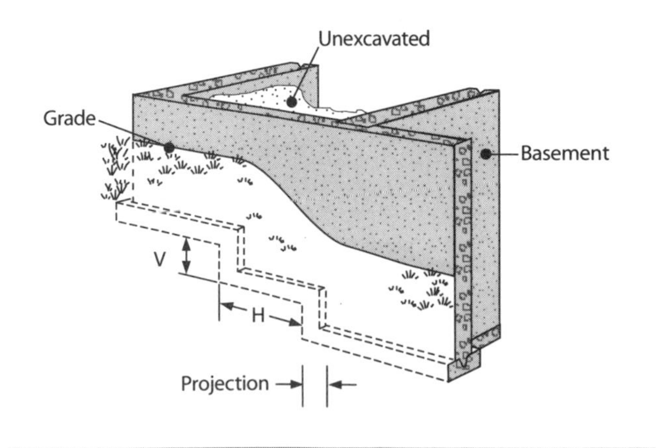

Figure 5-4
Stepped foundation on sloping ground

Proportions for Stepped Foundations

Follow along on Figure 5-4 to see how to plan a stepped foundation:

- *H* is the horizontal part of a single step in a stepped foundation. The minimum length for this part of a stepped foundation is 2 feet. The minimum thickness is 8 inches.
- *V* is the vertical part of the step or the rise. Its height is never more than 3/4 of the length of *H*. For example: if *H* is 4 feet, *V* can't be more than 3 feet, and 2 feet is better.
- *Projection* is the distance the footing projects beyond the foundation wall.

Masonry Materials

Concrete block is the most widely-used material for building foundations. Some of the properties that make concrete block such a popular material for building foundations are its strength, good insulating qualities, and low cost.

You can get lightweight or heavyweight concrete blocks. From the mason's point of view, the most important characteristics about concrete blocks are their loadbearing capacity, shape, compressive strength, and moisture content.

Let's look at the different types of blocks, and when to use them. A *solid loadbearing* block's core (the empty space or spaces inside a block) is less than 25 percent of its total area. Use these blocks for heavy-duty loadbearing walls.

The core of a *hollow loadbearing* block is more than 25 percent (usually about 40 percent) of the block's total area. Use them for exterior or interior loadbearing walls.

A *hollow nonloadbearing* block is the kind you'll use most often. It combines light weight with high compressive strength. Still, they're not the right block for every purpose. For example, never use hollow nonloadbearing blocks for any part of a building that's expected to carry a load. That's why these blocks are called nonloadbearing. On most of your jobs, you'll have many more nonloadbearing walls, partitions and screens than loadbearing walls.

Grades of Concrete Block

The American Society for Testing and Materials (ASTM) defines two grades of concrete block. A block's grade is based on its compressive strength measured in pounds of pressure per square inch, or psi.

Grade N blocks will hold up under as much as 800 psi without cracking or crumbling. You need this kind of strength if winters in your area have repeated freeze/thaw/freeze cycles. Use these blocks for all exterior walls, both above and below grade. Grade N concrete blocks resist the special kinds of stresses that go with this kind of weather.

Grade S blocks are rated at 600 psi. Use these only for above-grade work, for example in interior partition walls. You can also use Grade S concrete block for some protected exterior walls.

Types of Concrete Block

The ASTM also sets standards for two types of concrete block. A block's type is determined by its moisture content. The two types of block are:

Type I blocks are manufactured in a climate-controlled area so that they have a known low moisture content. This special treatment makes them more expensive. Use them only in very dry climates.

Most concrete blocks are *Type II* and they're the type you'll use most of the time. They aren't made under controlled-humidity conditions, so they don't have a specific or known moisture content.

Figure 5-5
Unload the block inside the foundation footing

Preparing to Build

Careful planning is important when you want to do high-quality work and make money at it. Study the blueprints or plans. Try to picture the step-by-step progress of a job when you're working on an estimate. Also picture what the job will look like when it's complete. Walking through a job in your mind like this makes it less likely you'll forget something important.

Delivery and On-Site Storage of Concrete Block

First of all, have the delivery truck park as close as possible to the footing. Then have the blocks unloaded into an area inside the footing. See Figure 5-5. Later on, you and your crew won't spend nearly as much time walking back and forth to pick up more block; you'll be laying them instead. That means you'll build faster and use fewer total manhours.

If you're not going to start laying the blocks right away, cover them to keep them dry. You can reuse that tar paper you covered the footing with (Figure 5-6). Concrete blocks are heavy enough already. When they're wet, they're a whole lot heavier. They're also slippery and hard to handle in this state. Trying to lay them in this condition is a waste of time and a fine recipe for

Figure 5-6
Cover blocks with recycled tar paper to protect them from rain

a nasty job-site accident. But that's not all. Wet blocks almost always sink in the mortar. But they don't sink evenly. You'll end up with a wall that's neither level or plumb. And wet blocks almost always cause mortar to smear.

Establish the Building Lines

We talked about batterboards — how to set them up and why you need them, in Chapter 1. You'll recall that the batterboard stringlines are guides to use when you're ready to fix the building lines. If you want to build straight foundation walls, you have to follow straight building lines. Follow these steps and that's what you'll have.

You start by finding and marking the exact corner points. Here's an easy way to do this:

1) Suspend a plumb bob from one of the points where the batterboard stringlines intersect.
2) Now lower the plumb bob toward the footing.
3) Stop when the tip of the plumb bob is close enough to use as a guide to mark the exact corner point on the top of the footing.

I like to use a lumber crayon to make my first mark here big, wide and easy to spot. Of course, this mark isn't very accurate. So I recommend making another, more precise mark with a pencil on top of the larger mark. Then drive a long masonry nail into the footing at the exact corner point. But don't drive the nail all the way into the footing. Always leave several inches sticking up so it'll be easy to spot.

Even after you start building the foundation, you need to be able to find these markers. Just cut away a corner from the first block you lay at each corner, as shown in Figure 5-7. What could be easier to do, or to remember, since the nail will be in the way?

After marking all the corners, take time to check your work. It'll only take you a few minutes to do. If there's a problem, you want to find it now. Fixing an out-of-square foundation after you've begun laying block is anything but easy. So here's a quick, accurate way to check your corner point setups.

All you do is measure the diagonals between the corners, as shown in Figure 5-8. If they're exactly equal, you've set the corner points correctly and the foundation you build will be square.

But, what if you measure the diagonals and they aren't exactly equal? It's time to do a little more measuring and comparing. This time, measure and compare opposite sides. They should be equal. If the opposite sides are equal but the diagonals aren't the same length, the footing's not square. That means one of the corners isn't really set at 90 degrees. Here's how to find out which corner's off and what you do to fix it.

1) Lay a large builder's square on the footing. Match the point (on the outside edge) of the square with your corner mark.

Figure 5-7
Corner of first block cut away to expose marker nail

Figure 5-8
Measuring the diagonals to make sure the foundation is square

2) Attach a line to the nail that's your corner mark.
3) Run this line, following the outside edge of the square, from the starting corner mark to the neighboring corner mark. Leave a gap of about 1/16 inch between the outer edge of the square and the line.
4) Now, stretch a second line from the starting corner mark to the other neighboring corner. The lines now form the two extended sides of the same right angle that the builder's square rests on.
5) If you've marked the corners correctly and the footing is square, then the second line (just like the first), will lie within 1/16 inch of the outside edge of the other arm of the square.

If the corner point is off a little, just move the nail you've used to mark the corner. Then recheck for squareness by measuring the diagonals again.

The real beauty of this check for squareness is that it's so accurate. Even if you're only a tiny bit out of square, unequal diagonals tip you off to it. Fix the problem right now, before you lay a single block, and you won't waste any time, labor or material.

Once you've located, marked and checked the corner points, use them to establish the wall lines. Set up and level the transit over one of the corner points. To set the first wall line, point the scope toward the far corner of the footing. Using a steel tape, measure the exact length of the wall between the corner markers, from the point of origin under the transit. Mark the footing at the point of the corner where the scope meets the tape. This point marks the outside of the wall.

To lay out your next wall line, turn a 90-degree angle with your transit and locate the next corner point in the scope's cross hairs. Have your assistant align the tape and mark out the second wall line. Set, measure and mark the third and fourth wall lines, following the same simple steps. After marking the fourth and final line, there's still one more important step. You can probably guess what it is. Measure those diagonals one last time, to make sure that the foundation's square.

Check Corner Elevations

Before you begin laying blocks you need to check and compare the corner elevations. Normally, all of the corners should have the same elevation. The first step is to set up and level your transit. I've found that the best position to work from is as near as possible to the center of the foundation. See Figure 5-9. Have your assistant hold the rod, or a rule, on the center point for each corner. As you take the sightings and record the elevations, make note of any differences. Later we'll talk about how to correct any problems you find. You'll do that at the same time as you're laying the first course.

Figure 5-9
Set up the level in the middle of the foundation or at a point where you can see all the corners

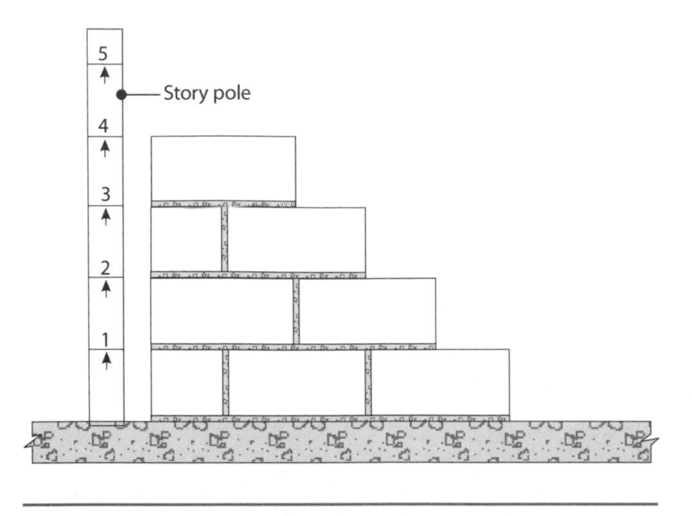

Figure 5-10
Number the story pole to show the courses of block

Making and Using a Story Pole

Masons use story poles to help them keep the different parts of the work at the right height. And if you have and use a story pole, your more expensive and fragile mason's rule stays clean and straight far longer.

Story poles are easy to make. You need a piece of good quality hardwood lumber, 8 to 10 feet long. A 1×4 or a 1 × 3 works best. Check one end of the board with a builder's square to see if it's square. Mark any excess and cut it off. Next, mark off 8-inch intervals on the pole, because that's the standard height (including the mortar bed) for one course of masonry. Take a look at the story pole shown in Figure 5-10. The numbers are course numbers and the oversize arrows show which end of the pole is up. Big, clear, bold markings like these are the best way to be sure that no one in your crew misreads the story pole.

When you're not using your story poles, store them in a dry and protected place. Proper storage is the best way to protect your poles both from accidents and warping. On large jobs you'll find that having more than one story pole is very helpful. Make all of your story poles, or at least several, at one time. That way they'll be more likely to match each other.

Here's an easy way to make sure that the story pole is held at the same height at each corner. It only works if all the corners are the same elevation, so be sure you've checked that first. Once that's done, make temporary platforms, with known elevations, to rest the story pole on. I do this by placing a piece of a hacksaw blade in the joint between the first and second courses at each corner. There's your platform, but don't forget to remove it when you're done using it.

Lay Out the Bond

Before laying out the bond, sweep away any dirt, dust or other debris from the footing. A clean footing will make sure your marks will be clear, unbroken, easy to see, and easy to read. Start by snapping a chalk line from corner to corner. I use blue chalk instead of the more typical yellow. I think it's easier to see blue chalk on the light concrete background.

To start laying out the bond, place a full block at any corner. Starting from this block, lay out the bond by marking the footing every 16 inches. The mason in Figure 5-11 is using a special mason's rule (it has a red mark every 16 inches) which makes it easier to mark the footing. This mason is using a folding wood rule, but you can also use steel tape measures marked the same way. If the structure is a modular design, laying out the bond will be a breeze. To see why, let's look at what modular design is.

Modular Construction

This is standardized construction based on a system developed by the American Standards Association (ASA). You'll find it used throughout the construction

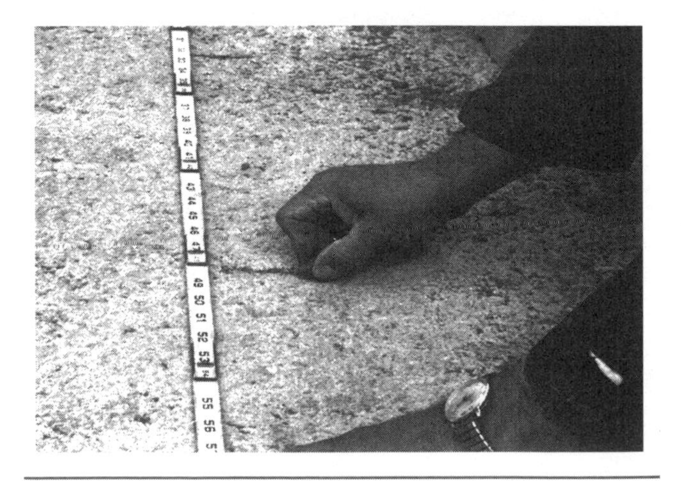

Figure 5-11
Mark the footing every 16 inches with a mason's rule to prepare for laying the bond

industry, from masons to rough framers. The base unit (and the starting point for all standards defining the system) is a 4-inch cube. That means all dimensions in any modular design are a multiple of 4 inches. Keep your eyes open for this tip-off whenever you're looking over a new set of plans. Do the plan dimensions fit the modular system? If the answer's yes, then the project's modular and that's good news for a mason. Here's why.

The ASA had one main goal in developing and introducing the modular system. This was to cut the costs of construction without cutting the level or quality of the craftsmanship. The modular system does exactly that, because:

- there's almost no on-site trimming of parts, which speeds up the work and saves labor
- few, if any, small pieces are used, which also speeds up the work and saves labor
- less material is wasted, saving you on the cost of the material

Most of the block sizes that you'll use regularly are modular. See Figure 5-12. Unit masonry that's not modular is usually clearly labeled nonmodular. Modular design makes it easier to lay out the bond because the walls usually work out in full bond. That means that each course starts with a full block and finishes with either another full block or a half block.

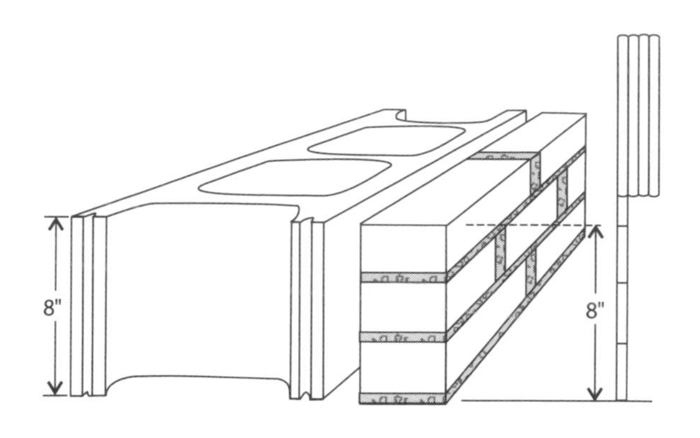

Figure 5-12
Standard modular block and brick with a rule marked to make it easier to place the brick

Problems Laying Out the Bond

Even modular designs have occasional kinks. Nonmodular designs, as you'd expect, are even more likely to have some quirks to deal with as you're laying out the bond. With that in mind, let's take a look at some of the problems that come up and what you do to solve them.

Bond lays out about 8 inches too short or too long

This is a nice, easy-to-solve problem. All you do is reverse the direction of the corner block at the next corner.

Bond lays out just a little short or long

Let's suppose you're using 16-inch concrete block to build a 48-inch-long wall. Both the wall (48 / 4 = 12) and the block (16 / 4 = 4) fit into modular standards. However, this combination of block and wall size won't lay out in full bond. Why? Because the last block (the closure block) in a course has no end joint. This is true no matter what size block you use or how long the wall is. The special problem in building a 48-inch-long wall using 16-inch blocks is that you'd only use three blocks plus two joints. Unfortunately, that only adds up to a total length of 47 5/8 inches, not 48 inches. Joint sizes shouldn't be opened wider than 1/2 inch. But in a short wall, with only a few joints, there's no other choice. The joints will have to be a little wide.

It's easy to make up that pesky 3/8 inch in a longer wall. All you do is either open or close up the joints a bit. In a wall with 17 blocks, if you close up the 16 joints by 1/8 inch, that makes the wall 2 inches shorter. Opening up 16 joints by 1/8 inch makes the wall 2 inches longer.

Bond lays out an inch or so short

Cut the *ears* (the extended ends on most stretcher blocks) off of one or a few blocks. By chiseling the ears off a block, you pick up about an inch. Of course, it takes time to chisel the ears off. Advance planning often does away with any need for these on-the-job fixes, and it's a better use of your time.

Plugs in walls

A *plug* is a piece of unit masonry that's smaller than a half unit. For example, a 4-inch piece from a 16-inch block is a plug. Don't use plugs; it's bad practice. It's also a sign of poor planning and poor-quality workmanship.

Here's how to avoid using a plug. Cut two or three 12-inch pieces from 16-inch blocks and lay them one after the other in each course. Group these pieces together, within the course and from one course to the next. The wall will look better and be stronger.

Bond alignment — jack-over-jack

A wall that's jack-over-jack has all, or a series of, its joints located directly above one another. Avoid ever laying block jack-over-jack. Such a wall is far more likely to crack.

Laying the First Course

After making sure the footing is clean, use a full bed of mortar for laying the first course of a foundation, as shown in Figure 5-13. A full bed of mortar helps keep water out and forms a solid bond between the foundation wall and the footing. But don't make the bed over 1 inch thick. An overly-thick bed of mortar doesn't hold the block as well. It will also take longer to stiffen.

If you find you need to compensate for a low spot in the footing, use small pieces of block embedded in the mortar to correct the elevation. If you find a high spot, cut the block to make the first course level. Make these corrections now and you'll save time and avoid problems later.

Check for Level

Be sure that the first course is laid level all the way around the footing. Here's an easy way to check this. Stretch a nylon mason's line between the two corners and pull it tight. In a long wall, no matter how tight you pull it, the line tends to sag. But a sag in the line is easy to correct if you have a trig.

A *trig* is a paper-thin metal clip about 4 inches long. It has a slot near one end to hold the mason's line. In Figure 5-14 you'll see a metal trig on top of a block, held down by the masonry scrap on top of it. If you don't have a metal trig close at hand, improvise! A matchbook cover or a stiff piece of paper will make a very serviceable trig. The block the trig sits on is called a *trig block*. The trig block must be level and plumb. With a builder's level you can set the trig block at the same elevation as the corner blocks. Then all you'll need to do is measure 8 inches up from the height of each succeeding trig block.

To check that the first course is level, stand at one corner and sight down the line and across the trig to the next corner. Problems like a hump or a wow in the wall are easy to see.

Stocking the Block

This is a little reshuffle of the dry block. Do it after the first course is in place, to save time and find out if you're short of material when there's still time to order more without holding up the work. Just count the number of

Figure 5-13
Make a clean footing with a good bed of mortar for the first course of block

Figure 5-14
A metal trig keeps the line from sagging or swaying

Figure 5-15
A 12-inch corner block

Figure 5-16
A 10-inch corner block

Figure 5-17
A 12-inch corner made of an 8-inch block and a concrete brick

blocks in the first course. Then multiply that number by the total number of remaining courses. The answer is the total number of blocks you need to finish the wall.

Stock the right number of blocks near the footing of the wall they'll go in. But before you move a block, decide where you'll stand while you work on the wall. Let's suppose it's a wall you'll build from the inside only. Then you'll want to stock as much of the block as you can inside the building lines. If there's not enough room for all of the block, leave a V-shaped opening to bring materials through, and fill it in later.

Use these same priorities to choose where the mixer should be set up. Putting it close to where you'll work means that everyone spends more time laying block and less time walking back and forth for fresh mortar.

Handle the blocks carefully anytime you move them. You'll waste less material and time. Here are a few points to keep in mind:

- ❖ Concrete blocks chip easily. Handle them carefully. Unless you're willing to write them off as a total loss, you have to patch the chips. That takes time, it's a boring job that no one wants, and the difference in finish makes it easy to see any patches.
- ❖ Don't let the blocks pick up mortar smears. Stains from mortar smears are almost impossible to remove. And don't assume that paint will cover them up. Most of the time, smears show right through.
- ❖ Blocks are almost always made wider on top than on the bottom. This makes the block easier to handle and lets it hold more mortar. Use the advantage you've been given, and always stock blocks with their wide, or top, sides up.

Lay the Corners

In this section we'll discuss corner construction, and some of the most-commonly-used specialty blocks (corner blocks).

Figure 5-15 shows a standard, L-shaped, 12-inch corner block. Notice that the orientation of the corner block changes in the second course. That way, the corner blocks not only continue the bond, they also avoid a jack-over-jack corner. Figure 5-16 shows a typical 10-inch L-shaped corner block. Its position also alternates with each course.

You can get both the 12- and the 10-inch L-shaped corner blocks from most concrete block manufacturers. These blocks are the most widely used and are also the easiest to use.

Figure 5-17 shows one way of turning a corner if you can't get 12-inch corner blocks. You can use a regular 8-inch block in combination with a concrete brick. In each course, the relative positions of the two will change. That way they won't come out jack-over-jack.

Figure 5-18 shows another solution to the same problem. This time the mason used a 12-inch-long piece of 4-inch block by placing it in alternating positions at the corner in each course. This combination of block works out in half bond.

Figure 5-19 shows a way to make a corner in a wall of 6-inch block. The mason used a 14-inch-long block at the corner in each course.

Figure 5-20 shows a way to build a corner in a wall made of 4-inch block. The mason used a 12-inch-long piece of 4-inch block at the corner in each course. This wall will lay out in half bond.

Figure 5-18
A 12-inch corner made of a 12-inch-long piece of 4-inch block

Figure 5-19
A 6-inch wall uses a 14-inch-long piece at the corner

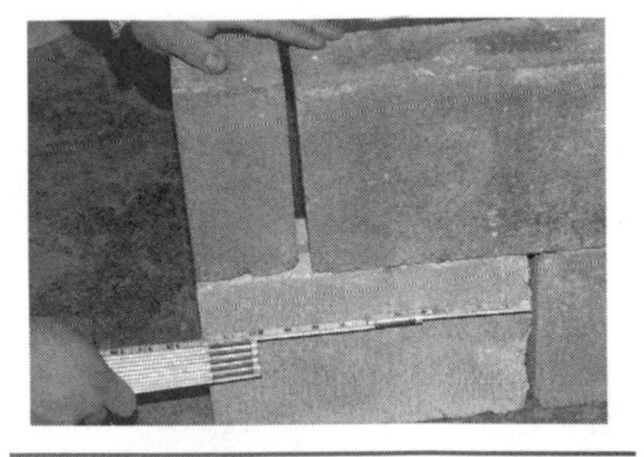

Figure 5-20
For a 4-inch wall, use a 12-inch-long piece at the corner

Figures 5-21 A and B show the first and the second course of a corner built from two sizes of block. One side of the right angle is an 8-inch block wall, the other a 12-inch block wall. Here's how to carry off this corner smoothly. In the first course, start the 12-inch wall at the corner by combining an 8-inch block with half of a 4-inch block. The 8-inch block in this group carries the bond through to the 8-inch wall. In the second course, place an 8-inch block at the corner to start the second course of the 8-inch wall. A 12-inch block, meanwhile, starts the second course of the 12-inch wall carrying the bond through. This 12-inch block overlays the 8 inch plus 4 inch combination used in the first course.

Lay the Closure Block

When you're ready to lay the last, or closure block, in a course, what do you do? Unlike all the other blocks in the course, this one can't be pushed against the block next to it. That means that you won't be able to compress the end joints. Instead, fill the ends of the block with mortar, as shown in Figure 5-22. You'll get a good bond this way. Besides that, it's a good habit to get into. Some building codes and inspectors insist on this method.

Pilasters

Typically, foundation walls are long and they're not tied to any partition walls or other supports. If these walls weren't reinforced somehow, they might eventually crack from the load of the surrounding backfill. A pilaster is one method of reinforcement. A pilaster is a square-sided column that's attached to a wall to reinforce and support the foundation walls.

Unreinforced masonry pilasters always project beyond the wall surface. How much and on which side varies from place to place. Check the local building code. Some codes, for example, require a minimum 4-inch projection from the face of the wall.

A First course

B Second course

Figure 5-21
Corner where an 8-inch wall meets a 12-inch wall

Figure 5-22
The closure block, with end joints filled in solid

A pilaster with a uniform cross section works just like a column or a vertical beam. If a pilaster's cross section tapers upward, it's called a buttress.

Pilaster placement depends on where the vertical loads are, where extra lateral support is needed, or a combination of these and other factors. If you don't place the pilasters correctly, the building won't be structurally sound. Leave pilaster design and placement to the project architect or structural engineer.

Figure 5-23 shows a 10-inch concrete block wall with an engaged (tied-in) pilaster. The block on the pilaster is a 10-inch jamb (or square) block. Right behind it there's a 6-inch block. The 6-inch block carries the bond through the pilaster. This pilaster projects 6 inches and all the projection is on the outside of the wall.

Figure 5-24 shows the same wall and pilaster after one more course of block has been added. In this course, the 6-inch and the 10-inch blocks simply trade places. The 6-inch block forms the projecting pilaster and the 10-inch behind it is laid in the normal position in the wall.

Masonry Retaining Walls

An important recent development in masonry is building retaining walls with unit block systems that you don't have to set in mortar. The blocks you use weigh from 30 to over 100 pounds each, so you have to handle them carefully. But installing them could hardly be easier. With most of these systems you won't need any new or specialized tools. In fact, with some systems, once the first course is in, you won't even need any hand tools! Building these retaining walls, as you can see, is fast work. Remember though, it all depends on getting the first course exactly right.

Figures 5-25 and 5-26 show examples of just two of these systems. The blocks are unusual because each has a lip along its lower back edge. Figure 5-25 shows the blocks upside-down so you can see the lip clearly. The

Figure 5-23
A 10-inch jamb or square block in the pilaster with a 6-inch block behind it

Figure 5-24
A 6-inch block in the pilaster with a 10-inch block behind it

Figure 5-25
Blocks turned upside-down to show lip

Figure 5-26
Retaining wall built with self-aligning mortarless unit-block system

lip of a block overlaps the backs of the blocks in the preceding course, making them self-aligning. This simple lip performs three functions:

1) It locks the individual block into the course.
2) It produces even exposure for all the blocks within the course.
3) It produces even exposures of blocks from one course to the next.

Figure 5-26 shows how this works. You can see the lip at work in the end block in the incomplete fourth course.

The running bond used in both examples is typical. Backfill this type of retaining wall as you go, filling in as many voids as possible. You'll find installation details sometimes vary between systems. Check with your masonry supplier. The basic methods and techniques, however, are similar and you'll quickly become familiar with them.

A retaining wall of this sort may not need a concrete footing. If you're not sure, check the architect's specs or local building codes. On jobs without a concrete footing, you'll still want to provide some kind of drainage for the wall. To do that, dig a trench following the planned route for the wall. Make the trench a bit wider than the wall will be and about 5 inches deep. Next, compact the soil in the bottom of the trench. Finally, fill in the trench with porous gravel, nearly to grade, or a depth totaling about 4 inches. If the site is particularly prone to wetness, consider also installing a perforated drain behind the wall.

Estimating Block for Concrete Block Walls

There are two basic kinds of estimating: quantity take-offs, and rule of thumb. Quantity take-off estimates are very precise and they cover everything. To assist those who use this method, the Portland Cement Association, the National Concrete Association, the Brick Institute of America and others have developed many reference charts. Large construction companies need this degree of accuracy. The smallest mistake multiplied hundreds of times creates a large error.

However, a small volume contractor usually doesn't need this level of precision. For most masons, rule-of-thumb estimating is the best way to go. First let's learn some basic rules. Then after a bit of practice, estimating concrete block walls will be quick and easy. Remember that most newer buildings are modular in design. In estimating, modular buildings will work out to some number of whole or half blocks.

Rule-of-Thumb Estimating for Concrete Block

1) Begin by figuring the total length of the wall in feet. This total may be the perimeter (the sum of the lengths) of the walls.
2) Multiply the total linear feet by 0.75 because three 16-inch blocks are needed to lay 4 feet of wall. The result is the number of blocks per course. If three 16-inch blocks equal 4 feet of wall, then 3/4 of a 16-inch block equals 1 foot of wall. Since it's easier to do the math with decimal numbers than fractions, change the fraction 3/4 to its decimal equivalent, 0.75.
3) Check the blueprints for the height of the walls. Divide wall height, in inches, by 8 inches. The result is the number of courses of masonry needed to build to that height.
4) Multiply the number of courses by the number of blocks per course. The result is the total number of blocks needed if the wall is solid.
5) Measure all the openings in the wall. Deduct the total number of blocks they eliminate. The result is the total number of blocks needed to build the wall.

Example

Here's an example of how this works with real numbers. Suppose Ms. Jones has hired your company to build a concrete block perimeter wall. The blueprints she supplied show an 8-foot wall around a rectangular area. The long sides measure 75 feet each. The short sides measure 50 feet each. Ms. Jones' blueprints also make it clear that none of the walls have any openings in them. Remember, this is just as example for estimating purposes. Don't worry about how people will get in and out. The estimate goes like this:

1) Add the sides to get the linear footage of the perimeter:

   75' + 50' + 75' + 50' = 250'

2) Multiply the linear footage by 0.75 to get the number of blocks in one course:

   250' × 0.75 = 187.50 blocks per course

3) Divide the wall height in inches by 8 to get the total number of courses needed to built to the correct height:

   96" ÷ 8 = 12 courses

4) Multiply the total number of courses by the total number of blocks in a course to get the total number of blocks needed:

   12 × 187.50 = 2,250 blocks

Pretty easy isn't it? But not every type of concrete block is covered by rule-of-thumb estimating. If you use specialty blocks, estimate them separately. Then don't forget to deduct them from the main estimate. Specialty blocks are separate, and often special-order materials. While you're ordering, it's smart to add a few spares. Think of the spares as an insurance policy.

In the next chapter, we'll take a close look at another type of foundation, the pier and grade beam foundation.

6

Pier and Grade Beam Foundations

We covered the most popular foundations, slab and block, in the last chapter. In this one, we'll learn about pier and grade beam foundations.

Pier Foundations

Pier foundations cost less to build than continuous wall foundations. Why? Simply because you don't use as much material to build them.

Pier foundations are more open than other types of foundations. This fact sometimes leads people, especially in windy areas, to believe that pier foundations are less safe. But a properly-designed and well-built pier foundation is as safe as any other type of foundation. It's true that because pier foundations are more open, resistance to wind uplift force is a design concern. Wind uplift damage to pier foundations does happen when they're poorly designed. Incorrect sizing of piers can result in piers that are undersized and too lightweight. Or incorrect anchoring of piers can cause a lessened ability to resist force of any kind (including wind uplift). With proper design, you can avoid both of these pitfalls.

Designing Piers for Foundations

It doesn't take a lot of math or complex tables and formulas to design adequate piers. Follow the steps outlined here and you'll design and build safe, stable piers for long-lasting and economical foundations. Let's start with your choice of materials.

You can build a standard pier foundation using any of the following materials:

- brick
- concrete
- stone rubble
- hollow core concrete block
- structural clay tile

Here's the first rule: If the plans call for slender piers (with a height more than 10 times their width), always make them using reinforced concrete.

Second, never build a masonry pier taller than 10 times its smallest dimension. Let's say you're building some piers out of modular 8 × 16 × 6 hollow core concrete blocks. Piers built with this size of block can't be taller than 60 inches. Now, let's see how this works for formed and poured unreinforced concrete piers. Suppose the forms are for rectangular piers measuring 5 inches by 6 inches. These piers can't be taller than 50 inches.

There's an additional rule for freestanding structural (supporting beams or girders) piers of hollow core concrete block or structural clay tile. If the piers are taller than six times the smallest dimension of the block, you must fill the hollow cores with either concrete or Type M or S mortar. For example, suppose you're using 8 × 16 × 6 blocks for piers that are 4 feet tall. The smallest dimension of the blocks is 6 inches, and the height is more than 36 inches (6 × 6 = 36). So you must fill the cores with concrete or mortar.

Placing Foundation Piers

How far apart you space the piers is mostly a question of the budget and the load the beams spanning the piers must carry. The closer you space the piers, the more piers you need and the higher your material costs. The farther apart the piers are, the heavier the spanning beams must be. In the end, building more piers costs less than buying longer, heavier beams. Here's a rule that helps to explain why:

The strength of a beam is an inverse function of the square of the span of the beam.

That means that a beam that spans 6 feet is four times stronger than a beam of the same thickness that spans 12 feet.

Reinforcing Piers

The size (diameter) and length of reinforcing material for a pier depends on the height of the pier. Piers are grouped by height into three size ranges. The reinforcing for all piers in a group follows the same guidelines. We'll look at each of the three size ranges later in this section. For a start though, let's look at the big picture. The following points are true for all reinforced piers.

You build reinforced piers from hollow core concrete blocks or poured concrete that's cast, on-site, in forms made of wood, steel and wood, or aluminum. Both kinds of piers are reinforced using the same material — pieces of steel rod with an open hook at either end. Figure 6-1 shows reinforcing running through a pier and footing. Here are the rules to follow:

- ❖ Embed reinforcing rods at least 5 inches in a pier footing.
- ❖ Don't place the rods more than 3 inches above the bottom of a pier footing.
- ❖ In both concrete block and poured concrete piers, face the rods' open hooks in, toward the center of the pier.

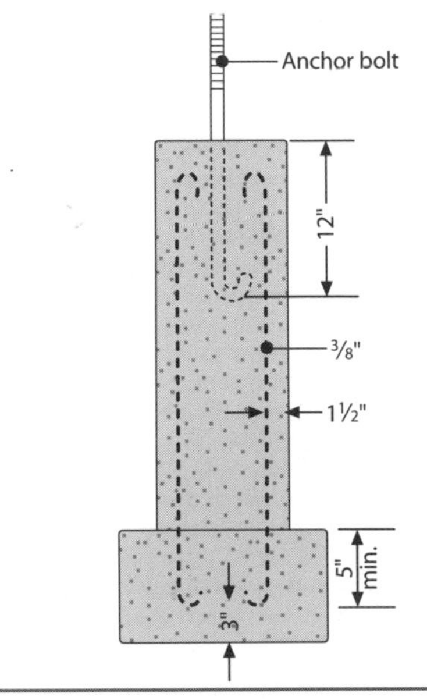

Figure 6-1
Reinforcing and anchoring short piers

- ❖ In hollow core concrete block piers, place the rods in the corners of the core, then fill the cores with concrete or mortar.
- ❖ In formed and poured concrete piers, place the rods near the corners of the form but leave a space that's at least 1½ inches wide between the rods and the forms, then pour the concrete as usual.
- ❖ Reinforcing rods in a pier must be at least 6 inches shorter than the total height of the pier and the footing, placed with equal spaces at the top and bottom. The rod's top should be 3 inches or more below the top of the pier. The rod's bottom should be 3 inches or more above the bottom of the pier footing.

Here are some tips on placing the reinforcement for piers from 3 to 4 feet high, the three size ranges of reinforced piers, including the correct rod diameter. Piers higher than 6 feet are used in major, large-scale projects. There are so many factors involved in the design of these piers and their reinforcing that it's a job better left to the structural engineer.

Piers 3 to 4 Feet High (Figure 6-1)

Use 3/8-inch-diameter steel rods that are at least 2 feet long to reinforce these piers.

Piers 4 to 6 Feet High (Figure 6-2)

Taller piers need more support, so you use ½-inch rods that are 6 inches shorter than the pier. These rods are heavier and harder to handle than 3/8-inch rods, so it's a good idea to use two shorter pieces of rod, spliced together, instead of one long, heavy piece. Here's how to do that, starting with the lower part of a pier:

1) Take four short pieces (or dowels) of ½-inch-diameter reinforcing rod, about 18 inches long.
2) Make sure one end of each rod ends in an open hook and the other is straight.
3) Embed the rods, hooked ends down, in the pier footing. Leave about 13 inches of the rod sticking up above the top of the footing to splice to matching pieces of rod later on. Be careful where you put the rods. You'll need to line them up correctly so you can splice them later.
4) For the upper part of the pier reinforcing, take four more pieces of ½-inch reinforcing rod (the length depends on the pier's total height). Have one end of each rod ending in an open hook and the other end straight.

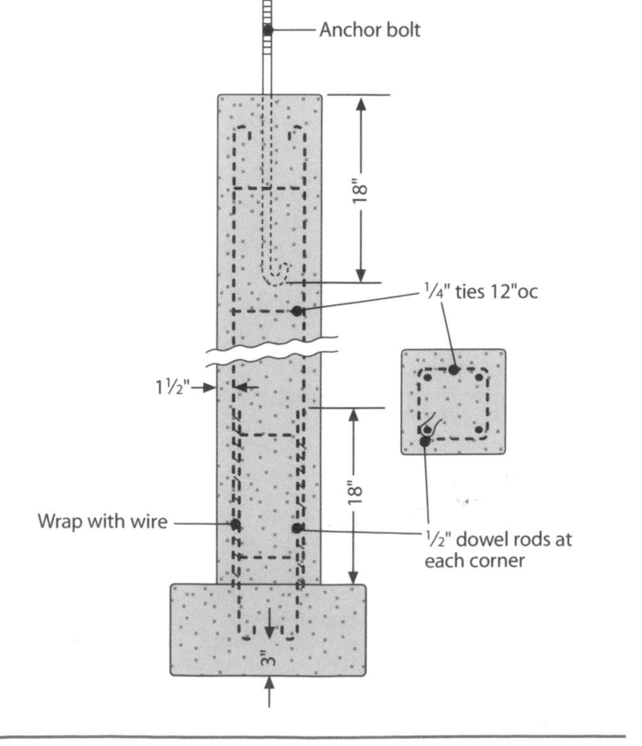

Figure 6-2
Reinforcing and anchoring large piers

5) Position the rods in the body of the pier, straight ends down. Splice the rods, following these rules:
   a) Overlap the two rods, in this case by at least 15 inches. The Uniform Building Code requires that the splice be at least 30 times the diameter of the rod (½ inch × 30 = 15 inches).
   b) Wrap the pairs of rod together with tie wire or weld them together.

Anchoring Piers to Structures

Anchor pier foundations to buildings with 5/8-inch-diameter anchor bolts that have one end with a hook and the other end threaded. Set anchor bolts with the hooked end down in the top of a pier. For formed and poured concrete piers, set anchor bolts 12 to 18 inches deep. For piers made with unit masonry, set bolts 3 feet deep with the threaded ends through the bottom plates or girders of the building. Use nuts to secure them.

Figure 6-3 shows how to anchor a wood post to a concrete pier. Use a pair of 3/8-inch by 1½-inch strap anchors that are long enough to embed at least 6 inches in the pier footing and still extend at least 12 inches above the top of the pier. Attach the strap anchors to the outside of the post with two lag screws in each.

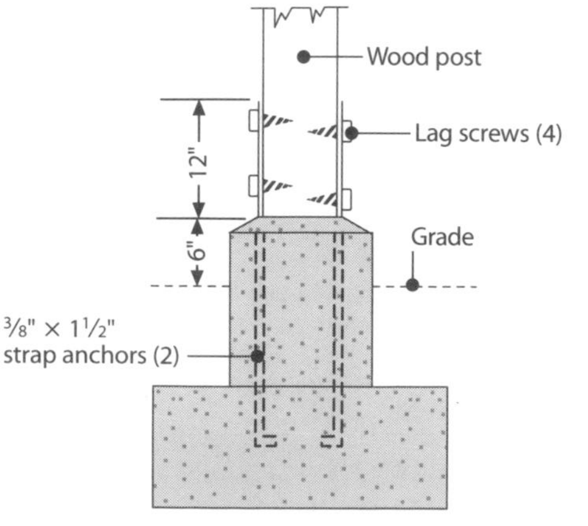

Figure 6-3
Anchoring wood posts to concrete

Grade Beams

A grade beam is a rectangular, reinforced concrete beam that serves as a type of continuous foundation. The beams rest on top of reinforced concrete piers, usually with 10-inch diameters. Figures 6-4 through 6-7 show some of the many types of grade beam construction.

A standard grade beam used in an average one-story house has a cross-section measuring about 8 by 12 inches. Dimensions for grade beams used in other buildings depend on the weight of the building. Only a structural engineer is qualified to design grade beams.

Rules for Grade Beam Foundation Piers

Reinforced concrete piers that support grade beams must be:

- 10 inches in diameter
- spaced 8 feet on center for a single-story building
- spaced 6 feet on center for a two-story building
- reinforced with 5/8-inch rods that extend into the grade beam, tying the two parts together

How to Build a Grade Beam Foundation

Grade beams are most useful on sites that have a soft soil overlaying another soil that's more compact. A grade beam foundation is about the quickest, easiest foundation you'll ever build. Here's all there is to it:

1) Drill holes through all soft soil down to more compact soil.
2) Space the holes 8 feet on center for a one-story building and 6 feet on center for a two-story structure.
3) Place a reinforced 10-inch diameter concrete pier in each hole.
4) Attach the reinforced concrete grade beam(s) to the piers. Site the grade beam so that about 8 inches is above grade and 8 inches below.

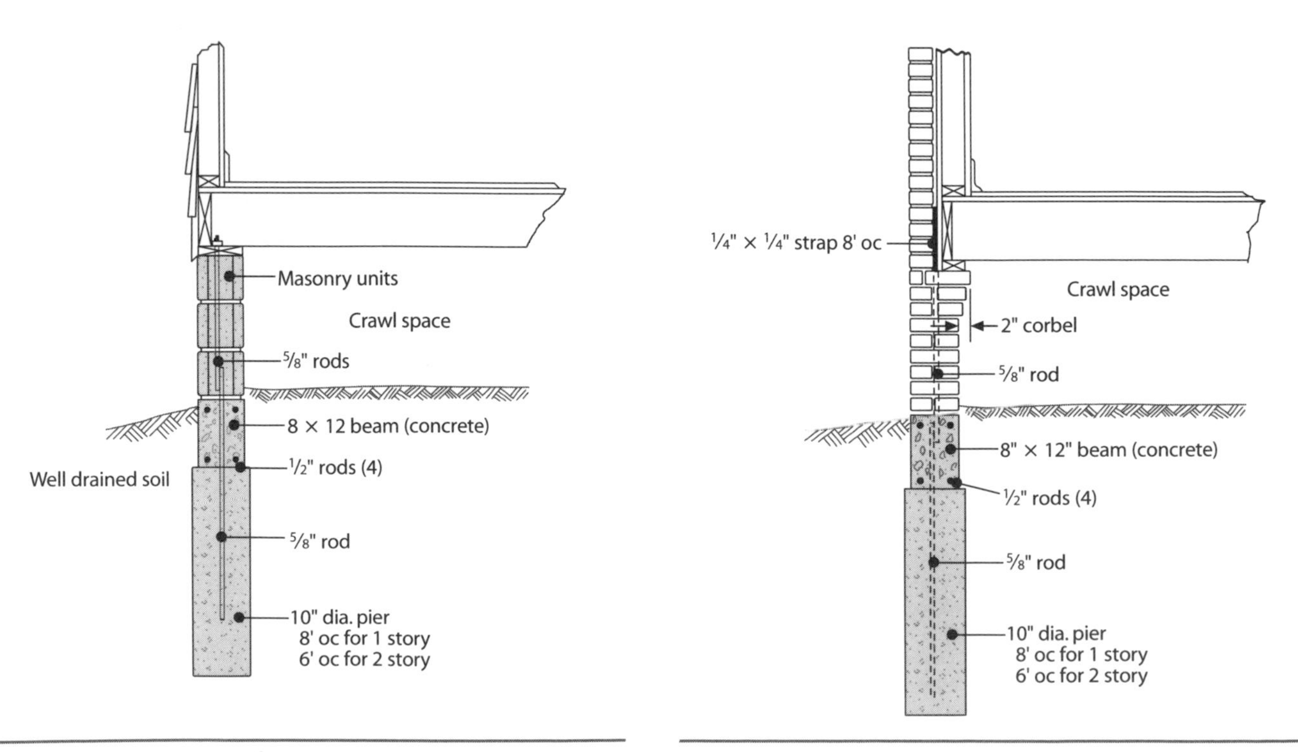

Figure 6-4
Grade beam for a frame house with crawl space

Figure 6-5
Grade beam for a brick veneer house with crawl space

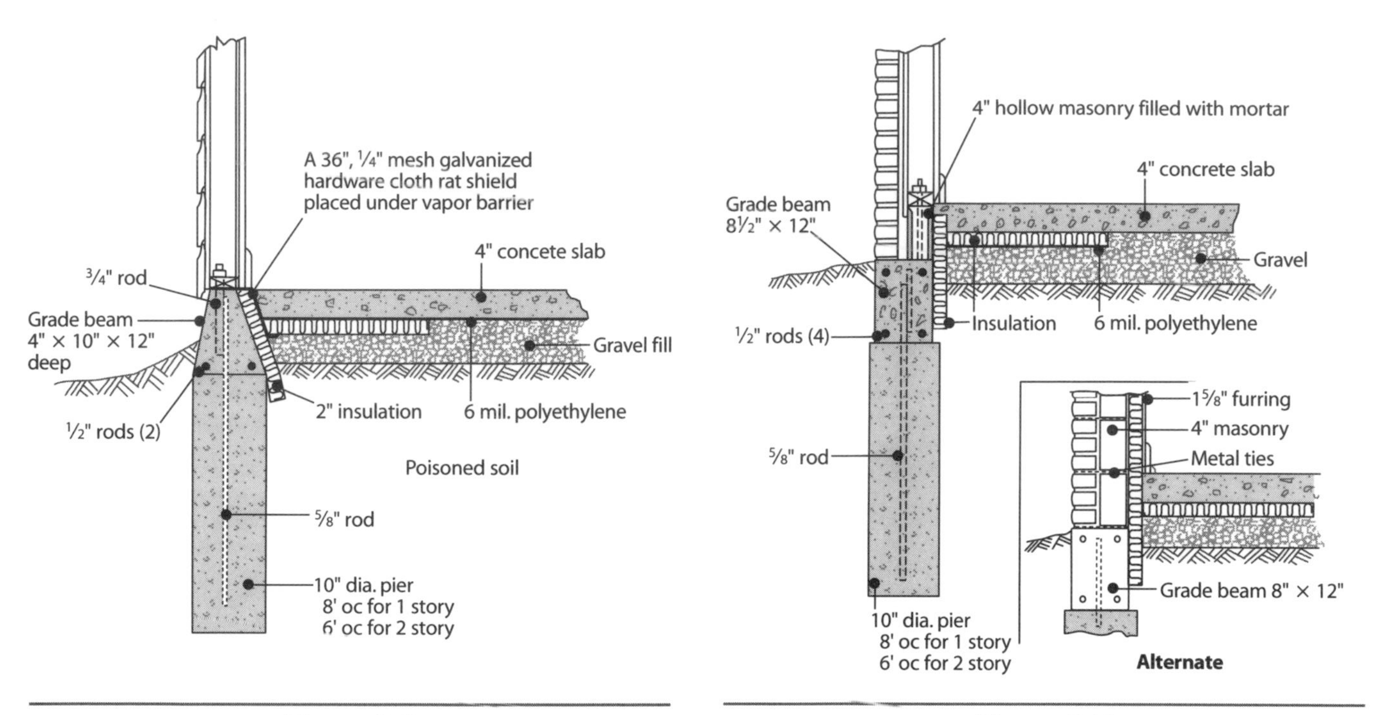

Figure 6-6
Grade beam for a frame house with slab floor

Figure 6-7
Grade beam for a brick veneer house with slab floor

5) Surround the below-grade part of the grade beam with a loose, porous fill such as gravel or cinders. The fill drains away any water that might collect beneath the beam.

Check Your Building Code for Grade Beam Restrictions

In many areas, grade beam construction isn't allowed by the building code. Before you agree to do any grade beam construction, make sure this type of construction is permitted. Grade beam construction isn't a good idea (and usually isn't allowed) in any area where frost heave is a problem. Obviously, if you build a grade beam foundation in an area where they're not allowed, you'll not only have to tear it out, but you probably won't get paid for either the construction or the demolition, *no matter what the owner or the architect or the plans say*. You're the builder. You're expected to know!

7

Concrete Form Construction

In today's time-is-money work environment, it's tempting to sacrifice accuracy for speed. But when you're building forms for concrete footings, take the time to work as accurately as possible. If the forms aren't level, square and in the center of the wall line, the errors will cause hours of additional work down the line. Workers will spend expensive time figuring out how to level the forms. It pays to do it right the first time. That's why many wall contractors prefer to pour their own footings.

Poured concrete walls are extremely expensive to replace. Spend the time you need during the preparation stage of wall construction. Here are three areas to watch out for:

1. Carefully examine the blueprints to make sure all the reinforcement is properly placed in the forms. Check whether the new epoxy-coated reinforcing is required.
2. Locate and align the anchor bolts correctly so the adjoining construction will line up. It's next to impossible to install anchors in the concrete after it's hardened.
3. Make sure dovetail anchor slots are vertical and in the correct alignment so the intersecting wall is properly anchored.

You can either use prefabricated forms or construct your own on the job site. If you're constructing your own, you'll need some very specific design information. A more detailed source is *Concrete & Formwork*, a manual published by Craftsman Book Company that gives everything you'll need to know about building your own forms. It's described in the order form bound into the back of this book.

In this chapter we'll look at the basic design criteria for concrete forms, including the materials and methods you use. We'll introduce two common prefabricated forming systems, the Jahn and the Steel-Ply forming systems.

You can pour concrete into a variety of forming materials. But whatever material you decide to use, it must be able to withstand a tremendous amount of pressure. Fluid concrete exerts about 150 pounds of pressure per square foot on the wall forms. And as the height of the form increases, the pressure increases. Later in the chapter you'll find tables to use in designing your forms to withstand the pressure.

Form Materials

Part of planning the forms is selecting the materials you'll use. We'll begin this chapter by taking a close look at form materials. If you decide to build your own forms, you'll probably use plywood for the construction material. You can use virtually any exterior-type

APA plywood, because all panels with that identification are manufactured with waterproof glue. But there's a plywood designed especially for concrete forms.

Plyform

Plyform is the recommended panel for most general forming uses. It's an exterior-type plywood composed of wood species and veneer grades that ensure quality and durability. All Plyform panels bear the APA trademark of APA — The Engineered Wood Association. It's available in two classifications: Plyform Class I and Plyform Class II. Of the two, Class I is the stronger and stiffer panel. Figure 7-1 is a grade-use guide for concrete forms.

Plywood is an engineered product, manufactured to exacting tolerances. A tolerance of plus 0.0 inch and a minus 1/16 inch is allowed on the specified width and/or length. The APA Standard requires panels to be square within 1/64 inch per linear foot for panels 4 feet by 4 feet or larger. Panels must be manufactured so that a straight line drawn from the corner to an adjacent corner will fall within 1/16 inch of the panel edge.

Sanded Plyform panels are manufactured with a thickness tolerance of 1/64 inch of the specified panel thickness for 3/4 inch and less, and plus or minus 3 percent of the specified thickness for panels thicker than 3/4 inch. Overlaid Plyform panels have a plus or minus tolerance of 1/32 inch for all thicknesses through 13/16 inch. Thicker panels have a tolerance of 5 percent over or under the specified thickness.

The tolerances and requirements set by the American Plywood Association ensure quality and minimize the time and labor required to build forms. But you've still got to be aware of these tolerances at the job site. In an extreme case, two 3/4-inch sanded panels, both within manufacturing tolerances, could form a joint with a 1/32-inch variation in surface level from panel to panel. Realigning the panels or adding shims are quick, easy solutions.

PLYFORM grades	Description	Typical trademarks	Veneer	
			Faces	Inner plies
B-B PLYFORM Class I & II** APA	Specifically manufactured for concrete forms. Many reuses. Smooth, solid surfaces. Mill-oiled unless otherwise specified.	APA PLYFORM B-B CLASS 1 EXTERIOR 000 PS 1-83	B	C
High Density Overlaid PLYFORM Class I & II* *APA	Hard, semiopaque resin-fiber overlay, heat-fused to panel faces. Smooth surface resists abrasion. Up to 200 reuses. Light oiling recommended between pours.	HDO PLYFORM 1 EXTRA 000 PS-1-83	B	C Plugged
STRUCTURAL I PLYFORM** APA	Especially designed for engineered applications. All Group 1 species. Stronger and stiffer than PLYFORM Class I and II. Recommended for high pressures where face grain is parallel to supports. Also available with High Density Overlay faces.	APA STRUCTURAL 1 PLYFORM B-B CLASS 1 EXTERIOR 000 PS 1-83	B	C or C Plugged

* Commonly available in 5/8" and 3/4" panel thicknesses (4' x 8' size)
** Check dealer for availability in your area

Figure 7-1
Grade-use guide for concrete forms

Unless otherwise specified, Plyform panels are sanded on both sides. They're face-oiled at the mill to reduce moisture penetration and to keep the concrete from sticking to the forms. You may have to use additional oil unless the mill-oiling is reasonably fresh when you first use the panels. Many users apply a top-quality edge sealer before the first pour. With proper care, Plyform panels often last through ten or more pours.

HDO Plyform

Instead of oiling at the mill, you can order Plyform with High Density Overlaid surface on each side. High Density Overlaid Plyform Class I and II meet the same general specifications as Plyform Class I and II. Both classes of HDO Plyform have a hard, smooth semi-opaque surface of thermosetting, resin-impregnated materials that form a durable, continuous bond with the plywood. Though the abrasion-resistant surface doesn't require oiling, many users wipe the panels lightly with oil or other release agents before each pour to ensure easy stripping. In my opinion, it's worth the trouble.

HDO Plyform is usually specified for the smoothest concrete finishes because the panel has a hard, smooth surface. Since both sides are moisture-resistant, you can use either side with equal effectiveness. With reasonable care, HDO Plyform can last for as many as 50 pours. Some concrete-forming specialists have used the same panels for as many as 200 pours or more with good results.

Structural I Plyform

This panel is stronger and stiffer than Plyform Class I or II and is designed specifically for engineered applications. HDO Structural Plyform contains all Group I wood species and is often recommended for high pressures where face grain is parallel to supports.

Related Grades

There are other plywood products designed for concrete forming, including special overlay panels and proprietary panels. They're designed to produce a smooth uniform concrete surface and are generally mill-treated with a form-release agent. Some proprietary panels are made of only Group I wood species and may have thicker face and back veneers than those normally used. This provides greater parallel-to-face grain strength and panel stiffness. Faces may be specially treated or release coated. Check with the manufacturer for design specifications and surface treatment recommendations.

Special Textures

Plyform is manufactured in many special surface textures, ranging from the polished high-density overlaid plywood to patterned board and batten siding panels. You can create virtually any pattern. Figure 7-2 shows two typical patterns.

A Board and batten texture

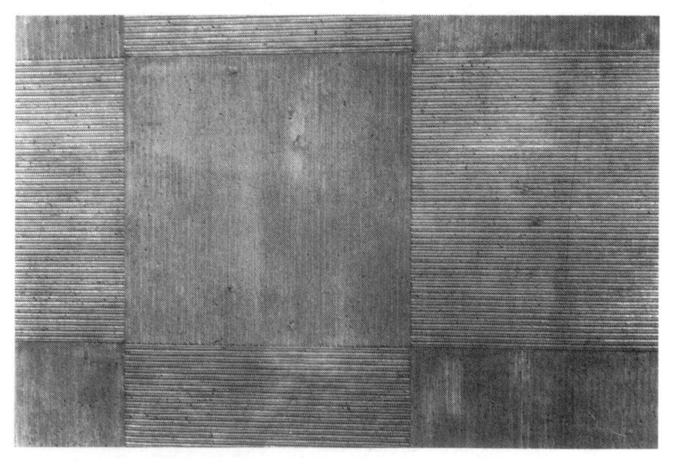

B Basketweave texture

Figure 7-2
Typical textures from plywood forms

There are two ways you can use textured plywood in form design. First, as a liner with a plywood backing, so the liner creates the texture but contributes little to the structure of the formwork. Second, as a basic forming panel. This works best in projects where the number of pours is limited, because the textured surface can increase the possibility of damaging the panels during stripping. Film coatings, like lacquer, polyurethane or epoxy, along with a release agent, can make the stripping easier.

Other Plywood Suitable for Concrete Forming

Though not manufactured specifically for concrete forming, you can use grades of plywood other than Plyform for various forming applications. Use the tables of allowable pressure, in the next section, for a good estimate of performance for sanded grades, like A-C Exterior APA and B-C Exterior APA, and unsanded grades, like C-C Exterior and C-D Interior with exterior glue, provided the face grain is *across* supports. For Group I sanded grades, use the table for Class I Plyform. For Group 2 sanded grades, use the table for Class II Plyform. For unsanded grades (Identification Index panels), use the Plyform Class I tables assuming 1/2-inch Plyform for 32/16 Identification Index panels, 5/8-inch for 42/20 and 3/4-inch for 48/24.

You can get textured plywood to create various patterns for architectural concrete. Many of these panels have some of the face ply removed in the texturing process. That reduces the strength and stiffness. Since textured plywood is available in a variety of patterns and wood species, I can't give exact factors for strength and stiffness reductions. For approximately equivalent strength, specify the desired grade in Group 1 or Group 2 species and the thickness based on Plyform Class II. When 3/8-inch textured plywood is used for a form liner, assume that the plywood backing must carry the entire load.

Tables for Form Design

Plyform is manufactured in various thicknesses, but it's good practice to base designs on 5/8-inch and 3/4-inch Plyform Class I, as they're the most commonly available. For large jobs or for those with special requirements, you may need other thicknesses. These could require a special order, so plan ahead.

The tables in this section will help you choose the right Plyform thickness for most applications. They're based on the plywood acting as a continuous beam which spans joists or studs. No blocking is assumed at the unsupported panel edge. To minimize differential deflection between adjacent panels, some form designers specify blocking at the unsupported edge, particularly when face grain is parallel to supports. You'll also find tables for choosing the proper size and spacing of joists, studs and walers.

Though many combinations of frame spacing and plywood thickness meet structural requirements, I recommend using only one thickness of plywood and varying the frame spacing for different pressures. If you have to use two layers of plywood, you can add the allowable pressure for each layer to find the total.

Concrete Pressures

The required plywood thickness, as well as size and spacing of framing, depends on the maximum concrete pressure. So the first step in form design is to determine the maximum concrete pressure on your job. This is based on such things as pour rate, job temperature, concrete slump, cement type, concrete density, method of vibration, and height of form.

Pressures on Column and Wall Forms

There are several methods of calculating pressures for wall and column forms. Figure 7-3 shows pressures for vibrating concrete at different pour rates and temperatures based on the recommendations of the American Concrete Institute. These values are for internal vibration of the concrete only. For external vibration, double the pressures shown. When concrete isn't vibrated, reduce the pressures in the table by 10 percent.

Concrete pressure is in direct proportion to its density. Pressures shown in Figure 7-3 are based on a density of 150 pounds per cubic foot (pcf). They're appropriate for the usual range of concrete poured. For other densities, adjust pressures proportionately.

Pour rate (ft/hr)	Pressures of vibrated concrete (pcf) (a), (b)			
	50° F		75° F	
	Columns	Walls	Columns	Walls
1	330	330	280	280
2	510	510	410	410
3	690	690	540	540
4	870	870	660	660
5	1050	1050	790	790
6	1230	1230	920	920
7	1410	1410	1050	1050
8	1590	1470	1180	1090
9	1770	1520	1310	1130
10	1950	1580	1440	1170

Notes (a) Maximum pressure need not exceed 150h, where h is maximum height of pour.
(b) Based on concrete with density of 150 pcf and 4" slump.

Figure 7-3
Concrete pressures for column and wall forms

Pressures on Slab Forms

Forms for concrete slabs must be able to support workmen and equipment as well as concrete. Figure 7-4 gives some pressures which represent the average when you're using either motorized or nonmotorized buggies for placing concrete. These pressures include the effects of concrete, buggies, and workmen.

Depth of slab (in)	Concrete pressure (psf)	
	Nonmotorized buggies (a)	Motorized buggies(b)
4	100	125
5	113	138
6	125	150
7	138	163
8	150	175
9	163	188
10	175	200

Notes (a) Includes 50 psf load for workers, equipment, impact, etc.
(b) Includes 75 psf load for workers, equipment, impact, etc.

Figure 7-4
Concrete pressures for slabs

Pressures on Curved Forms

Plyform can also be used for curved forms. The radii in Figure 7-5 are approximate minimums for mill-run panels of the thicknesses shown when bent dry. Shorter radii can be developed by selecting panels that are free of knots and short grain, and by wetting or steaming the panels. An occasional panel may develop weaknesses at these radii.

Plywood thickness (in)	Across the grain (ft)	Parallel to grain (ft)
1/4	2	5
5/16	2	6
3/8	3	8
1/2	6	12
5/8	8	16
3/4	12	20

Figure 7-5
Minimum bending radii

Allowable Pressures on Plyform

Allowable pressures on Plyform Class I, Class II and Structural I are shown in Figure 7-6. Use these tables to design architectural concrete forms where appearance is important. Calculations for the pressures shown in these tables were based on a deflection limitation of 1/360 of the span.

Selecting the Framing for Wall Forms

Design the lumber studs and double walers for the Plyform you've selected. Maximum concrete pressure is 540 psf.

Designing the Studs

Since the plywood must be supported at 12 inches on center, space studs 12 inches on center. Assume the load carried by each stud equals the concrete

Support spacing (in)	Plywood thickness (in)					
	1/2	5/8	3/4	7/8	1	1 1/8
A Plyform Class I						
	Face grain across supports					
4	3265	4095	5005	5225	5650	6290
8	970	1300	1650	2005	2175	2420
12	410	575	735	890	1190	1370
16	175	270	370	475	645	750
20	100	160	225	295	410	490
24	—	—	120	160	230	280
32	—	—	—	—	105	130
36	—	—	—	—	—	115
	Face grain parallel to supports					
4	1860	2350	2910	3450	4615	5455
8	605	905	1120	1325	1775	2100
12	215	360	670	820	1100	1300
16	—	150	300	480	725	895
20	—	105	210	290	400	495
24	—	—	110	180	255	320
B Plyform Class II						
	Face grain across supports					
4	2675	3570	4515	4595	4925	5475
8	670	890	1135	1380	1885	2105
12	295	395	505	615	840	965
16	150	225	285	345	470	545
20	—	135	195	240	325	375
24	—	—	105	140	205	240
32	—	—	—	—	—	115
	Face grain parallel to supports					
4	1610	2235	2765	3280	4370	5165
8	455	800	1065	1260	1680	1985
12	130	255	485	745	1040	1230
16	—	105	210	335	505	670
20	—	—	150	240	355	490
24	—	—	—	125	195	265
C Structural I Plyform						
	Face grain across supports					
4	3925	5240	6175	6490	6535	7240
8	980	1310	1680	2060	2515	2785
12	415	580	745	915	1335	1540
16	175	270	380	485	725	845
20	100	160	230	305	465	550
24	—	—	120	165	260	315
32	—	—	—	—	115	145
36	—	—	—	—	—	125
	Face grain parallel to supports					
4	2520	3185	3940	5110	6255	7395
8	830	1225	1515	1965	2405	2845
12	255	425	825	1215	1490	1760
16	105	108	360	570	865	1145
20	—	125	255	400	555	685
24	—	—	130	215	335	435

Figure 7-6

Allowable pressures for Plyform for architectural uses (continuous across two or more spans)

Equivalent uniform load (1lb/ft)	Continuous over 2 or 3 supports (1 or 2 spans)								Continuous over 4 or more supports (3 or more spans)							
	Nominal size								Nominal size							
	2x4	2x6	2x8	2x10	2x12	4x4	4x6	4x8	2x4	2x6	2x8	2x10	2x12	4x4	4x6	4x8
200	49	72	91	111	129	68	101	123	53	78	98	120	139	81	118	144
400	35	51	64	79	91	53	78	102	38	55	70	85	98	57	84	111
600	28	41	53	64	74	43	63	84	31	45	57	69	80	47	68	90
800	25	36	45	56	64	38	55	72	27	39	49	60	70	41	59	78
1000	22	32	41	50	58	34	49	65	24	35	44	54	62	36	53	70
1200	20	29	37	45	53	31	45	59	22	32	40	49	57	33	48	64
1400	19	27	34	42	49	28	41	55	20	29	37	45	53	31	45	59
1600	17	25	32	39	46	27	39	51	19	27	35	42	49	29	42	55
1800	16	24	30	37	43	25	37	48	18	26	33	40	46	27	40	52
2000	16	23	29	35	41	24	35	46	17	25	31	38	44	26	37	49
2200	15	22	27	34	39	23	33	44	16	23	30	36	42	24	36	47
2400	14	21	26	32	37	22	32	42	15	22	28	35	40	23	34	45
2600	14	20	25	31	36	21	30	40	15	22	27	33	39	22	33	43
2800	13	19	24	30	34	20	29	39	14	21	26	32	37	22	32	42
3000	13	19	23	29	33	19	28	37	14	20	25	31	36	21	31	40
3200	12	18	23	28	32	19	27	36	13	19	25	30	35	20	30	39
3400	12	17	22	27	31	18	27	35	13	19	24	29	34	20	29	38
3600	12	17	21	26	30	18	26	34	13	18	23	28	33	19	28	37
3800	11	16	21	25	30	17	25	33	12	18	23	28	32	19	27	36
4000	11	16	20	25	39	17	25	32	12	17	22	27	31	18	27	35
4500	10	15	19	23	27	16	23	30	11	16	21	25	29	17	25	33
5000	10	14	18	22	26	15	22	29	11	16	20	24	28	16	24	31

Notes Spans are based on the 1991 NDS allowable stress values. Spans are based on dry, single-member allowable stresses multiplied by a 1.25 duration-of-load factor for 7-day loads. Deflection is limited to 1/360 of the span with 1/4" maximum. Spans are measured center-to-center on the supports. Lumber with no end splits or checks is assumed. Width of supporting members (e.g. wales) assumed to be 3 1/2" net (double 2x lumber plus 1/2" for tie).

Figure 7-7A

Maximum spans in inches for Douglas-Fir Larch or Southern Pine No. 2 framing

pressure multiplied by the stud spacing in feet. (This method is applicable to most framing systems. It assumes the maximum concrete pressure is constant over the entire form. Actual distribution is more nearly trapezoidal or triangular.)

$$540\ psf \times \frac{12}{12}\ ft = 540\ lb/ft$$

Figures 7-7 A and B show the maximum spans for lumber framing. Assuming you're using 2 x 6 Douglas-fir studs continuous over three supports (two spans), Figure 7-7A indicates a 51-inch span for 400 pounds per foot and a 41-inch span for 600 pounds per foot. Interpolate between these spans for a load of 540 pounds per foot.

$$\frac{540 - 400}{600 - 400} \times (51 - 41) = \frac{140}{200} \times 10 = 7\ in$$

For 540 pounds per foot, the span is 51 inches – 7 inches = 44 inches.

The 2 x 6 studs must be supported at 44 inches on center. Assume this support is provided by double 2 x 6 walers spaced 44 inches on center.

Designing the Double Walers

The load carried by the double walers equals the maximum concrete pressure multiplied by the waler spacing in feet, or:

$$540\ psf \times \frac{44}{12}\ ft = 1980\ lb/ft$$

Equivalent uniform load (1lb/ft)	Continuous over 2 or 3 supports (1 or 2 spans)								Continuous over 4 or more supports (3 or more spans)							
	Nominal size								Nominal size							
	2x4	2x6	2x8	2x10	2x12	4x4	4x6	4x8	2x4	2x6	2x8	2x10	2x12	4x4	4x6	4x8
200	48	71	90	110	127	64	96	117	52	77	97	118	137	78	112	137
400	34	50	63	77	90	51	76	99	37	54	69	84	97	56	83	109
600	28	41	52	63	73	43	62	82	30	44	56	68	79	46	67	89
800	24	35	45	55	64	37	54	71	26	38	48	59	69	40	58	77
1000	22	32	40	49	57	33	48	64	23	34	43	53	61	36	52	69
1200	20	29	37	45	52	30	44	58	21	31	40	48	56	33	48	63
1400	18	27	34	41	48	28	41	54	20	29	37	45	52	30	44	58
1600	17	25	32	39	45	26	38	50	18	27	34	42	49	28	41	54
1800	16	24	30	37	42	25	36	48	17	26	32	39	46	27	39	51
2000	15	22	28	35	40	23	34	45	17	24	31	37	43	25	37	49
2200	15	21	27	33	38	22	33	43	16	23	29	36	41	24	35	46
2400	14	20	26	32	37	21	31	41	15	22	28	34	40	23	34	44
2600	13	20	25	30	35	21	30	40	15	21	27	33	38	22	32	42
2800	13	19	24	29	34	20	29	38	14	20	26	32	37	21	31	40
3000	12	18	23	28	33	19	28	37	14	20	25	31	35	21	30	39
3200	12	18	22	27	32	18	27	36	13	19	24	30	34	20	29	38
3400	12	17	22	27	31	18	26	35	13	19	23	29	33	19	28	36
3600	11	17	21	26	30	17	25	34	12	18	23	28	32	19	28	35
3800	11	16	21	25	29	17	25	33	12	18	22	27	31	18	27	35
4000	11	16	20	24	28	17	24	32	12	17	22	26	31	18	26	34
4500	10	15	19	23	27	16	23	30	11	16	20	25	29	17	25	32
5000	10	14	18	22	25	15	22	29	10	15	19	24	27	16	23	31

Notes Spans are based on the 1991 NDS allowable stress values. Spans are based on dry, single-member allowable stresses multiplied by a 1.25 duration-of-load factor for 7-day loads. Deflection is limited to 1/360 of the span with 1/4" maximum. Spans are measured center-to-center on the supports. Lumber with no end splits or checks is assumed. Width of supporting members (e.g. wales) assumed to be 3½" net (double 2x lumber plus ½" for tie).

Figure 7-7B

Maximum spans in inches for Hem-Fir No. 2 framing

Since the walers are doubled, each waler carries half of the total, or 990 pounds per foot. Assuming 2 x 6 walers continuous over four or more supports, the table shows a 35-inch span for 1000 pounds per foot and 32-inch span for 1200 pounds per foot. Interpolate to find that 2 x 6s can span 32.5 inches for 990 pounds per foot. Support 2 x 6s at 32.5 inches on center with form ties. (Place bottom waler about 8 inches from the bottom of the form.)

Figuring the Load on Ties

The load on each tie equals the load on the double walers multiplied by the tie spacing in feet, or 1980 pounds per foot x $^{32.5}/_{12}$= 5362.5.

If the allowable load on the tie is less than 5000 pounds, decrease tie spacing accordingly. For example, a tie with 5000 pounds allowable load should be spaced no more than:

$$\frac{5000}{1980} \times 12\ in = 30\ in$$

Other Loads on Forms

You must also brace concrete forms against lateral loads due to wind or any other construction loads. Lateral loads are at least 10 pounds per square foot for wind load, or greater in some local building codes. In all cases, design forms over 8 feet high to carry at least 100 pounds per linear foot applied at the top of the form.

Design wall forms to withstand wind pressure applied from either side. Inclined wood braces can be designed to take both tension and compression. With careful design, you may need braces on only one side. Guy-wire bracing, on the other hand, can take only tensile loads. If you use it, you'll need bracing on both sides of the forms.

In general, wind bracing also resists uplift forces on the forms, provided the forms are vertical. If forms are inclined, uplift forces may be significant. Special tiedowns and anchorages may be required in some cases.

Building the Forms

You can build forms from plywood and 2 x 4s, or buy manufactured forms. Most concrete contractors only work with prebuilt forms. Some won't even take jobs that require job-built forms. They're more expensive and don't last as long as the factory-built. In this chapter, we'll look at both kinds of forms.

Factory forms go together like an erector set. If they're set on a fairly level surface, they fit together easily. As you put the forms together, you'll know if the footing has high or low spots because the pin holes won't match. The concrete will usually fill in any low spots, but the forms will ride on the high spots. If the footing isn't level, you'll probably have to shoot a level line with a transit. Mark it with nails in the plywood forms to show where to screed the top of the concrete in the forms. If the footing is level, you can avoid this whole operation.

Laying out walls is one of the most important parts of the job. You can lay out both concrete and block walls with either a transit or a tape measure. For most small jobs, a tape measure is adequate, since the excavation is open and you can use the tape in all directions. Whichever method you use, be sure to measure the diagonals to check for square. If the diagonals are the same, the walls will be square. But make sure the people holding the tape at each end are starting at the same side of the wall layout — either from the inside or the outside of the corners.

After establishing the corners, use a chalk line to show the location of the bottom of the forms. Some contractors will nail a shoe or plate to the footing to hold the bottom of the forms in line. Others just use the chalk line to line up the forms.

To begin a plywood-formed concrete foundation or wall, set and brace one corner first (Figure 7-8). Then place the succeeding panels and add bracing to hold them in place. Figure 7-9 shows the walers and strongbacks in place. Make sure everything is held together properly. Have the crew go around and hit

Figure 7-8
The corners are the first part of a plywood-formed wall

Figure 7-9
Plywood concrete form braced by walers and strongbacks

every pin a second time after the setup is complete. This might prevent a wall blowing out. Sometimes a pin can slide in above or below the tie. If only one pin that holds a tie is missing, the whole wall can rupture from the pressure in a domino effect.

If a foundation will have an opening for a garage door or access door, you can provide a bulkhead that will help the carpentry crew later. Install a treated board the thickness of the wall, with a nailed-on keyway made of a sawed-in-half 2 x 4, to the inside of the form. Install it with 16d nails. This will serve as both a bulkhead and a nailing surface for the door frame.

When installing window openings, I advise using steel window bucks. You can buy them from form manufacturers. Usually four 16- x 32-inch and four 20- x 32-inch forms are enough. Place them in the form locations according to the prints or the owner's specifications. They provide a sill in the concrete. Then the window frame is an integral part of the foundation. The glass is added later. On modular and premanufactured homes, there have to be openings for the units to be set in. The owner should specify these openings.

If the plans call for odd-shaped walls such as 45-degree turns, contact your form supplier. Chances are they'll rent or sell you an adjustable corner form. These have a hinge running the height of the form. Usually the inside corners are 6 inches wide and the outside corners are 2 inches wide. You can use them for a number of different angles.

Carefully examine the blueprints to make sure you're placing the correct size of reinforcing in the right location. Carefully locate and align the anchor bolts so the adjoining structural members will line up. Remember, it's next to impossible to install some anchor bolts in concrete after the concrete has hardened.

Prefabricated Forms

For years concrete forms were made strictly of wood and were built on the job. In most construction today, contractors use prefabricated forms built of wood, steel and wood, or aluminum. These forms come in many sizes for assembly on the job site.

Formwork labor and material often make up 30 to 50 percent of the total concrete wall costs, with labor averaging 2 to 3 times the material cost. To stay in business, you've got to select the least expensive forming system. Higher labor productivity means lower labor costs and higher profit. It also means faster construction cycles, a better chance of meeting schedules, lower overhead per job, less chance of delaying other trades, and more jobs handled per year.

The Jahn Forming System

Though there are many different brands of concrete form accessories, the Jahn forming system is probably the most common. It uses $^{5}/_{8}$-inch or $^{3}/_{4}$-inch plywood. Although $^{5}/_{8}$-inch is slightly lower in cost, $^{3}/_{4}$-inch is stronger, lasts longer, and uses fewer ties and brackets. I like saving money as much as the next guy, but I've found $^{3}/_{4}$-inch is more economical in the long run. A $^{5}/_{8}$-inch eccentric take-up on "A" or "C" brackets allows you to use a $4^{3}/_{4}$-inch wall tie with either $^{3}/_{4}$-inch or $^{5}/_{8}$-inch plywood.

The only preparation required with the Jahn forming system is gang-drilling the plywood, as shown in Figure 7-10. Drill holes $^{1}/_{8}$ inch larger than the snap tie head. This usually takes a $^{5}/_{8}$-inch diameter drill bit. The snap tie spacing layout you use will depend on the temperature, rate of pour, plywood thickness, and nature of the job. Figure 7-11 shows typical snap tie layouts.

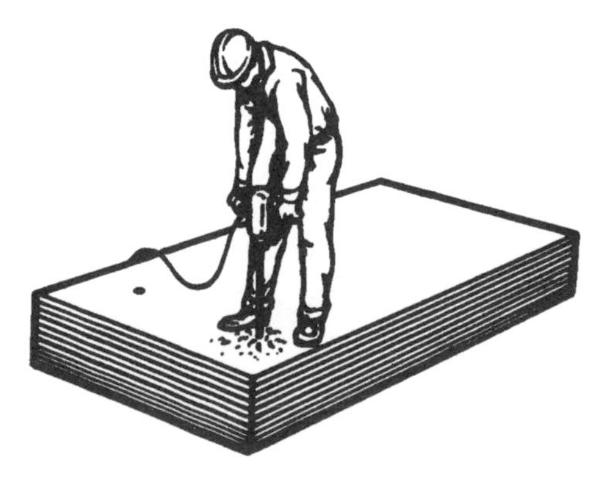

Figure 7-10
Gang drilling the plywood

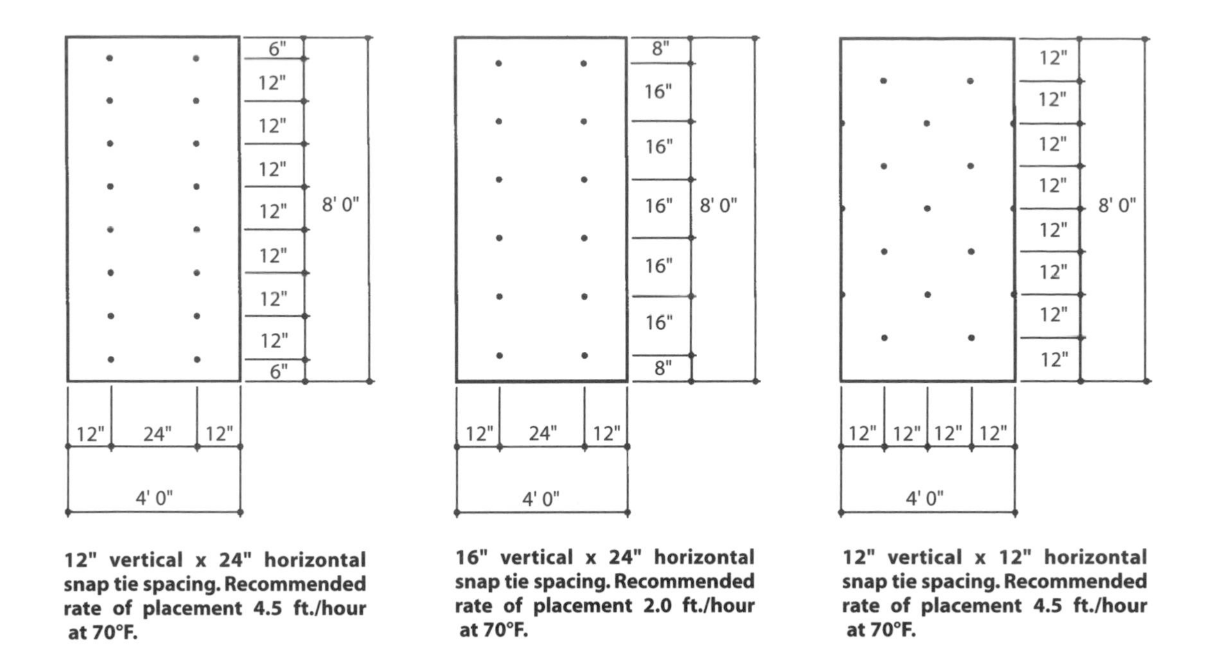

12" vertical x 24" horizontal snap tie spacing. Recommended rate of placement 4.5 ft./hour at 70°F.

16" vertical x 24" horizontal snap tie spacing. Recommended rate of placement 2.0 ft./hour at 70°F.

12" vertical x 12" horizontal snap tie spacing. Recommended rate of placement 4.5 ft./hour at 70°F.

Notes The above recommendations are based on the use of ¾" Plyform Class I, and 2 x 4 S4S studs (Douglas Fir-Larch, Southern Pine or equal having a minimum allowable fibre stress of 1,200 psi). Design is based on all formwork members being continuous over 4 or more supports.

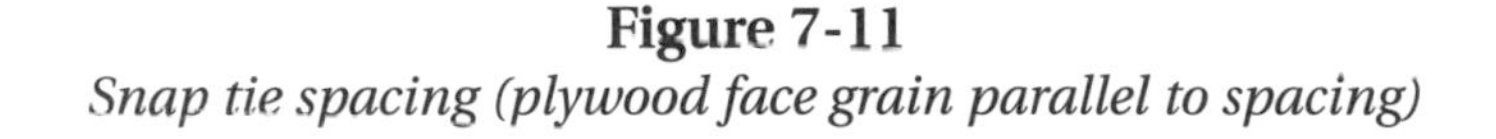

Figure 7-11
Snap tie spacing (plywood face grain parallel to spacing)

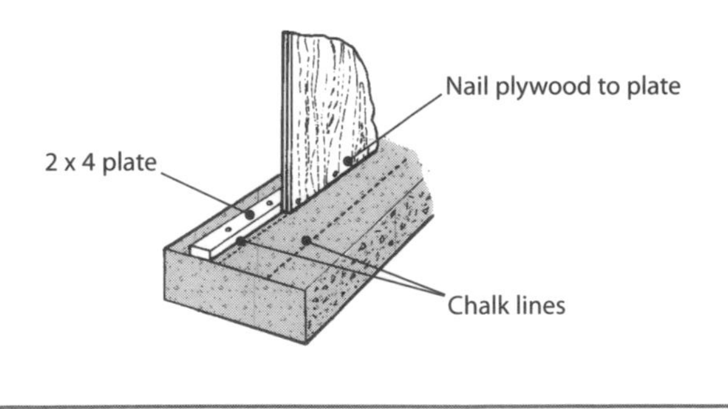

Figure 7-12
Footing plates

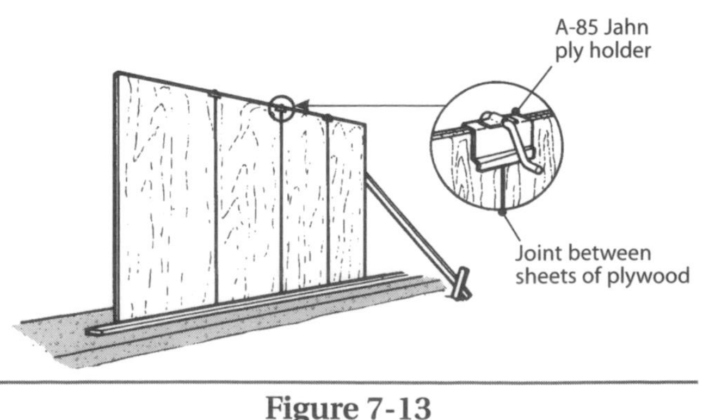

Figure 7-13
Panel erection

The Footing Plates and Panels

Good forming starts with good level footings. Snap a chalk line behind the plywood at the wall line, then nail down a 2 x 4 plate (Figure 7-12).

Figure 7-13 shows the correct positioning of the panels. Put up the first sheet of plywood, plumb it and nail it to the plate. Brace it temporarily. Then erect additional sheets, nailing them to the plate and holding them temporarily with Jahn ply holders. Make sure all vertical joints are tight and snug, and that the panels are level at the top if they're going to be stacked.

Installing the Brackets and Walers

Insert the end of the snap tie through the hole in the plywood. Jahn recommends the 4¾-inch L&W A-S snap tie, 2,250 pounds SWL with the "A" bracket, ⅝- or ¾-inch plywood and 2 x 4 walers. Two men can do this job quickly and efficiently if one inserts the ties and the other installs the brackets. (See Figure 7-14.)

Figure 7-14
Installing the snap ties and "A" brackets

Figure 7-15 shows how to install the "A" bracket both with and without the walers in place. Working from top to bottom, install the walers in the "A" brackets, tightening as you go (Figure 7-16). Make sure the waler joints occur at the brackets. Check the panels for alignment and plumb.

You can construct most of the second wall from the outside. As shown in Figure 7-17, this can best be done by two men slipping the plywood over the tie ends. Starting at the bottom, move the panel from side-to-side or up-and down to align the holes in the plywood with the tie ends.

There's no special treatment needed for inside corners. Just alternate the walers as shown in Figure 7-18 A. To reduce the number of panels you have to cut, start the corner on the inside wall with a full 4-foot plywood panel. When you erect the outside wall, install full panels in line with the inside panels, and fill out the outside corner with filler panels cut the width of the wall thickness plus the thickness of the plywood. See Figure 7-18 B.

Use the Jahn Cornerlock to secure the walers to the outside corners. The cam action of the Cornerlock draws the 2 x 4 walers together and eliminates costly

A Without waler in place

Place "A" bracket on tie end by slipping slotted bracket body over tie end as shown.

Slip eccentric **loosely** over tie end so tie is just in slot as shown.

Drop 2 x 4 wale in place as shown. Seat in place with a hammer blow, if necessary.

B With waler in place

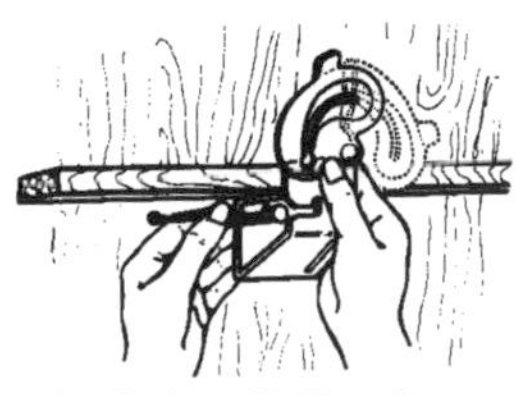

Slip back slot of "A" bracket over tie directly behind tie head.

Push bracket back toward plywood until tie head emerges through round hole in front slot.

Swing the eccentric over tie end and tighten.

Figure 7-15
Installing the "A" brackets

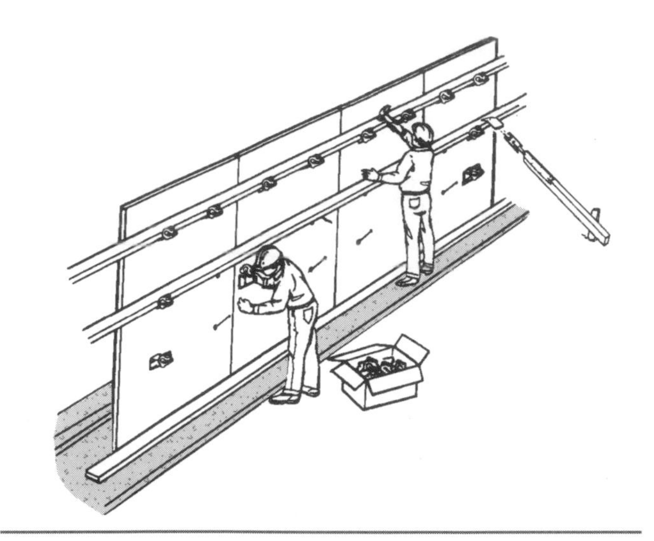

Figure 7-16
Installing the wales

Figure 7-17
Erecting the inside wall panels

overlapping, blocking and nailing. Place one waler flush at the corner and let the other run free. Slip the Cornerlock into place with its handle perpendicular to the waler. Drive two nails through the holes on the clamp and pull the handle around 90 degrees (Figure 7-19). The result will be a snug, tight outside corner.

You can install vertical strongbacks to align the forms and to tie stacked panels together. Use loose double 2 x 4s for the strongbacks, along with "C" brackets and 8¼-inch snap ties or 4¾-inch ties with a tie extender. Normal spacing for the strongbacks is 8 feet on center. See Figure 7-20.

Stacking the Forms

If you're going to stack the forms, first install joint covers. One method is to drill ⅝-inch holes 1⅛-inch down from the top edge of the plywood. Install snap ties, "A" brackets and walers, then the second panel of plywood. Nail the top sheet of plywood to the waler. Or you can install the snap ties in the joint between panels, with double walers and "C" brackets. Another alternative is to nail 4 x 4 walers to the lower plywood panel, held in place with strongbacks, before adding the second layer of plywood.

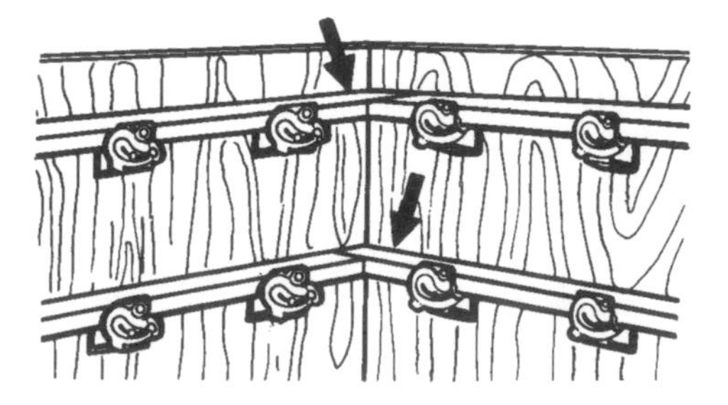

A Inside corner forming

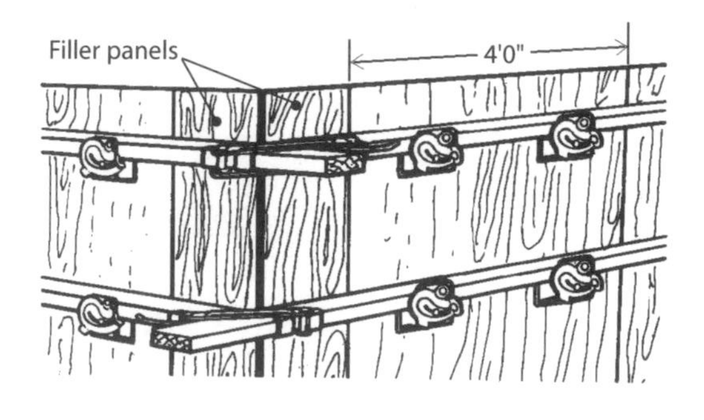

B Outside corner forming

Figure 7-18
Corner forming

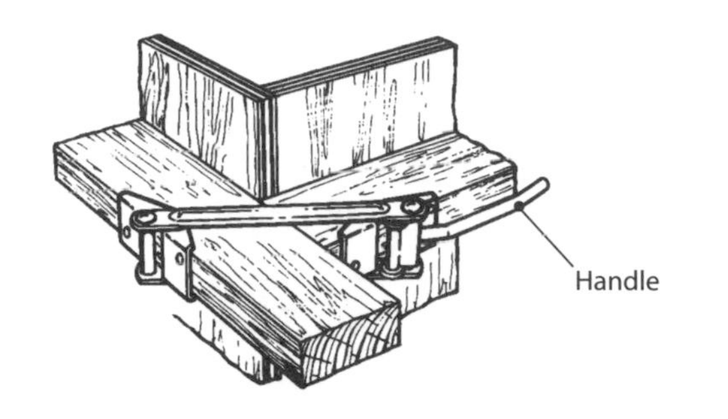

Figure 7-19
Installing the Cornerlock

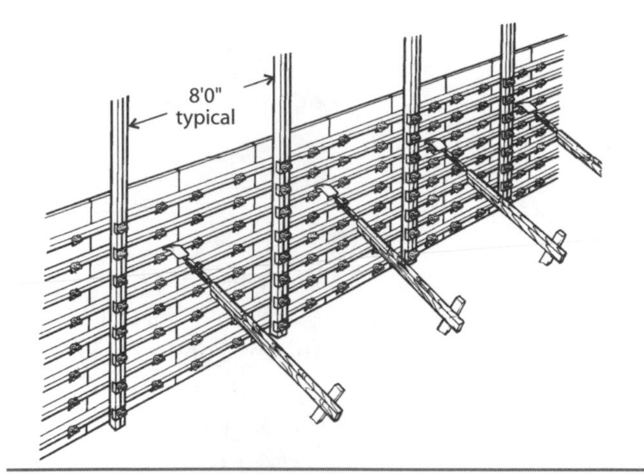

Figure 7-20
Installing the strongbacks

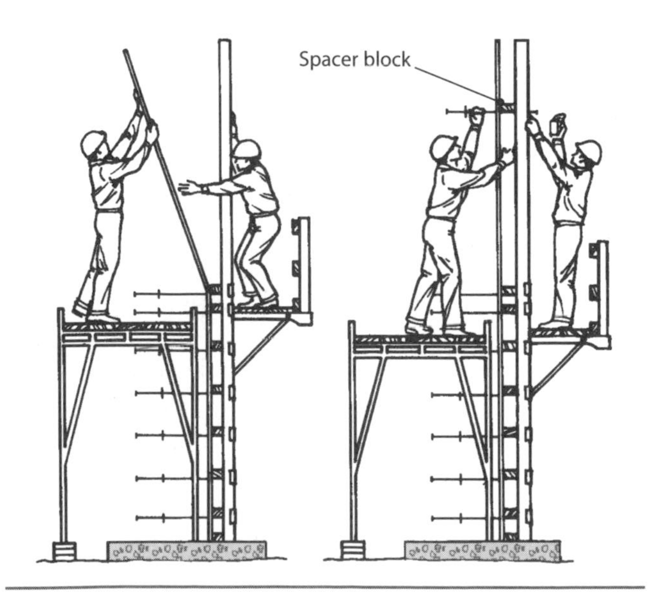

Figure 7-21
Installing the second lift of plywood

Then lift the layer of plywood and set it into position (Figure 7-21). Nail the bottom of the sheet to the joint cover waler while holding it in place with a short 2 x 4 spacer block, snap tie and "C" bracket. Set the additional panels, nailing them to the joint cover waler and securing them to the adjoining panel with a ply holder. Finally, install snap ties, brackets and walers, working from bottom to top.

Three-way Wall Forming

Figure 7-22 shows a three-way wall configuration, with horizontal walers attached with a combination of "A" and "C" brackets. Double the walers opposite the intersecting wall.

The Steel-Ply System

The Symons Company makes over 75 different panels and fillers for their Steel-Ply system, along with the hardware to put them together. With such a wide variety to choose from, you can form practically any dimension by combining panel and filler sizes and erecting them vertically or horizontally. A 2- by 8-foot panel weighs 75 pounds. You can set the panels directly on the concrete footing, using a chalk line as a guide, or on lumber plates. Plates are better, because they provide a positive on-line wall pattern and level out any rough areas on the footing.

When starting, use an outside corner piece and two fillers, as shown in Figure 7-23. The outside corner pieces come in 3-, 4-, 5-, 6- and 8-foot lengths and are 2½ inches by 2½ inches. There are slots in both sides for the wedge bolts. The slots are 6 inches on center. The special configuration of the wedge bolts (Figure 7-24) allows them to be used in many different types of form construction. Two identical wedge bolts function as a lock-bolt set, one as a connecting bolt and the other as a clamping wedge. You can get an adequate wedge bolt connection by pushing down on the head of the vertical wedge bolt with one hand while striking the head of the lateral wedge bolt with a hammer.

Set the corner piece and two fillers on a lumber plate or concrete footing, working along a chalk line. Connect the fillers to the outside corner by wedge-bolting through the corner piece and the filler side

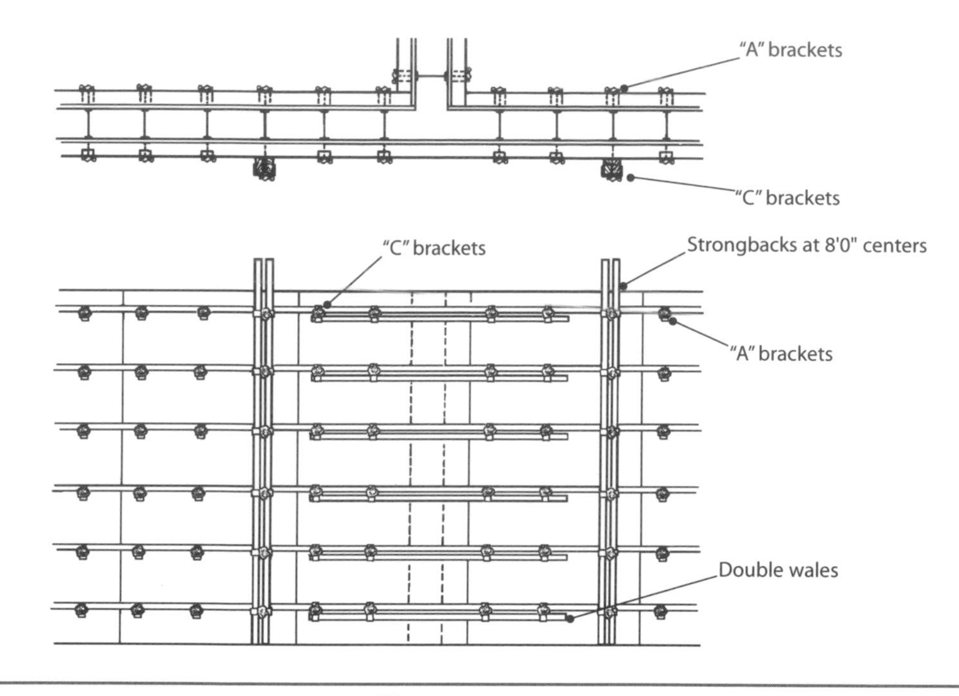

Figure 7-22
Forming a three-way wall

rails adjacent to the cross members. Tap the vertical wedge bolt with a hammer to secure the connection.

Place a panel next to the filler. Insert a tie through the dado slot. Now insert a wedge bolt horizontally through the filler slot, tie loop, and panel slot. Insert a second wedge bolt vertically and tap with a hammer to secure the filler, tie, and panel. As an alternative method, set both sides of the wall, wedge-bolting adjacent panels at the top and bottom. Then insert and wedge-bolt the ties using other slots.

Several factors determine the proper tie selection: spacing, safety, lateral concrete pressure, type of formwork (vertical two-sided, battered wall or unopposed single side), and the concrete finish desired.

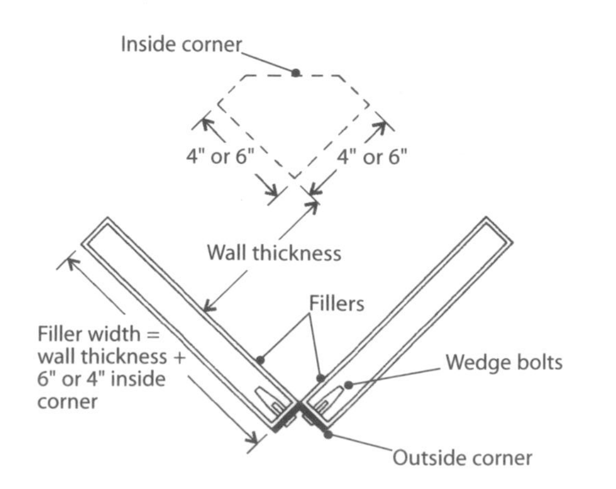

Figure 7-23
Plan of corner with Steel-Ply system

Figure 7-24
The wedge bolt shape allows their use in a wide range of form construction

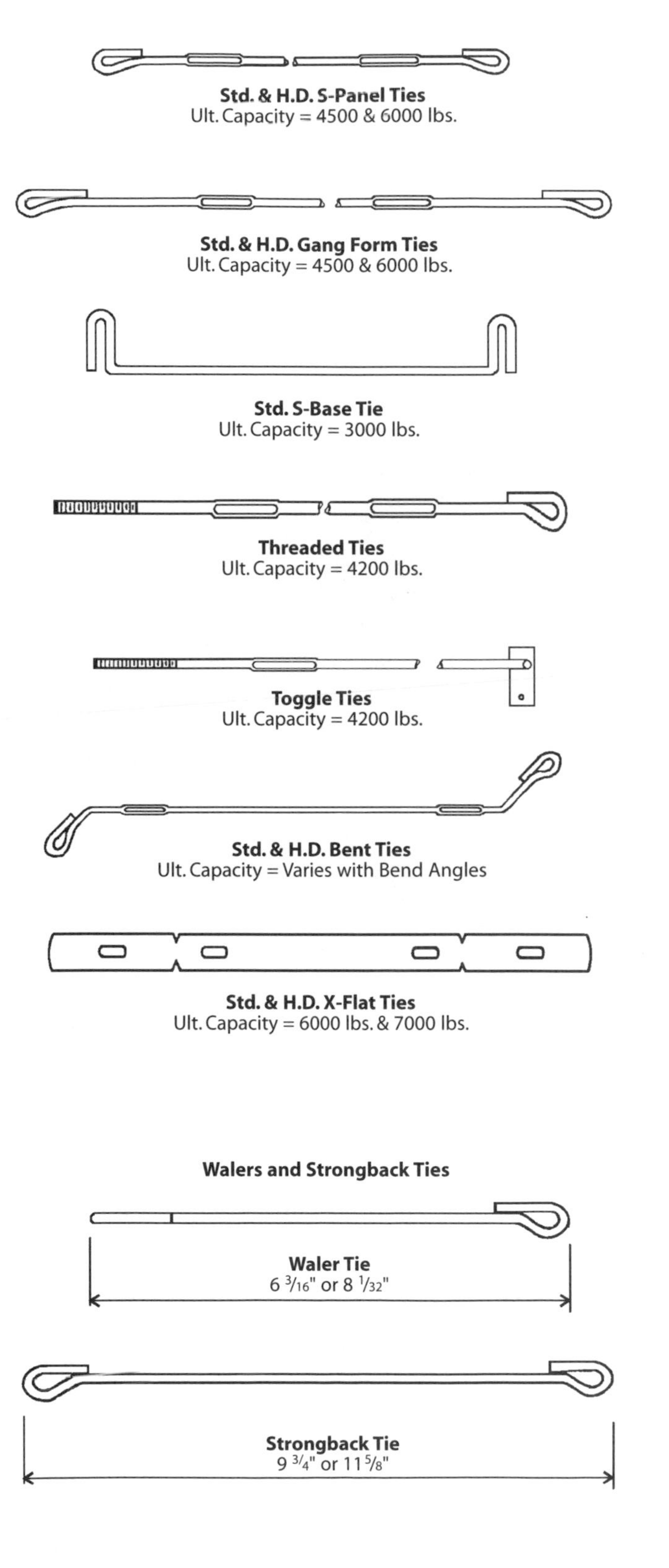

Figure 7-25
Symons ties

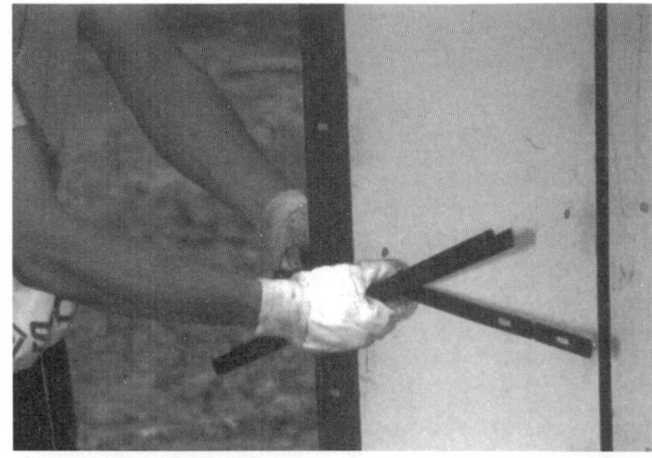

Figure 7-26
Installing the flat ties

Various ties are shown in Figure 7-25. Figure 7-26 shows the flat ties being installed. Figure 7-27 lists safe load ratings of wire and flat ties.

After setting 8 feet to 10 feet of panels, install walers using a one-piece waler clamp (Figure 7-28). Position single or double 2 x 4s and tap the locking wedge with a hammer. With this type of fabricated form, less material and labor are used for erection. Also, there is no measuring, sawing, or nailing. Use walers only on

Wire or flat tie	Ultimate load (lb)	Rating according to factor of safety	
		1.5 (lb)	2.0 (lb)
Standard duty wire tie	4,500	3,000	2,250
Standard duty threaded tie [1,2]	4,200	2,800	2,100
Standard duty S-base tie	3,000	2,000	1,500
Heavy-duty wire tie	6,000	4,000	3,000
Standard duty flat tie	6,000	4,000	3,000
Heavy-duty flat tie	7,000	4,500	3,500
Heavy-duty adjustable flat tie	7,000	4,500	3,500
Toggle tie (1)	4,200	2,800	2,100

[1] Tie capacity is dependent on adequate anchorage.
[2] When anchored with threaded inserts in 3500 psi concrete, an ultimate load of 4000 lb and a safety factor of 4.1 is recommended.

Figure 7-27
Safe load rating of Symons wire and flat ties

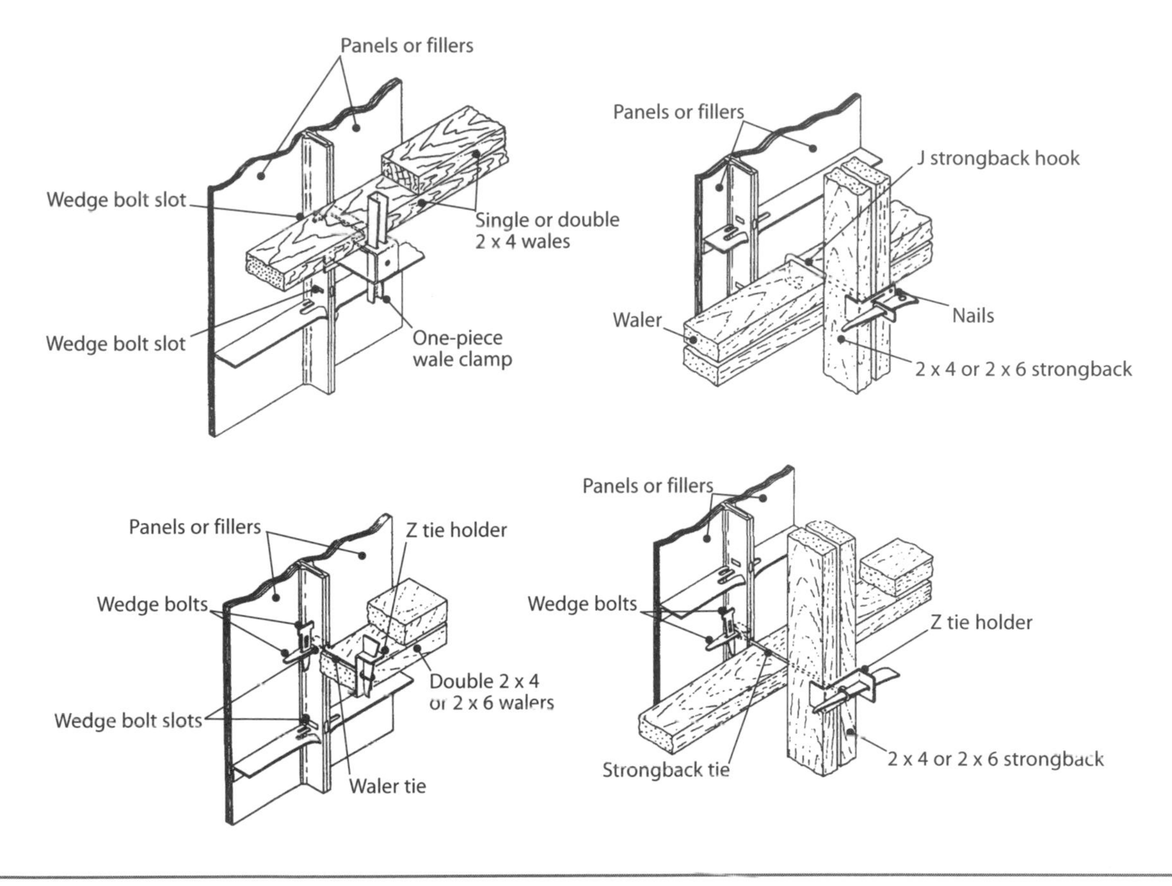

Figure 7-28
Waler and strongback clamps and ties

one side of the form. Waling is for horizontal panel alignment only and is not a structural part of the formwork. Job-built plywood forms use up to five rows of walers on each side of the wall.

Plumb and brace to vertical alignment. Only one side of the formwork needs to be braced, thus saving time. By using the steel brace plate, turn-buckle and steel stakes, you eliminate the need to cut, fit and nail together scrap lumber each time. This type of bracing can be used repeatedly.

Then start with an inside corner and set opposing panels of identical width so the ties are aligned. Slightly deflect the ties for clearance and rotate the panel into position. Secure the corner, ties and panel with wedge bolts.

Install the scaffold brackets after the first tier is set, using wedge bolts, not form ties (Figure 7-29). Starting at a corner, set the next tier using the same hardware

Figure 7-29
Installing the scaffold bracket

Figure 7-30
Stripping the forms

and techniques as the lower tier. Wedge-bolt the upper tier to the lower through panel end rails, using two connections through corner fillers and one in each other panel. Align and install the walers and strongbacks.

To strip the forms, tap the wedge bolts loose to remove braces, strongbacks, walers and panel connections. If a form coating was used, the forms will remove easily, leaving a smooth clean finish. See Figure 7-30.

Forming Columns and Pilasters

With prefabricated forms, you don't need walers, braces or ties on most jobs. You can combine inside and outside corners, panels and fillers to form just about any pilaster. Pilaster forms are nonsymmetrical. Wedge-bolt one edge to the wall form for 2-, 4-, 6-, 8-, 10- or 12-inch depths. Simply turn the pilaster form end to end for 1-, 3-, 6-, 7-, 9-, or 11-inch depths. One or more panels or fillers easily form any pilaster width. A 24-inch-wide pilaster needs no ties if you use a 30-inch column form opposite the pilaster form. Look at Figure 7-31.

Form Stripping and Maintenance

If you use manufactured forms, they should last up to 20 years if you take care of them. Make sure your crew understands that taking care of forms will make more money for everyone. When stripping the forms, don't use metal stripping bars or pries on plywood. Use wood wedges, tapping gradually when necessary. The strength, light weight and large panel size of the plywood helps reduce stripping time. Cross-laminated construction resists edge splitting.

Cleaning

As soon as possible after removal, inspect plywood forms for wear. Clean and repair them before storing them. Use a hardwood wedge and a stiff fiber brush for cleaning. Light tapping with a hammer will generally remove a hard scale of concrete. Then spot-prime, refinish and lightly oil before reusing. If the forms are oiled adequately, the concrete won't stick unless the forms are left on too long.

On prefabricated forms, you can reverse plywood panels after they get ragged. Patch the tie holes with metal plates, plugs or plastic materials. Remove nails and fill the holes with patching plaster, plastic wood, or other suitable materials. You may be able to get another five years from the panels.

Coatings and Parting Agents

Protective sealant coatings and parting agents for plywood increase form life and aid in stripping. Mill-oiled Plyform panels may require only a light coating of form oil between uses. For regular plywood, apply a liberal amount of oil a few days before the plywood is used, then wipe to leave a thin film. Adding a little diesel fuel to the form oil makes it easier to spray. But check the specifications before using any oil or compound on the forms.

You can get specially-coated panels with long-lasting finishes that make stripping easier and reduce maintenance costs. Care in handling will provide a maximum number of reuses.

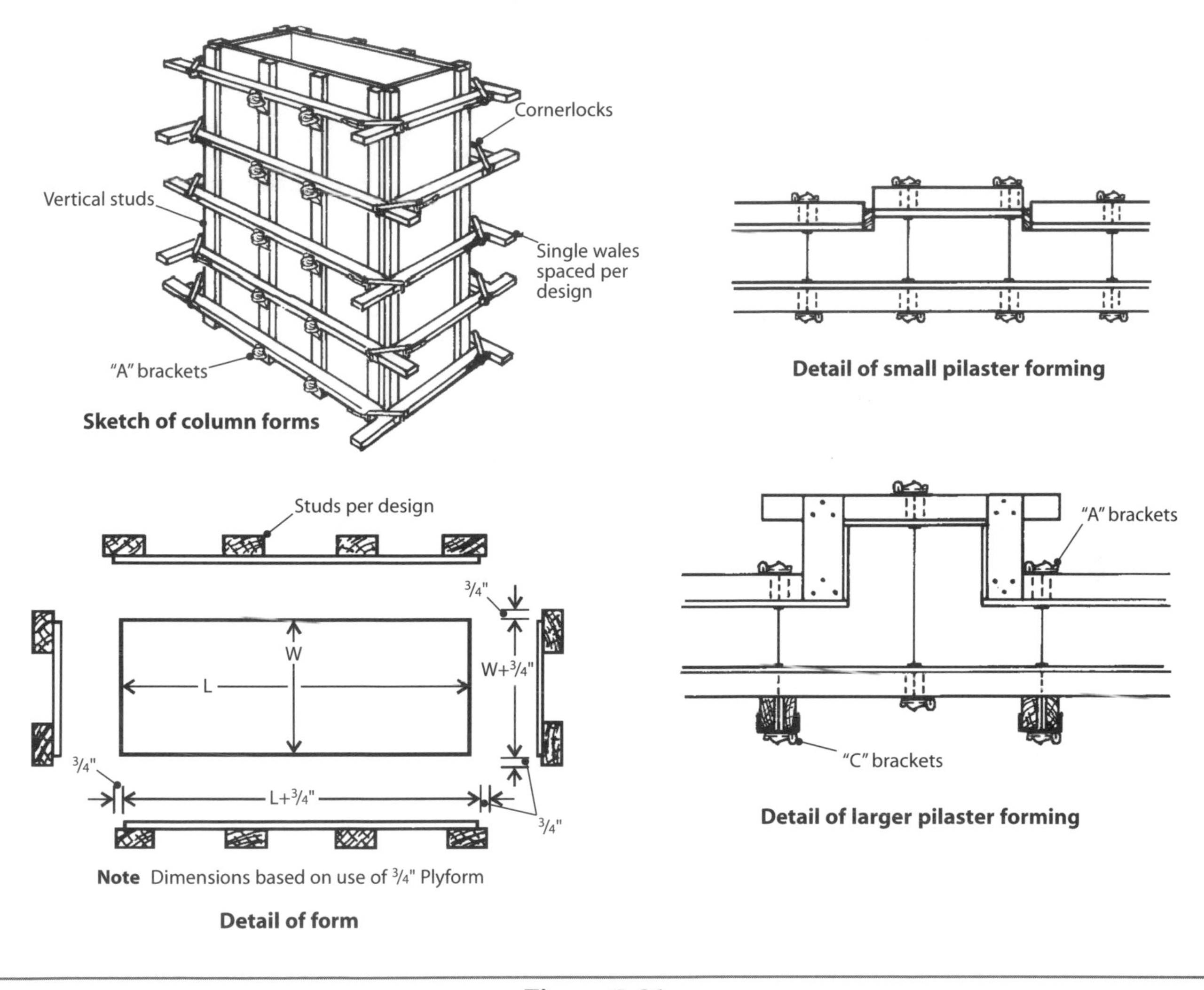

Figure 7-31
Column and pilaster forming ideas

Hairline cracks or splits may occur in the face ply. These "checks" may be more pronounced after repeated use of the form. Checks don't mean the plywood is delaminating. A thorough program of form maintenance, including careful storage to assure slow drying, will minimize face checking.

Plywood Form Coatings

Lacquers, resin, or plastic-base compounds and similar field coatings sometimes are used to form a hard, dry waterproof film on the plywood surface. In most cases, field-applied coatings reduce the need for oiling between pours. The performance level of the coating is generally rated somewhere between B-B Plyform and High Density Overlaid plywood. Many contractors report obtaining significantly greater reuse than with the B-B Plyform, but generally fewer than with HDO plywood.

Mill-coated products of various kinds are available in addition to mill-oiled Plyform. Some plywood manufacturers suggest no oiling with their proprietary concrete forming products, and claim exceptional concrete finishes and a large number of reuses. In any event, select a release agent based on the product's influence on the finished surface of the concrete. For example, you don't want to use release agents that include waxes or silicones when the concrete will be painted.

Handling and Storage

Handle the forms carefully to prevent panel chipping, denting and corner damage. Panels should never be dropped. Carefully lay the forms flat, face-to-face and back-to-back, for hauling. You can solid-stack them or stack them in small packages with their faces together. This slows the drying rate and minimizes face checking.

Using plywood stack-handling equipment and small trailers for hauling and storing panels between jobs will reduce handling time and minimize damage. If possible, store the plywood panels inside a building or a shed, or cover them loosely to allow air circulation without heat buildup. When panels are no longer suitable for formwork, you can use them for subflooring or wall and roof sheathing if they're in good enough condition.

Besides making them last longer, proper care of your forms can prevent some common problems with the surface of the concrete — dusting and staining.

Dusting

You may find surface dusting on concrete poured against a variety of forming materials, including plywood. There appears to be no single reason for this. Dusting has been traced to many possible causes, including excess oil, dirt, dew, smog, and unusually hot, dry weather, as well as chemical reactions between the form surface and the concrete.

Several methods have been successful in dealing with this problem. Some of these include proper form storage (cool, dry conditions) and cleanliness (avoiding needless exposure to dust, oil, and weathering). If dusting does occur, a water spray may help speed surface hardening. Cure the affected areas for a few days with water in a spray that's fine enough not to erode the soft surface. Other concrete specialists have recommended surface treatment solutions such as magnesium fluosilicate or sodium silicate.

Staining

Occasionally, a reddish or pinkish discoloration may appear when the concrete has been poured against High-Density Overlaid plywood forms. The stain, a fugitive dye, is temporary and usually disappears with exposure to sunlight and air. Where sunlight can't reach the stain, natural bleaching takes longer. Household bleaching agents such as Clorox or Purex (5 percent solutions of sodium hypochlorite), followed by clear-water flushing, have been effective in hastening stain removal.

When castor oil or other vegetable oil is used on the Plyform, it may cause turkey-red staining. Using mineral oil instead is the most direct solution. If you must use castor oil, test the concrete mix to determine whether it will cause staining.

8

Poured Concrete

❖❖

Once the forms are completed and inspected, it's time to order the concrete. And to make sure you can get what you want when you need it, it's important to establish a good working relationship with the concrete suppliers. When it works well, you both make money. But remember that you're not their only customer. Schedule ahead on large pours to make sure the supplier has the trucks available when you're ready for them.

Also listen to your suppliers' advice on estimating and technical questions, including admixtures. If your location is distant from the plant, you may need a retarder. Occasionally drivers will make suggestions before they unload. They deliver concrete every day and have probably seen every problem you can imagine (and some you can't). And if they're helpful, a tip may ensure that their help continues.

Always consider the difficulties your site may pose for a truck that weighs up to 100,000 pounds, fully loaded. The rule is that once you've ordered the concrete and it's in the mixer, it's yours — whether you can unload it or not.

Some concrete companies have large front-unloading mixers available to reach inaccessible spots. Make sure the supplier knows ahead of time if that's what you need. If your site is flat enough and clear enough for the truck, the driver can probably maneuver the chute and deliver the mix, even around a wall.

Common Concrete Delivery Problems

Here are some common problems with concrete deliveries. If any of these apply on your site, address them before ordering the concrete:

- Small roads to the site may have short bridges that aren't rated for the weight of the truck. Check ahead of time, and map out an alternate route for the driver to take if it's a problem.
- Check the height of overhead wires. Most front-end loaders are over 13 feet high, and rear discharge trucks are about 12 feet. If you're lucky, you'll just have to repair any damaged lines. But if it's an electrical line, someone could get seriously injured — or worse.
- Are there tree branches in the way? Get permission and clear away any obstructions.
- Consider the access roads and driveways to the site. Remember, concrete trucks have a large turning radius, and are much wider than your pickup. They tend to ride off the edge and can overturn if there's a significant drop-off. And if you let these trucks park in the wrong place at the job site, they may crush or seriously damage underground pipes or tanks. You're responsible for the truck once it leaves the curb line. If it gets stuck, you pay to have it towed out. But if it's still full, you'll probably have to unload it first — wherever it is.

- ❖ Try to arrange a place for the driver to clean out his chutes. Extra water on the site is always a good idea, especially in hot weather.

As part of your preparation for the pour, make sure you have extra equipment on the site. You'll be glad you have an extra wheelbarrow on hand if the first one blows a tire. And if you're doing a large pour, organize the work so the trucks can unload as soon as possible, to prevent the load setting up in the truck.

One of your biggest problems could be running short of concrete. I recommend overestimating a little. It's less expensive to throw away half a yard than to have to send for half a yard. You can get a short load, but you'll pay dearly for it. And the delay in getting the additional concrete can cause cracking in the finished pour.

The Importance of Water

One of the causes of poor concrete is excessive water in the mix. Many contractors add extra water to make it easier to shoot the concrete. This can come back to haunt you if the job fails and you have to redo it later.

You need water in concrete or it won't harden. It combines with the cement to form a paste that binds the aggregates together, and plays a large part in the curing process. Too much water dilutes the paste, which make the concrete weaker. That can lead to any of the following problems:

1) Dusting. As the concrete hardens, water bleeds to the surface. If there's too much water in the mix, it carries fine aggregates to the surface, which causes dusting and crazing. Dusting can also be caused by troweling when there's too much water on the surface, and inadequate curing. But don't spread dry concrete on the surface of fresh concrete for any reason. If surface water doesn't evaporate, use a garden hose as a drag hose to pull off the water.
2) Scaling. A good way to reduce scaling is to use air-entrained concrete. Too much water defeats the purpose of the air-entraining agents because slumps greater than 5 inches usually decrease the air content. Freeze/thaw cycles and salt or deicing agents placed on the surface can also cause scaling. To reduce scaling, don't trowel when there's water on the surface, and begin curing as soon as possible, using wet burlap, waterproof covering or ponded water.
3) Rough surfaces and sand streaking. As the extra water bleeds to the surface or the sides of the forms, it washes out the cement paste, leaving a streaked, unattractive surface. It's more difficult to trowel the concrete surface.
4) Too much water tends to increase shrinkage and create a lower tensile strength. These contribute to cracking in foundation walls.
5) Concrete that's too wet takes longer to set up. For contractors, this means higher labor costs, and possibly overtime.

Here are some rules of thumb for adding water to the concrete mix:

- ❖ Adding 1 gallon of water to 1 yard of concrete will increase the slump 1 inch.
- ❖ For each gallon you add, you decrease the rating. So if you add 3 gallons extra to a 5-gallon mix, you decrease the strength from 5,000 psi to 2,000 psi. And you increase the potential for the concrete to crack.

Most workable concrete mixes contain more water than needed to make the concrete hard. About half the water in a typical mix is there to make it more workable. Finding the balance is the trick. You'll be more successful with concrete if you place it as dry as possible. Use the minimum amount of water you need to make it workable.

Placing the Concrete

Always place concrete as close as possible to its final position. Don't try to flow or push it too far. That can cause honeycombing and segregation of the aggregates, which weakens the concrete.

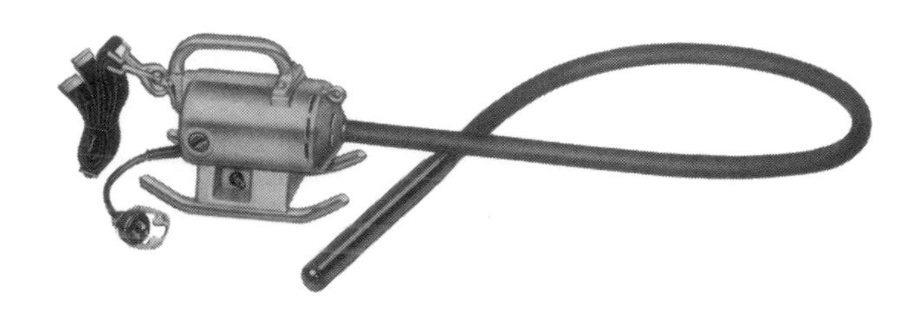

Figure 8-1
Electric concrete vibrator

Spade or vibrate the concrete as it enters the forms. Figure 8-1 shows a typical concrete vibrator. This machine has a flexible shaft that moves and settles concrete within forms. You can rent one at an equipment rental yard or from a construction/building supply store. It's worth the cost. You're less likely to overwork the concrete if you use a concrete vibrator.

But there are pros and cons to vibrating. If you pour concrete with a 4- to 6-inch slump, light vibrating will settle the mix and help prevent honeycombing. Take care, though. Too much vibrating can cause excess pressure on the forms, settle the aggregates, or move the fine aggregate and cement to the top of the wall, weakening it. If you use a vibrator to flow concrete around a wall form, you may just be moving the finer parts around the wall.

Pour concrete walls with as low a slump as possible. I've seen contractors soup up concrete to make it flow around the forms to the other side. When the forms are removed, the wall looks fine. But the concrete that has flowed around the wall for some distance will be considerably weaker.

If you're mixing concrete on site, place it within 20 minutes after you finish mixing it. In warm weather, concrete can begin its initial set in about 20 minutes. If you have to disturb the concrete after its initial set, it'll lose strength. Never rewet and remix concrete that sets up before you place it in the forms. Throw it out and mix a new batch.

Slab-On-Ground Construction

We covered the perimeter construction of slabs-on-ground in Chapter 4 on footings. This chapter covers how to pour the rest of the slab. The slab-on-ground is constructed in warmer climates, and also used to make interior floors such as basements or slab floors.

Reinforcing is generally placed in the top half of a concrete slab, as shown in Figure 8-2. Some contractors pour a layer of concrete, drop the mesh on top of it, and pour the rest over the first layer immediately. Others prefer to lay the mesh on the ground. Then, while pouring the concrete, they pull the mesh up into the slab using special hooks. Consult local codes to see if they specify the procedure for placing the wire mesh reinforcement.

The Superior Screed Joint, developed by the Superior Concrete Accessories Company, makes concrete pouring easier and saves both time and money. The Superior screed key joint shown in Figure 8-3 is a 24-gauge galvanized metal contraction joint that acts

Figure 8-2
Placement of wire mesh reinforcement

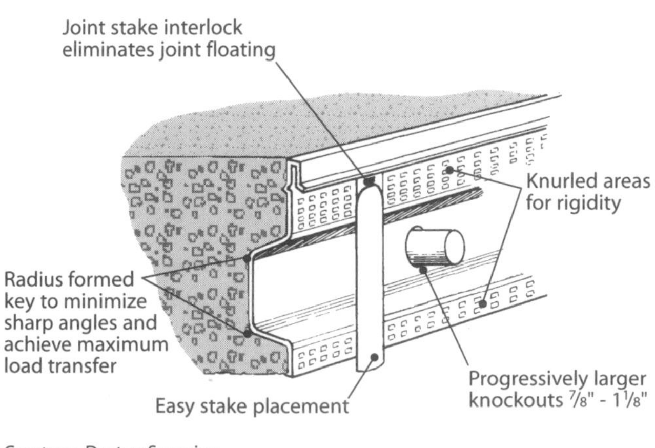

Courtesy: Dayton Superior

Figure 8-3
Superior screed Key-Loc® joint

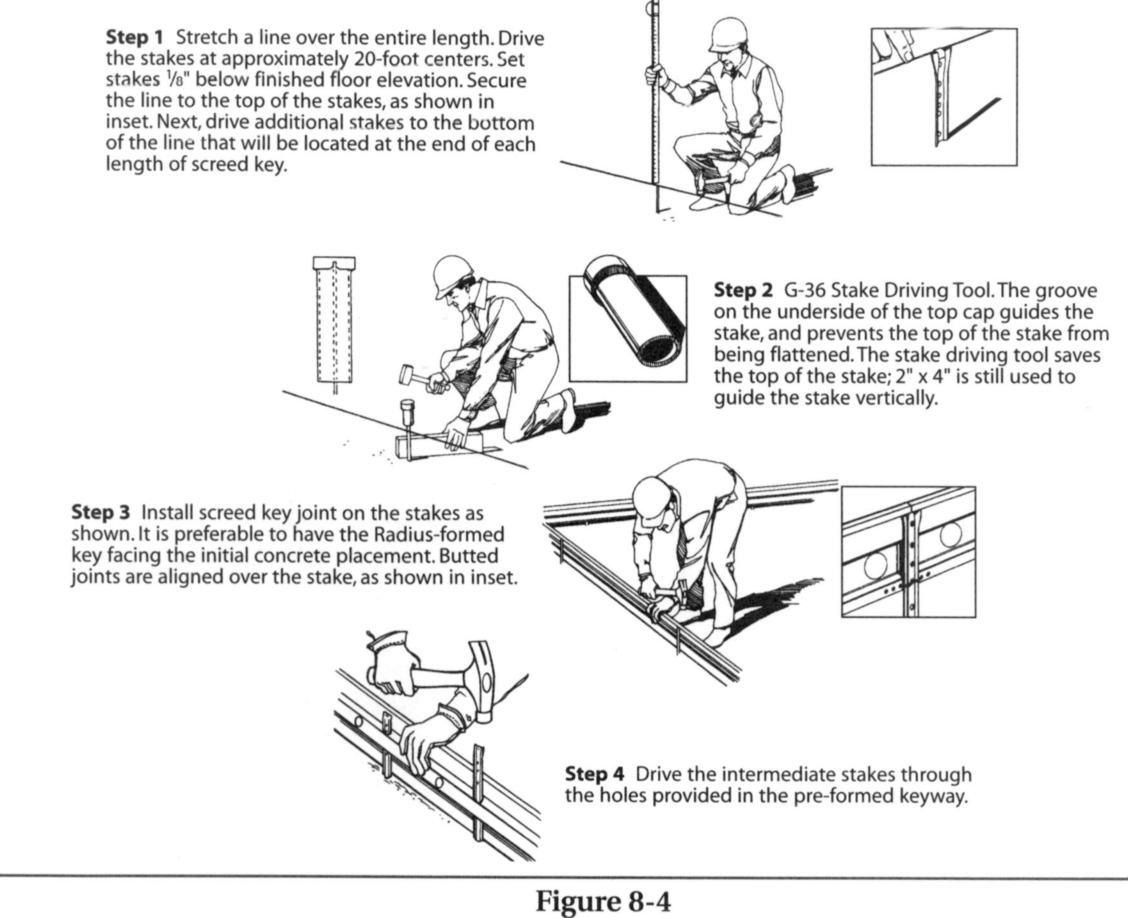

Figure 8-4
Installing the Superior screed key joint

as a screed and is left in the slab; no form removal, joint repair or filling is necessary. It comes in 10-foot lengths for slabs from 4 to 12 inches thick. The square top screed edge minimizes spalling at the joint. A built-in engineered key joint provides good load transfer. And the key has rounded edges to eliminate sharp-angle fracture points in the slab. The dowel knockouts, to allow placement of the reinforcement, are on 6-inch centers. Figure 8-4 shows how to install one of these screed joints.

You also need a keyway when you're pouring slabs at different times, and to bond masonry walls to the footing. You can construct keyways for footings by suspending 1-by or 2-by boards from the spreaders in the forms. Or Dayton Superior has a nail-on metal keyway that's more economical than job-constructed keyways (Figure 8-5). They're quick and simple to install, and create a straighter keyway.

Finishing the Slab

No matter how well the pour goes, proper finishing is essential for a durable, attractive slab.

Tools for Finishing Slabs

On large concrete pours, I recommend using a power-driven screed that lets you screed up to 10,000 square feet per hour. This power screed will strike off, compact and float the slab in one pass, leaving a semi-finished surface. For slabs 16 feet to 24 feet wide, you can use two equally-spaced units. Rent this equipment for large pours if you don't own one.

The Rollerbug is a valuable tool for floating and finishing. Contractors who have used the Rollerbug like the surface it leaves. It's ideal for all finishes —

Step 5 Tin snips or a metal cutting saw may be used when it is necessary to cut to length, or to trim around conduit, pipe, etc.

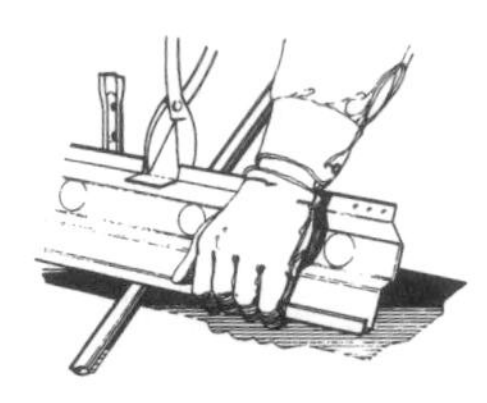

Step 6 Where joints meet at right angles, load key joint may be trimmed to fit as shown. Note the stake supports are close to the joints.

Step 7 1⅛" knockouts are supplied as shown on 6" centers where dowelling is specified. When screed key joint is used as a bulkhead for a construction joint, the knockout tab would be bent back into the pour at a 45° angle as shown.

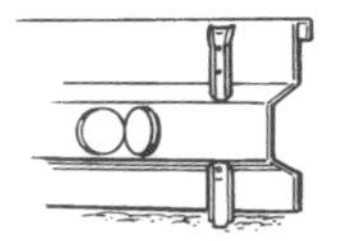

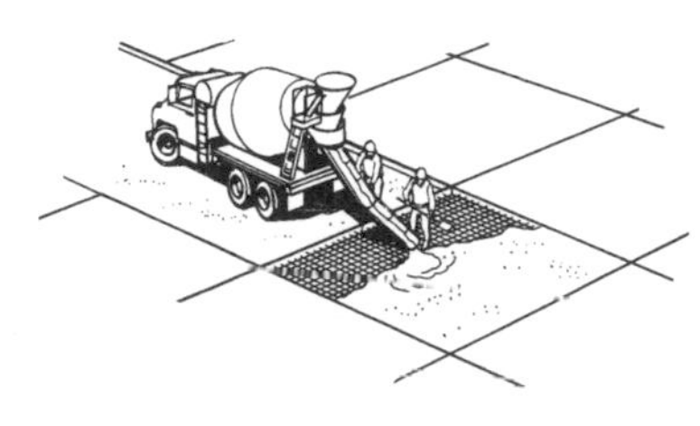

Step 8 Entire slabs may be poured at once, by leaving out 10-foot sections of screed key joint temporarily until the concrete trucks can pull ahead. Good construction practices dictate that slab reinforcement be properly supported with bar supports after the truck has pulled on through.

Step 9 The more common method is to pour concrete in strip fashion as shown. When a strip is poured and finished, there are no added and costly steps of cutting, joint treatment, form stripping, edge damage or concrete surface damage.

Courtesy: Dayton Superior

Figure 8-4 (Continued)
Installing the Superior screed key joint

exposed aggregate, float, broom or trowel. And it's excellent for depressing exposed aggregate. The rollers are the only moving parts. Maintenance is simple — just clean the rollers the same way you clean other cement tools.

The Jitterbug tamper brings fine material and fat to the surface for fast, easy finishing. It also settles larger aggregate to consolidate the slab for greater strength.

The rugged power trowel shown in Figure 8-6 is designed for the roughest concrete work you have to do. It helps achieve a perfect finish even on large projects. This machine has a deadman's control lever that stops the blades the moment the lever is released.

The pony trowel shown in Figure 8-7 is a low-cost easy-to-use trowel that can float up to 2,000 square feet per hour. You have to hook a slow-speed heavy-duty drill to the trowel to power it. The drill should be ½ inch or larger with a chuck speed of 500 rpm or less.

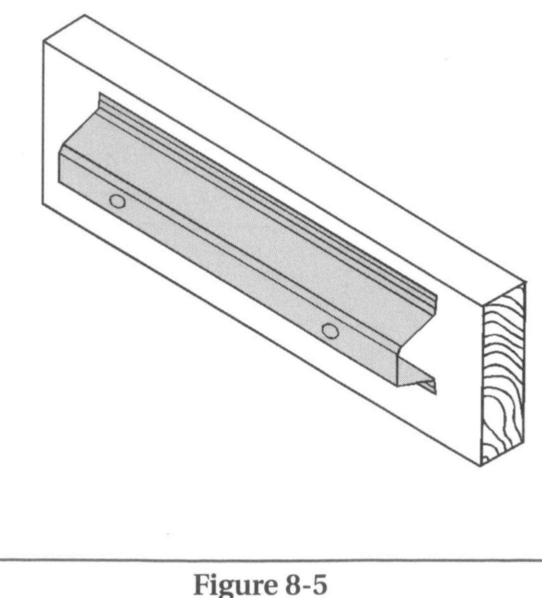

Figure 8-5
Nail-on keyway

Figure 8-6
Power-driven trowel

Figure 8-7
Pony trowel

Curing Concrete

Concrete hardens due to hydration — a chemical reaction when you combine water and cement. It begins on the surface, as the outer portion of the concrete hydrates and forms a concrete gel. As water from below the surface continues to soak through this gel, the hydration continues. Concrete reaches most of its strength within the first 28 days. But as long as moisture is present, the hydration can continue for many years. Nothing can stop the hydration process except a lack of moisture or freezing temperatures.

For the best curing, keep the concrete damp and about 70 degrees F. If it dries out too fast in its initial curing stage, the quality will be poor forever. Good quality concrete that's poured and cured properly is impervious to moisture. That's your goal. If water can't get in, it can't cause problems. If water can enter, it will subject the concrete to cracking during the freeze-thaw cycles.

Controlling Cracking in Concrete

Practically all concrete will expand or contract with:

- ❖ Changes in temperature
- ❖ Unequal loads placed on the surface
- ❖ Changes in moisture under the surface
- ❖ Heaving or settling of the subgrade
- ❖ Freezing and thawing of the subgrade

Freshly-poured concrete contains more water than it actually needs to hydrate. When this extra water starts to evaporate, the concrete develops tensile stresses. Cracks develop when the tensile stress is greater than the tensile strength of the concrete. See Figure 8-8.

Shrinkage Cracking

Slabs shrink differently depending on their thickness and shape. Here are some of the different conditions that will affect cracking:

- ❖ Humidity changes during the drying process
- ❖ Water/cement ratio
- ❖ Temperature
- ❖ Condition of the subbase

After you place any concrete it will change in volume as it loses moisture. This happens faster in fresh concrete than in concrete that's begun to harden. In fact, up to 80 percent of water loss due to evaporation occurs within the first 24 hours. Cracking due to shrinkage is typically classified as either *plastic* or *drying* shrinkage. Plastic shrinkage occurs before the concrete has reached initial set. Drying shrinkage occurs after the set. *Early* shrinkage is the combination of plastic shrinkage and drying shrinkage that occurs during that first day.

Concrete that isn't restrained won't crack due to shrinkage. But most concrete is restrained if only because there's a difference in shrinkage between the surface and the underlying layer.

There's a fiberglass additive you can add to help minimize shrinkage. But concrete will still shrink some, and probably crack. It's up to you to control the cracking so it doesn't happen randomly. It's always a good idea to put control joints and isolation joints in a concrete surface to predetermine the crack location.

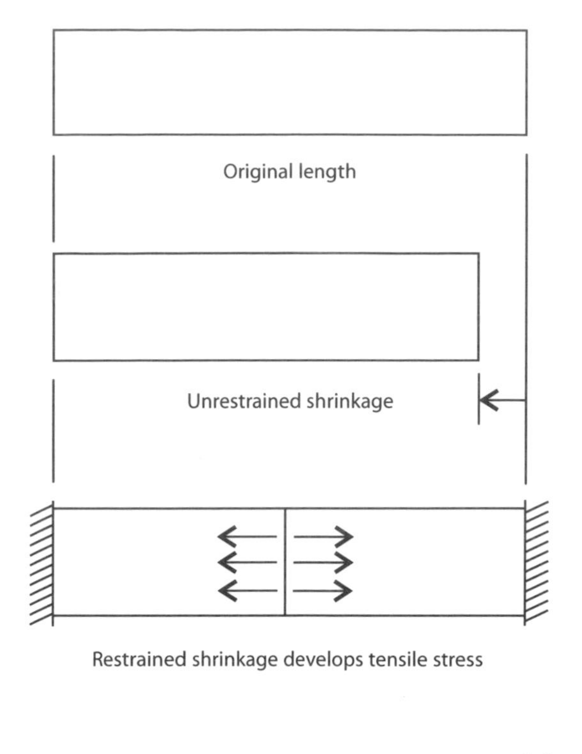

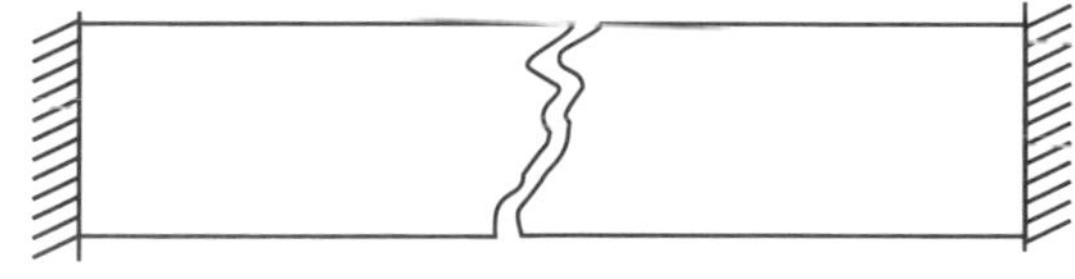

If tensile stress is greater than tensile strength, concrete cracks

Figure 8-8
Cracking of concrete due to shrinkage

Control Joints

Cut control joints into the surface of concrete as soon after pouring as you can. That makes it easier for the joint tool to force the larger aggregates down in the joint. In small slabs, cut a control joint with a groover, using a board or other straightedge laid across the fresh concrete as a guide. Some groovers are made with a deeper bit than others. Thin deep-bit groovers cut a thin, deep groove to keep spike heels from catching in the walks. They're available with up to 2-inch deep bits.

For most jobs, use a groover like the one in Figure 8-9 with a 3/4- to 1-inch cutter, depending on the depth of the slab. Cut the joint 1/4 to 1/5 the thickness of the slab. You can use a margin tool or a bricklayer's trowel initially to make the joint deeper and then finish it off with the groover.

Another way to control the cracking is to cut a joint in the concrete with a diamond blade when the concrete is about 8 hours old. In most cases, you can cut the joint about 1/3 the thickness of the slab. Place the joints at points of stress.

In a typical sidewalk, you usually place control joints 4 to 5 feet apart. On a large walk, place the joints so they cut the walk in squares. For large slabs, locate joints every 10 feet. If a slab is shaped irregularly, put joints at the juncture of an offset. Figure 8-10 shows control joints intersecting at the spot where a steel column will go.

Figure 8-9
Groover

Figure 8-10
Control joints intersect at location for a steel column

Another less costly method is to use the Form-A-Key by Dayton Superior. To install the key, run a string line along the slab where you want the concrete to crack. Insert the E-Z Crack material vertically into the slab until it's flush with the top of the slab (Figure 8-11). Be careful not to overlap the ends. E-Z Crack stays rigid and straight when you're installing it.

Other Joints

You can install an *isolation joint* that lets a slab and adjoining surfaces move separately so cracks won't occur. Put an isolation joint wherever a slab meets walls, columns and footings. You can make isolation joints from ½-inch-thick rigid asphalt-impregnated insulation felt, building paper (tar paper) or polyethylene materials. Make the joint watertight by applying a strip of caulk or tar along the joint. Figure 8-12 shows an isolation joint and the perimeter insulation. The laser level in the picture is now commonplace on large construction jobs because of its extreme accuracy.

Another type of joint is the *construction joint*. You make true construction joints by using steel reinforcing between a previously-poured section and a new

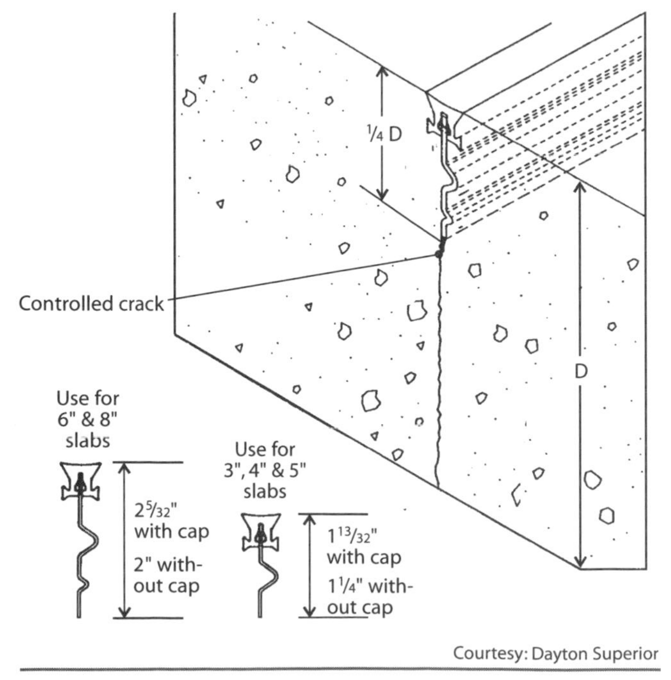

Figure 8-11
E-Z Cracks induce cracks in concrete slabs on grade

Figure 8-12
Isolation joint where slab meets the wall

pour. Sometimes you'll need a bonding agent to help bond the steel to the concrete. Be careful to keep the old surface clean and damp.

There's also a *combination control and construction joint* that runs all the way through a slab. That way the two surfaces aren't bonded. Usually you make this type of joint with a key that interlocks the slabs, but allows them to move independently. The key joints shown back in Figure 8-3 are one example.

Pouring Concrete in Cold Weather

Temperature has the greatest effect on the rate of hydration. Concrete poured at near-freezing temperature won't gain much strength in its early stages. But if the concrete isn't allowed to freeze, it will eventually develop the required strength over time.

Sometimes you can't avoid pouring concrete in cold weather. Where I live, if I didn't want to pour concrete in cold weather, I'd be working only six months out of the year. But there are precautions you can take. Plan ahead for special heaters, covering materials and admixtures. You can use accelerators or high-early-strength concrete. Use Type III portland cement or Type I with calcium chloride dissolved in the mixing water at the rate of 2 pounds per bag of cement.

When air temperature is below 40 degrees F:

1) Heat the sand, gravel and water to just below 150 degrees F. Heat the sand and gravel in separate piles over culvert pipe or an improvised firebox. Stir and rake frequently for even heating.
2) Remove snow and ice from the forms.
3) Place the concrete immediately after mixing.
4) Try to retain as much heat as possible. Cover the concrete with canvas, straw or hay for four or five days. Or provide enclosures that can be heated before, during, and after the concrete is poured.
5) Don't place concrete over frozen ground. Cure concrete at least 48 hours before allowing it to freeze. It's even better to prevent freezing for four to five days.
6) Remove forms only after sufficient curing. Pour hot water on the concrete to see if it has frozen. If the concrete has properly cured, there will be no effect, but if the concrete has frozen when setting it will soften up on the surface. If that happens, you'll have to replace it.

Pouring Concrete in Hot Weather

Hot weather can create as many problems as cold weather. The main problem is rapid evaporation of the water in the concrete. It's essential that hydration take place naturally, and not too quickly. If concrete dries too fast, it will be weak and develop cracks. And if the surface is too dry, it's hard to finish.

Moisture in concrete evaporates faster when the temperature is high, the humidity is low, or the wind is blowing. Combine these factors and the effect increases. Here's how to fight the problem.

- ❖ Pour early or late in the day when the temperature is lower.
- ❖ Put up barriers to keep wind from blowing across the surface.
- ❖ Have water available to spray the forms and subbase, and to moisten the pour after the initial set.
- ❖ Cover the surface with polyethylene or other suitable material to slow evaporation.

Pouring Sidewalks and Driveways

Sidewalks and driveways are usually the last masonry jobs you do on a project. Many contractors schedule this work last because the final grade has been established and other building materials have been used up or removed from the site by this time.

A sidewalk can be just about any design, shape, color or texture. Generally, sidewalks are made 4 feet wide except for service walks, which are generally 3 feet wide. Always use air-entrained concrete for a sidewalk.

Figure 8-13
A base of crushed stone helps drain excess water

Figure 8-14
A typical chute for placing concrete

Pouring and Tamping

Before pouring, clear the ground of any roots, large stones or other debris. Then compact the ground under the slab by tamping. This helps keep a slab from settling and cracking. Use a power-driven tamper to help compact the base of large slabs or long sidewalks. You can rent a tamper or most any large power tool used in concrete work. If you don't use a tool very often, it's usually cheaper to rent than to buy.

If there is no moisture problem, compact the surface and pour the concrete on the ground. In areas where you suspect dampness, excavate the area a few inches below the bottom level of the slab and place a fill of coarse gravel or crushed stone to help drain excess water, as shown in Figure 8-13.

For a sidewalk, make the forms of 2 x 4 lumber, and stake the outside at least every 3 feet using duplex-headed nails. A sidewalk generally doesn't need reinforcement because of the control joints or expansion joints.

Do the actual pouring of the concrete as near as possible to where it's needed. Use a chute like the one in Figure 8-14, or use old coal chutes. You can also make chutes from wood. Just don't let the concrete fall too far. Use a Georgia buggy like the one in Figure 8-15 instead. These motorized buckets carry much more than a worker can with a wheelbarrow.

Screed the concrete as it's poured into the forms, as shown in Figure 8-16. Be careful not to leave any concrete on the top edge of the forms. The screed will

Figure 8-15
A Georgia buggy

Figure 8-16
Screed the concrete as it is poured into the forms

Figure 8-17
Bullfloating concrete

Figure 8-18
Use a hand float to smooth the surface

run over any dried lumps of concrete, producing humps in the concrete.

After the concrete is screeded, use a bullfloat on the surface to smooth out the screed marks. The bullfloat also makes finishing the surface easier by causing the aggregate to come to the surface of the concrete. Run the bullfloat at a 90-degree angle to the direction of the screed. Push the float across the slab with its front slightly raised, and pull it back across with its back raised to keep it from digging into the wet concrete. Look at Figure 8-17.

When you've finished bullfloating and the concrete has started to set (about 20 minutes under normal conditions), use a hand float to further smooth the surface. Move the float in a 180-degree arc across the slab, raising the edge slightly to prevent digging in. Note in Figure 8-18 the expansion joint material that has been set into the wet concrete. It will be pushed down into the concrete until it's just below the surface.

Figure 8-19
After the surface of the concrete is floated, finish the edges with an edger

Sloping and Edging

Slope sidewalks 1/8 to 1/4 inch to one side or the other to drain off water. Round the edges of the walk to dress it up and keep the concrete from spalling.

After the surface is floated, give the edges a once-over with an edger, as shown in Figure 8-19. Concrete edgers of stainless steel and bronze are made in many sizes. Sidewalk edgers are usually 6 inches long and 2 3/4 inches wide. There are combination edgers such as the edger-groover shown in Figure 8-20. They help

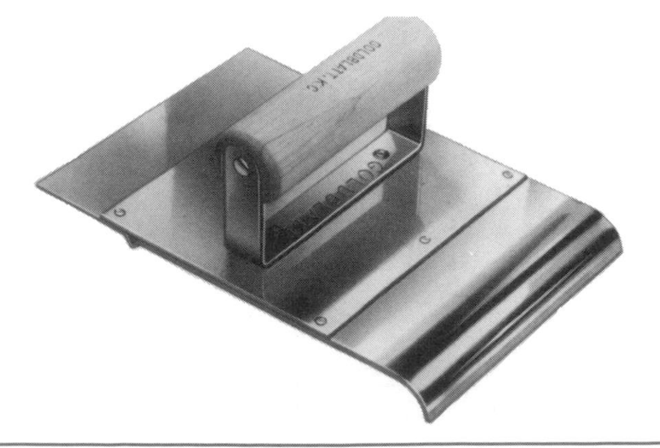

Figure 8-20
Edger-groover

cut costs and working time when you pour a walk and curb together. The heavy-gauge edger forms the top edge of the curb and cuts a groove about 1/4 inch deep and 3/8 inch wide at the top with one pass along the slab.

Finish the edge soon after pouring the slab, since the form material makes the concrete dry out faster than the rest of the slab. You'll have to do the edging two or three times before the job is completed, but the first time is the most important because it smoothes the aggregate while it's still soft.

Once a walk is finished and the area cleaned up, cover the slab to help the concrete cure. Let concrete cure at least six days before using it for heavy traffic. Various materials you can use to cover concrete are burlap, wet sand (after the concrete has set), straw or plastic. Plastic is the quickest and most economical. But be careful — if you put it on too soon, the plastic will make marks in concrete and discolor it slightly. Place rocks, boards or other heavy objects on the edges of the plastic to keep it from blowing away, as shown in Figure 8-21.

Driveways

Since they support heavy vehicles, make the concrete for driveways about 6 inches thick. That means you'll need thicker forms than the ones you used for sidewalks. Reinforce concrete for driveways with wire mesh or with rods crossing each other in the slab. Driveways are usually at least 10 feet wide and double driveways are usually 20 feet wide.

Much of the information about sidewalks is true for driveways. If there's a problem with dampness, excavate the subgrade at least 4 inches and fill with crushed stone or gravel. A driveway should have a slope from one side to the other of about 1/8 inch per foot for surface water to drain. Put expansion joints 10 feet or less apart. If a driveway is adjacent to the foundation, install isolation joint material between them. Make concrete for driveways with air-entrained cement.

Precast Concrete Slab Sidewalks

You can buy precast concrete slabs for sub-slabs and to span openings. Precast slabs are very heavy and you'll need a machine to remove them and to place them in the correct spot. Figure 8-22 shows a precast slab being placed over an opening where a sidewalk will be located. Figure 8-23 shows the slabs in place. It's very important to get the slabs level with each other. They can have a pitch away from a building for runoff, but make sure the edges are level with each other.

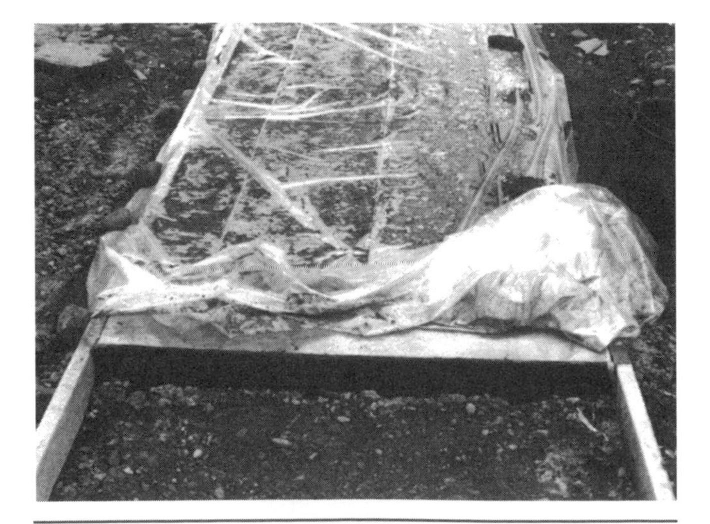

Figure 8-21
Cover the finished walk to promote curing

Figure 8-22
Placing precast slabs

Figure 8-23
Precast slabs in place

Figure 8-24
Grouted precast panels

Figure 8-25
Precast slabs coated with liquid and elastomeric membranes

After the slabs are in place, grout them with a mixture of portland cement and a grouting aid such as Interplast-N. That's an expanding grouting aid that doesn't contain calcium chloride, nitrates or other chemicals that might corrode steel. This type of grouting aid increases fluidity and produces a slow, controlled expansion before hardening.

Be sure your forms are tight and well fitted in areas to be grouted that need forms. When using Interplast-N grout, restrain the expansion of the grout to produce the highest possible density, bond and strength. Use top forms where there are open areas.

Place the grout in the joints within one hour after you mix the grout. Figure 8-24 shows precast panels with grout added to the joints between the slabs. After the precast slabs are in place and grouted, coat them with Hydrocide liquid membrane to a thickness of about 55 mils. Place a layer of elastomeric membrane on top of the liquid membrane.

Finally, pour the top or finished slab of concrete. Figure 8-25 shows a top layer of concrete. You can also cover a precast slab with paving brick or some other finish material.

9

Reinforced Masonry Construction

Back in Chapter 5, on block foundations and retaining walls, we covered the basics of masonry construction. Look there for information on laying the bond and constructing corners, for example. In this chapter we'll cover structural unit masonry walls and requirements for reinforcement. We'll also look at the important subject of planning for doors and other openings in the walls.

Masonry Walls and Partitions

Masonry walls must be strong enough to withstand all vertical loads and horizontal forces, plus allowance for any unusual off-center loads. The minimum nominal thickness of a wall depends on whether it's load-bearing or not, and on the height between its supporting diaphragms. Don't use cavity walls unless each wythe is individually designed as an independent structural wall.

Reinforced Grouted Masonry

Reinforced grouted masonry has concrete, or in some instances mortar, poured or pumped into the cavities in the masonry wall. The cavity could be the hollow cells of block masonry or the space between two masonry walls. See Figure 9-1. Of course the concrete or mortar has to be a mix that will flow into the cavities. The fluid consistency of grout determines how easily it will pour. It's important to strictly adhere to the architect's specifications for the grout material.

Placing the Reinforcing

Most single width walls are grouted with rods placed both vertically and horizontally. See Figures 9-2 and 9-3. When you're going to grout a wall, the reinforcing rods in the footing must project into the exact center of the wall location. They shouldn't protrude more than a few feet, or they'll make it more difficult to construct the wall. As construction continues, you can weld new sections to the reinforcing rods or use form wire to tie new sections together. Your masons must be able to work easily around the reinforcement. In single width block walls, the mason has to lift the masonry units over the rods, so take this into consideration with placing rods. See Figure 9-2.

The horizontal rods are placed in the top course of blocks after they're laid. You can order special blocks to accommodate horizontal reinforcing rods. But some contractors prefer to have the mason make a channel for the reinforcing as they work. They can use a masonry saw or simply knock out the top half of the block's webs to lay the rods on. For maximum lateral strength, place mortar on all top surfaces, including

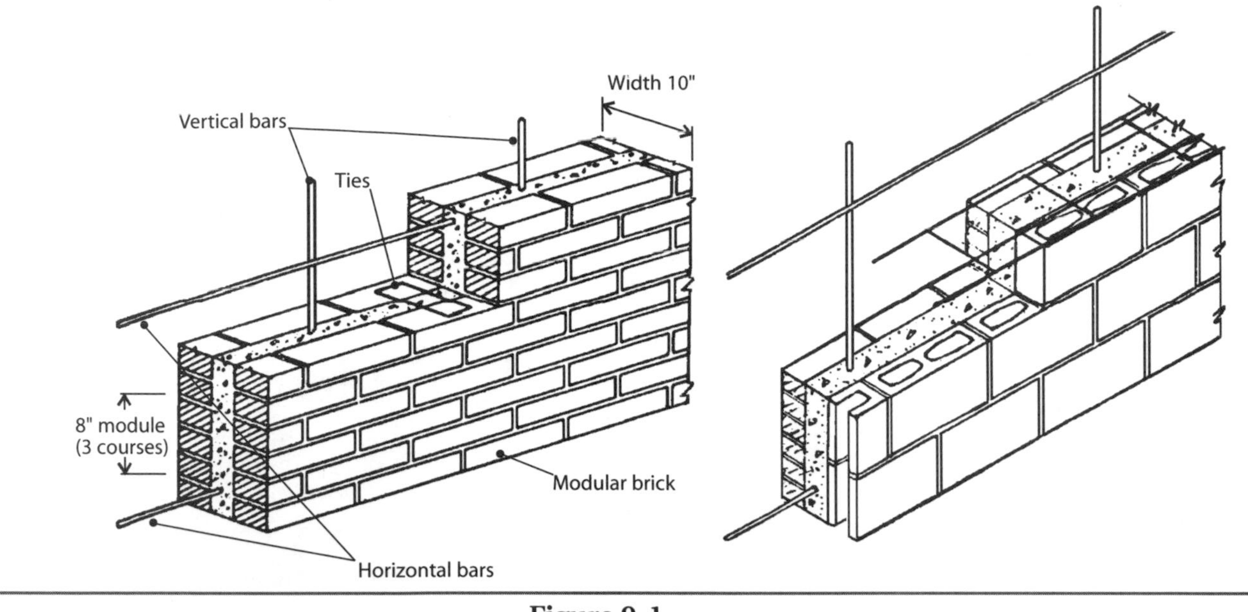

Figure 9-1
Reinforced grouted masonry

cross webs, for concrete blocks laid in a wall that's going to be grouted.

Bond beams or reinforced concrete masonry lintels are masonry units with the webs cuts out or left out at the factory. See Figure 9-4. After laying concrete blocks, the masons place reinforcement and grout in the opening of the block. See Figure 9-3 B and C. Often bond beams are incorporated in walls as part of a continuous bond beam course around the building. This helps to further distribute shrinkage and temperature stresses in the masonry above openings. The major advantage of masonry bond beams over steel in the elimination of the differential movement between the steel and the masonry during temperature changes.

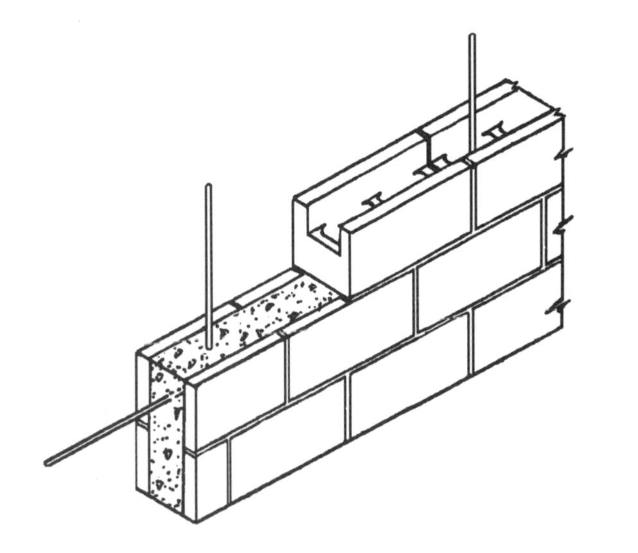

Figure 9-2
Reinforced filled-cell masonry

Include Cleanouts

When laying the first course of blocks, include cleanout holes to allow removal of debris from the wall center. If possible, leave out every other unit. If you place bracing such as a wood block or brick vertically to support the second course, work can continue. Then spread sand in the clean cavity after the first courses are laid. This prevents mortar droppings from sticking to the footing.

During wall construction, masons should keep the vertical alignment of the inside of the wall clean and smooth so the grout will flow freely. Clipped brick headers (cut with a trowel or hammer) shouldn't protrude into the cavity and cause an obstruction.

The masons should place all masonry anchors or ties and reinforcing wire as they're laying the wall. Make sure they tie the reinforcing rods to spacers to keep them in the center of the cavities. If possible use all corrosion-resistant anchors and ties.

It's unavoidable that some mortar will fall into the cavity as the masons build the wall. Here's one possible solution for that problem. You can place small boards supported on wires in the wall. That lets the masons pull up the board frequently to remove any mortar droppings.

After the wall is finished, it's a simple matter to remove the blocking at the bottom of the wall and clean out the sand and other debris in the cavity. Then the mason fits the missing masonry unit into the cleanout opening. After the final units are in place, wait until the new masonry cures before grouting. This usually takes three days.

Pouring the Grout

For most single-width masonry block walls, you'll pour the grout after the wall's 4 feet high. Pour or pump the grout in slow continuous pours or lifts to prevent dry seams. If work has to stop for some reason, don't wait more than one hour before continuing the lift.

Grouting is done in what is called *lifts*. For cavity walls, you'll usually use low-lift grouting as the wall is laid. The masons will construct the wall so the courses on each side are close to being the same height. In modular construction this will be every 8 to 16 inches. Make sure your masons know how to lay units for grouted walls. First, they should *shove* (fill and push together) all mortar joints. Second, they should completely fill and press in the end joints in the last unit laid. Since it takes time to lay successive courses in the wall, you're not likely to make successive pours too close together. There should be at least 15 minutes between pours. Leave each lift slightly below the top of the last courses laid so the next lift won't be at the mortar joint.

High lifts are usually limited to 4 feet at a time. If you use high-lift grouting, you'll have to let the walls cure at least three days — more if the humidity is high. All masonry should be laid in running bond with full end and bed joints. It's also best not to furrow the bed joints. Place metal ties between the walls according to

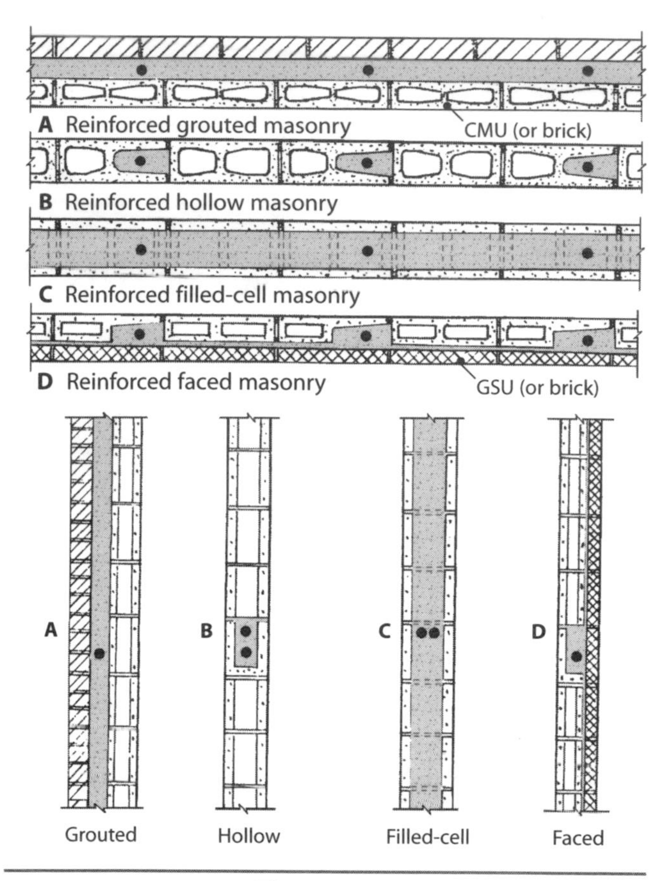

Figure 9-3
Basic types of reinforced masonry

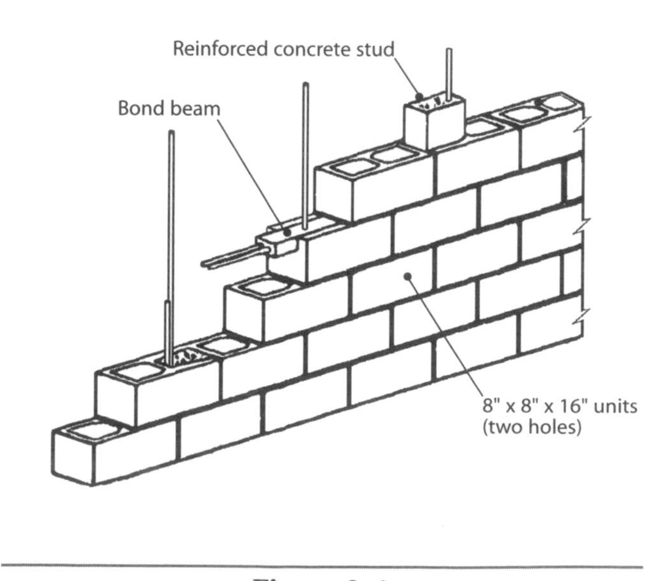

Figure 9-4
Reinforced hollow masonry

Figure 9-5
A concrete pump will force grout to inaccessible locations

Figure 9-6
A simple hopper to guide grout or insulation into the shell

local building codes. Build up both walls to the same height to make it easier to pour the grout. One wall shouldn't be more than 8 inches higher than the other during the pour. Build vertical barriers in the walls about every 20 feet.

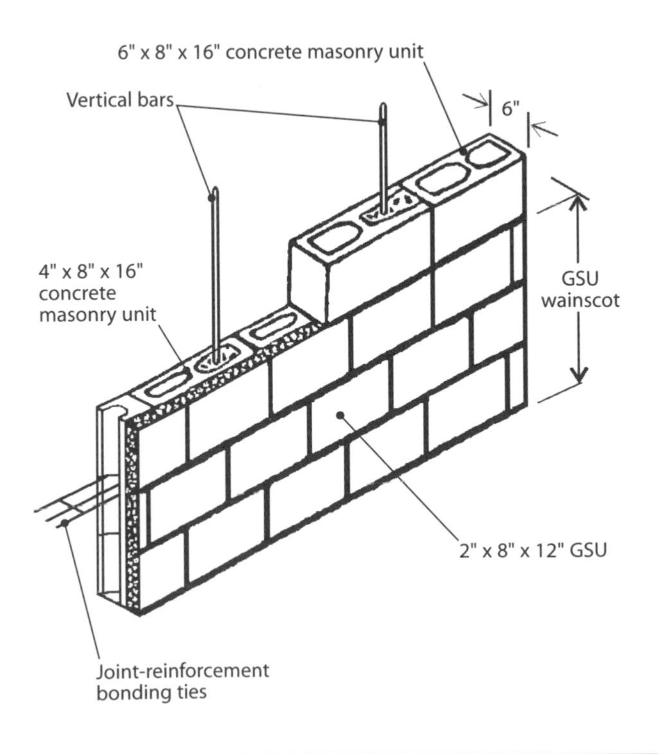

Figure 9-7
Reinforced faced masonry

Lifts should be done slowly to prevent rupturing joints or causing bulges in the wall. As a safety precaution, consider bracing a wall before high-lift grouting. Allow at least one hour lapse between pours so excess water can bleed into blocks or through the wall, and the grout can settle. Figure 9-5 shows a concrete pump you can use to move the grout to the top of the wall. Figure 9-6 shows a simple homemade hopper that's designed for pouring insulation into walls, but you could use it to pour stiff grout. Take every precaution to prevent grout from running down the face of the walls. It's a minor disaster to get grout on porous masonry surfaces. The cleanup takes a long time.

Reinforced Faced Masonry

Reinforced faced masonry is made with two widths of masonry units, with a structural-bonded facing and backing of different materials. See Figure 9-3 D. An example is hollow masonry units (CMU) faced with glazed structural units. The facing of GSU is usually laid in running bond or structural strength, and anchored to hollow masonry units with joint reinforcement such as Dur-O-Wal. See Figure 9-7. There are different types of Dur-O-Wal reinforcement for different types of applications.

Openings in Masonry Walls

Before you lay out the coursing in a masonry wall, you must plan ahead for the openings. For example, you want to have the courses over the top of the door frame laid out in full units. So first lay out the brick dry along the opening and starting at the other side of the opening with a full or half unit for correct bond.

Door Frames

Metal door frames or *bucks* are designed to fit coursing of both brick and block walls. Figure 9-8 shows the standard door frame sizes. The X's in the figure show the availability of a particular size. Mark the coursing in pencil on the door edge. Use these guide marks to build the wall on both sides and you won't have to make a cut at the top later. On some layouts you'll have to start the first course of all walls with a 4-inch starter block.

Figure 9-9 illustrates door frames using the butted method. Use this method when the wall is wider than the frame, so the wall is laid against or "butted" to the frame. You can see the different types of masonry

	Height			
	1³/₈" x 18 Ga	1³/₈" x 16 Ga	1³/₄" x 16 Ga	1³/₄" x 14 Ga
	6'8"	6'8"	6'8"	6'8"
	7'0"	7'0"	7'0"	7'0"
	7'2"	7'2"	7'2"	7'2"
	7'10"	7'10"	7'10"	7'10"
Width	8'0"	8'0"	8'0"	8'0"
2'0"	X	X	X	X
2'4"	X	X	X	X
2'6"	X	X	X	X
2'8"	X	X	X	X
2'10"	X	X	X	X
3'0"	X	X	X	X
3'4"	X	X	X	X
3'6"	X	X	X	X
3'8"	X	X	X	X
3'10"	X	X	X	X
4'0"	X	X	X	X

Figure 9-8
Standard frame sizes

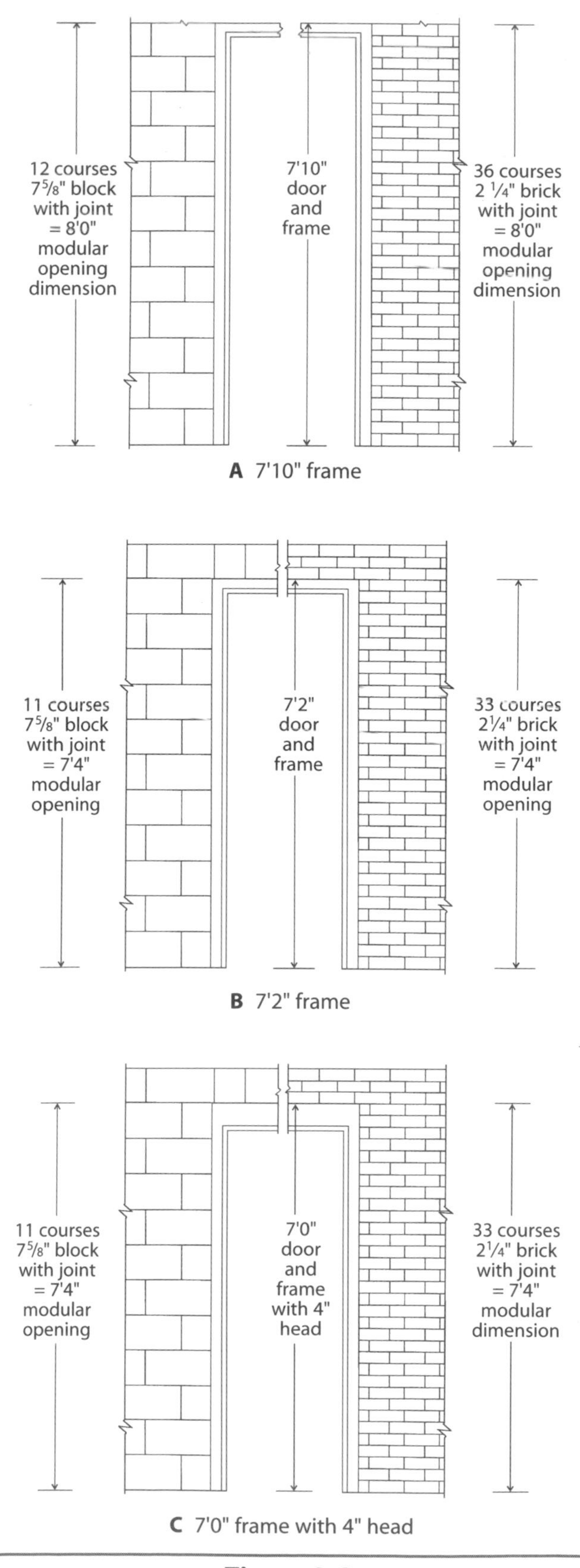

Figure 9-9
Door frames — butted masonry

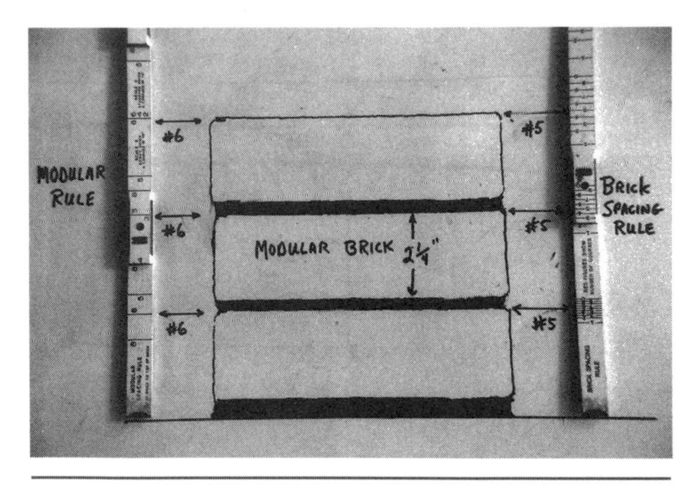

Figure 9-10
Modular and brick spacing rules

materials placed against the door frames. Mortar is placed in the interior of the frame to help anchor the frame to the wall. Courses of block and brick are laid out to allow full units to be laid over the frame. While this isn't always possible, it's better for both looks and strength if you can do it. A mason can use either the spacing rule or the modular rule to lay out course markings on the frame. A simple adjustment to the first course could avoid having to make a cut over the door. Figure 9-10 shows the modular and brick spacing rules.

Figure 9-11 shows jamb details for a butted installation, while Figure 9-12 has the head details.

Figure 9-13 illustrates a wraparound installation. It shows how the different masonry materials run into the frame. This is the method of choice when the frame is thick enough to encompass the masonry wall. The top course of masonry may or may not come out even with the top of the frame. However it's easier to lay the units if they work out in full sizes. Therefore the mason should also lay out the height of the coursing with a modular or spacing rule to allow the work to come out in full units if possible.

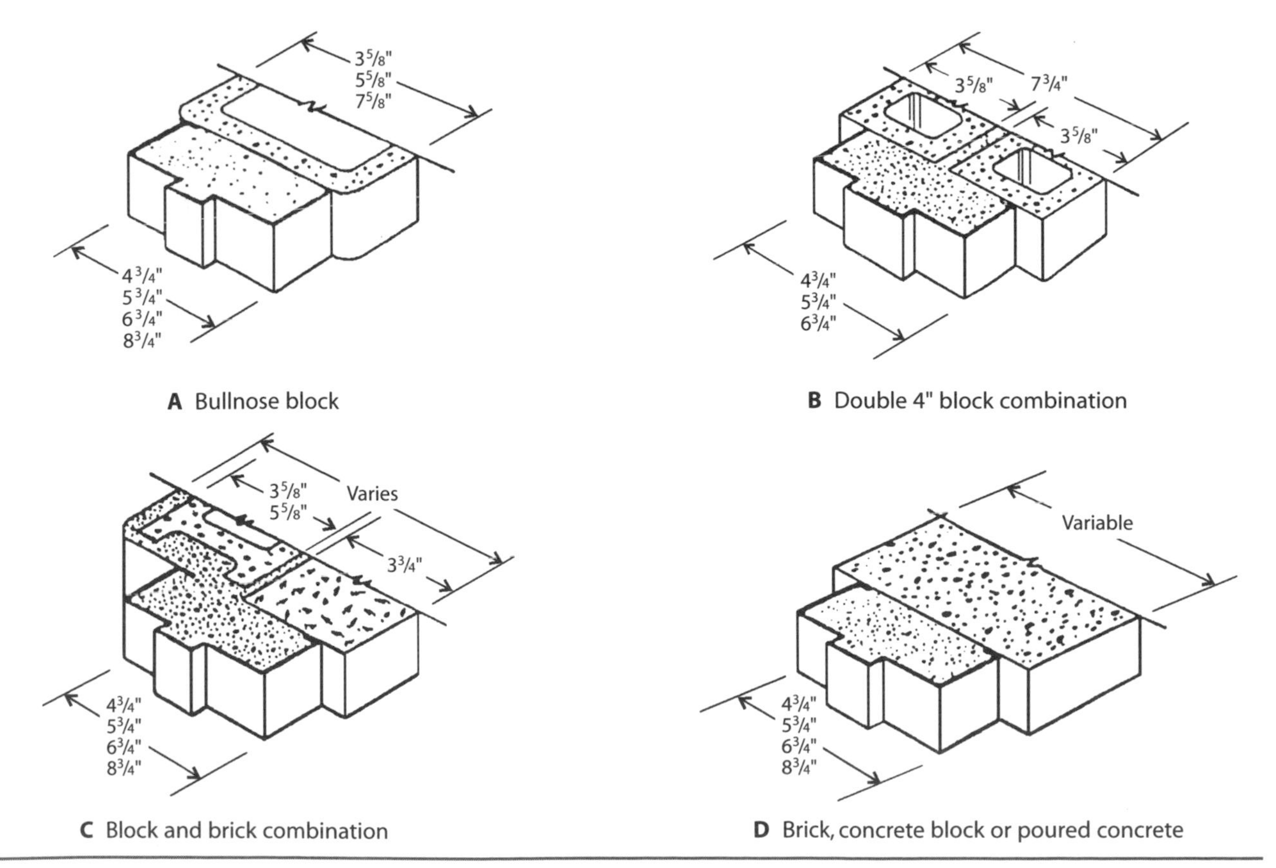

Figure 9-11
Jamb details — butted installation

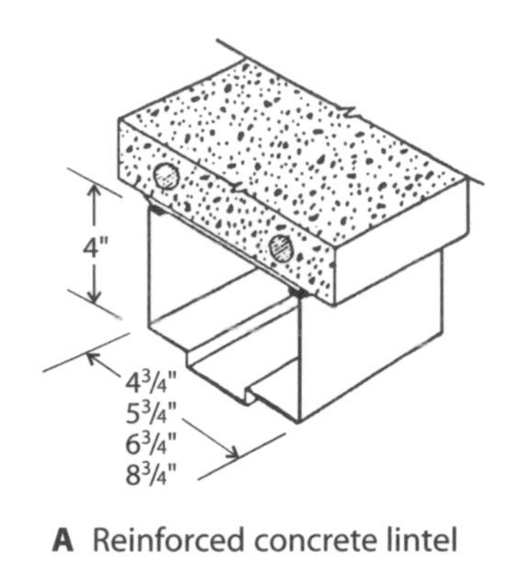

A Reinforced concrete lintel

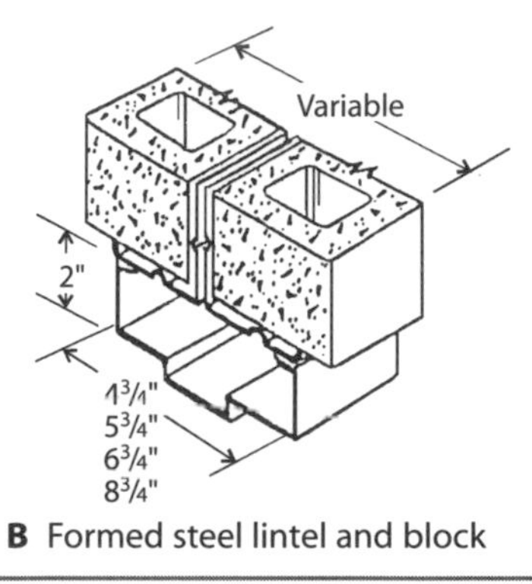

B Formed steel lintel and block

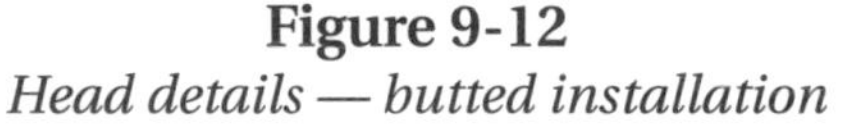

Figure 9-12
Head details — butted installation

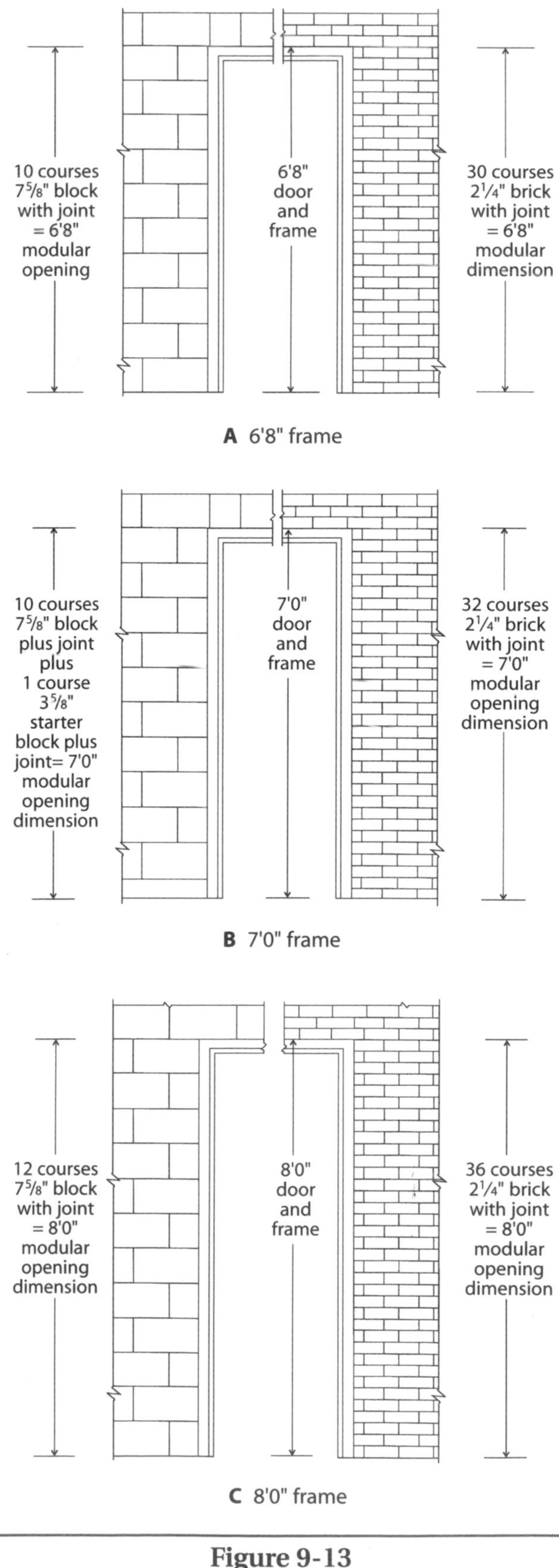

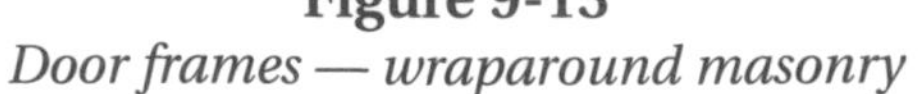

Figure 9-13
Door frames — wraparound masonry

Figure 9-14 shows the jamb details for a wraparound installation. Figure 9-15 has the head details. Figure 9-15 A is the combination of a wall inside the frame and another wall passing outside the frame on an angle iron with a masonry collar joint between them. Section B shows the masonry inside the frame with a steel angle set on the frame for additional support.

Making a Story Pole for Doors and Windows

If you're an experienced mason you may not need a story pole for basic work, such as building corners. But every mason should use a story pole to record heights for all the sills and lintels for door and window openings. The accuracy you gain in your work more than makes up for the time it takes to make a story pole. For this story pole, mark and label the points where brick or block pass beneath sills or over lintels. Then, near the bottom of the story pole, make a big, bold arrow and label it "UP." Figure 9-16 shows a story

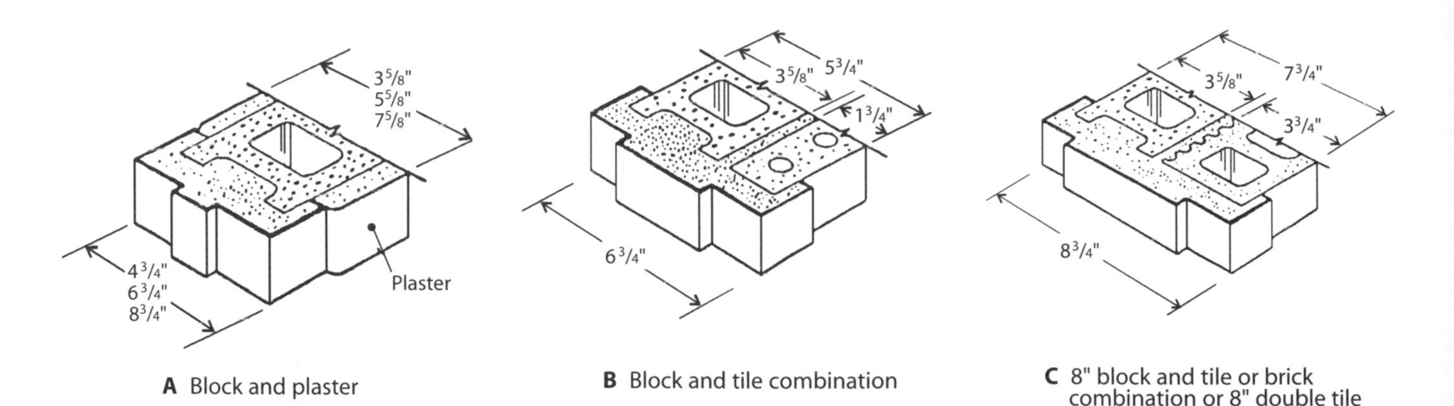

Figure 9-14
Jamb details — wraparound installation

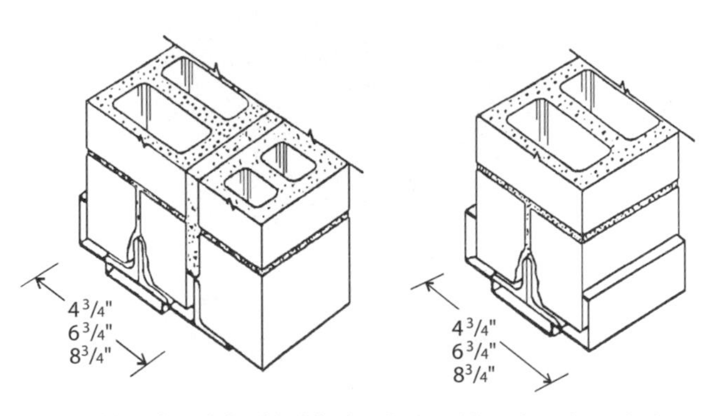

Figure 9-15
Head details — wraparound installation

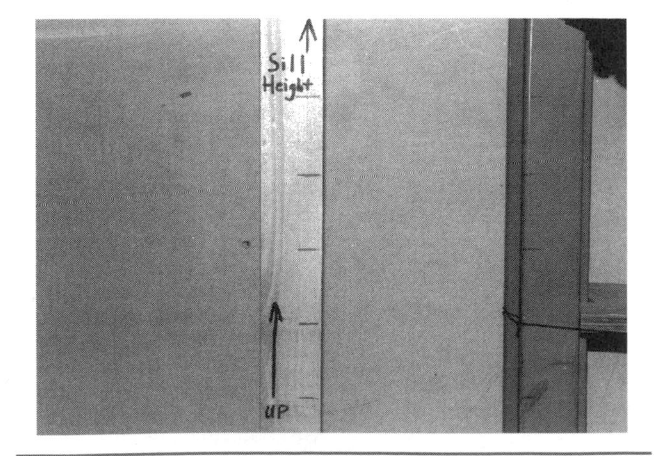

Figure 9-16
Story pole for door and window openings

pole like this. There's no better insurance against errors and confusion.

Using a modular rule, check whether the brickwork works out in full courses at the sills. Brick window sills, for example take 4½-inch spacing. So if you're working in brick, the story pole's first course mark is 4½ inches down from the sill. If you're lucky, the spacing (there are six to choose from on a modular rule) works out very close for the brick you're using. Starting from the sill line, mark off each course on the story pole. Once you've finished that, check the coursing from the sills up. If you're still lucky, you'll find that the same course number works here.

If you can't get any course number on the modular rule to work out, try a mason's spacing rule instead. This rule also has course height markings, some of which may not be on the modular rule. Make any slight adjustments needed, check them with either rule, and your story pole's complete.

It'll only take you a moment to write the course numbers on the story pole. Having this information at hand may avoid a wall with a hog. That's a wall that has more courses at one end than the other. As a backup, make a copy of the story pole. If your crew is large enough to split up and work on the wall from both ends at once, a duplicate story pole is a must.

Anchoring Door Frames

Always anchor the door frames well. Anchors help keep the frame from moving within the wall. Figure 9-17 A shows the wire anchor method. These wire ties

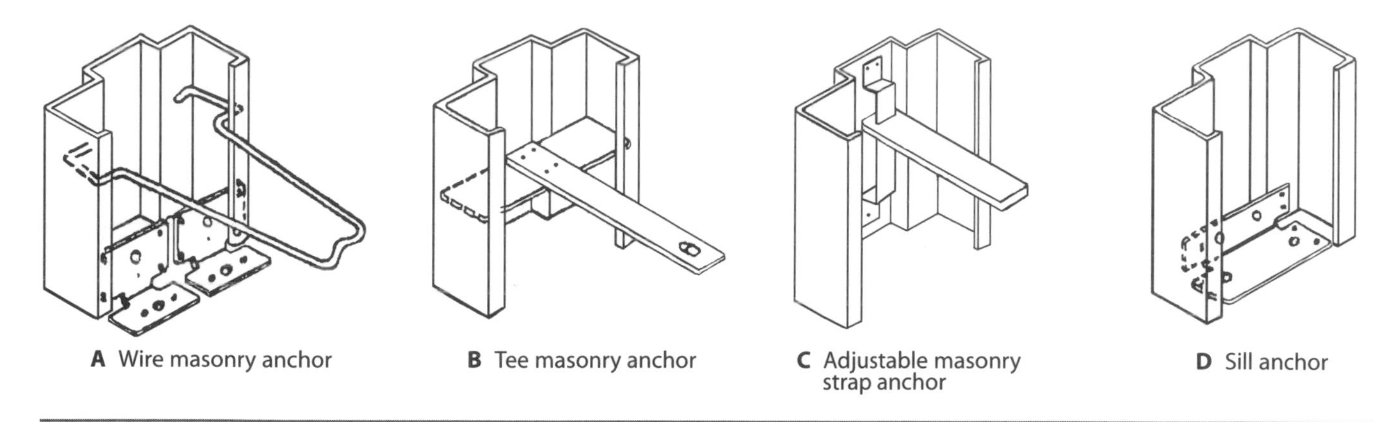

Figure 9-17
Masonry anchors for door frames

are placed as specified by the architect or engineer, but they're usually in every other course. The tie is placed on top of a course of masonry and the door frame is filled with mortar. You can also see the bottom anchor of the frame. These are usually applied to the frame with screws and then attached to the floor with masonry anchors. Figure 9-17 B is the tee-shaped anchor. It's applied the same way as the wire anchor. Section C of the figure is a factory-installed strap tie that can be adjusted up or down to fit in a masonry joint. Section D shows the floor anchor that comes welded to the frame for attachment to the floor with masonry anchors.

Install temporary spreaders at the top and center of the door. They keep the door frame from moving inward during construction. Never lay masonry up against only one side of a door. To keep the frame square and true, build both of the walls that bracket a door frame at the same time. And make sure all the masonry you lay against the metal frame is flush and straight. Your reputation depends on details like this.

Lintels

A lintel is a horizontal structural member placed over an opening in a wall to carry the weight above it. Analyzing loads and stresses on lintels involves complex calculations. A project architect or structural engineer should design and choose suitable lintels.

Most lintels are one of three types: reinforced masonry, precast concrete, or shaped structural steel. Let's look at each of them, highlighting the advantages as well as any disadvantages.

Reinforced Masonry Lintels

These are the most widely-used type of lintel (Figure 9-12 A). They have many advantages, including:

- Lower material costs (the only steel is the reinforcing)
- Lower labor costs (there's less heavy steel to handle)
- Lower maintenance costs (there's no need for regular repainting)
- Greater fire safety (fireproofing is built in)
- Less chance of cracking (the lintel has the same thermal expansion coefficient as the wall)

Precast Concrete Lintels

This type of lintel is both economical and versatile. Figure 9-18 shows precast lintels. You can get a close match between walls and precast concrete lintels by coloring and texturing the lintel. And you can order odd or unusual sizes made to your specifications. Precast concrete lintels usually arrive marked with their location and size, with the top labeled. Use this information to properly handle, store, and place the

Figure 9-18
Precast concrete lintels

lintels. A final caution: Be sure to store precast concrete lintels topside up, never on their sides. Figure 9-19 shows the safe loads for modular precast concrete lintels.

Clear span	Lintel length	Uniform load (lbs/ft) 4" x 8" 27½ lb/ft	6" x 8" 41 lb/ft
2'0"	3'0"	1600	2400
2'4"	3'6"	1300	2000
2'8"	4'0"	1100	1700
3'0"	4'0"	950	1500
3'4"	4'6"	850	1300
3'8"	5'0"	750	1200
4'0"	5'0"	700	1000
4'4"	5'6"	600	950
4'8"	6'0"	575	900
5'0"	6'6"	525	800
5'4"	6'6"	500	700
5'8"	7'0"	450	600
6'0"	7'6"	425	550
6'4"	8'0"	400	500
6'8"	8'0"	375	450
7'0"	8'6"	350	400
8'0"	9'6"	300	300

Note Load estimates are using 3000 psi concrete mix. Loads shown are superimposed loads. Load range may vary with strength of concrete. Consult your manufacturer.

Figure 9-19
Safe loads for modular precast concrete lintels

Formed Steel Angles

These angles (called lintels) span the opening and rest approximately 8 inches on the masonry on each side of the opening. See Figure 9-12 B.

Of the three materials, steel is by far the most expensive, but that cost can be offset because they speed up the construction. There is a disadvantage, however. A steel lintel installed in a masonry wall doesn't expand and contract at the same rate as the wall. Over time, the stresses caused by this difference produce cracks in the masonry. That's why they're used mostly in interior walls where temperature changes are minimal.

Pockets

The block wall in Figure 9-20 shows pockets made in masonry walls. Pockets give plumbers the access they need to complete their rough-in work. You have to build pockets in exactly the correct place. And they have to be plumb for you to build the intersecting wall correctly. The best way to do this is to use a masonry saw to cut the blocks for pocket openings. As the enclosing walls are built, the mason lays masonry units into the openings, tying the walls together.

Figure 9-20
Pockets for plumbing rough-in

10

Reinforcing and Crack Control in Masonry Walls

One problem with masonry walls is the potential for cracking. There are several possible causes, including:

- The load on the wall
- Changes in the materials caused by temperature or moisture fluctuations
- Movement in the foundation
- Pressure from earth piled up against the wall, from water or from wind
- Deflection in the beams or slabs supporting the wall
- Movement in the building components connected to the wall

There are three ways you can control cracking in a masonry wall. First, you can use products that reduce the amount of moisture getting into the wall. Second, you can add control joints that let the wall move. Finally, you can use reinforcement in the wall to make it stronger and more crack-resistant between the control joints. And with proper design, you can use a combination of these that should virtually eliminate any risk of cracking in a wall. Let's take a closer look at control joints and joint reinforcement.

Control Joints

A control joint is a division in masonry designed to prevent cracks in a masonry wall when it's forced to move. For a control joint to form a true stress-relieving joint it must be:

- Cut through the masonry wall completely from the top to bottom
- Structurally sound to provide lateral stability
- Self-sealing or caulked to keep moisture out

Since a control joint is a weakened plane in a wall, it must be both strong and free to move longitudinally at the same time. One material which has these features is *Rapid Control Joint,* a patented product of the Dur-O-Wal Company. Figure 10-1 shows this product and some recommended applications.

Locating Control Joints

You need to work out where to put the control joints early on a job. You can't add, remove or relocate them once the structural design is complete. Here are some typical locations for control joints:

- Major changes in wall height or thickness
- Construction joints in foundation, roof and in floors

Rapid® Poly-Joint — P.V.C. Compound

Available in Regular, Tee and Wide Flange designs. Joint is designed for use with standard sash block. PVC material conforms to ASTM D2287 type PVC 654-4 with a Durometer hardness of 85±5 when tested in accordance with ASTM D2240. Cold crack brittleness –35°F (–37°C).

Courtesy: DUR-O-WAL, Inc.

Table 13a: Shear Strength

Control Joint Type	Average Load per 8" of Joint (lbs)	Shear Strength (psi)
Wide Flange	2283 (10.2kn)	490 (3.4mPa)
Regular	2831 (12.7kn)	566 (3.9mPa)

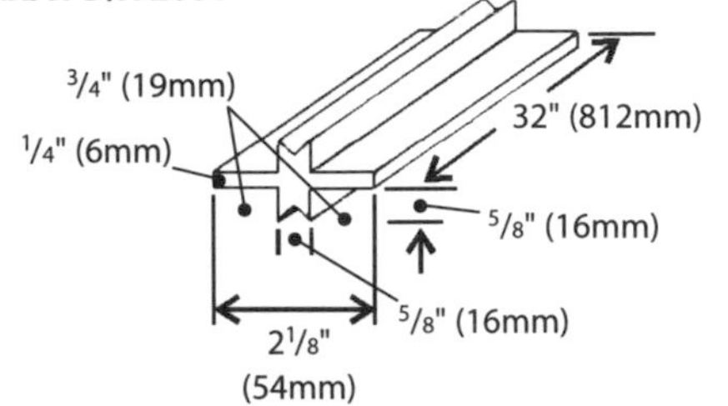

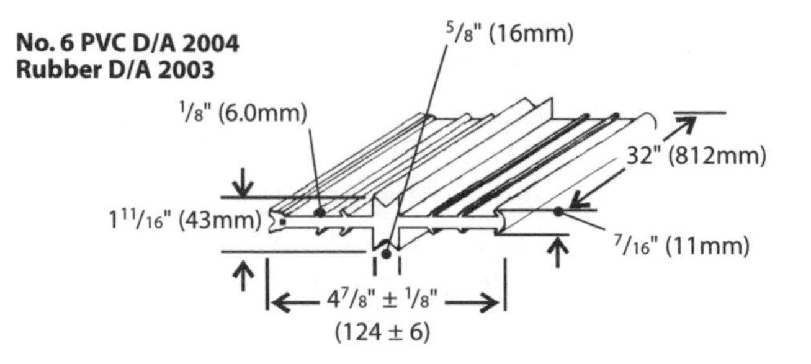

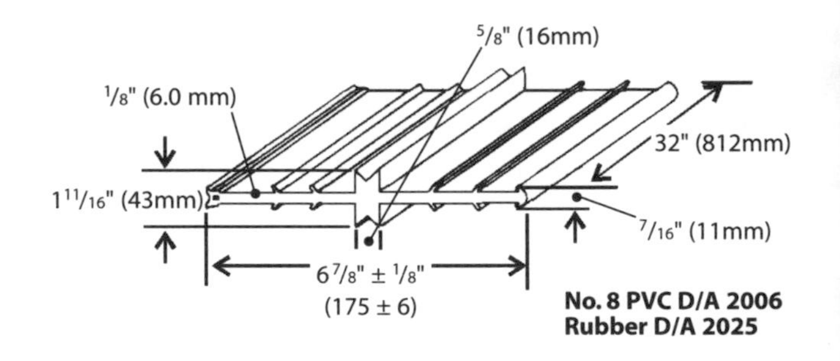

No. 8 PVC D/A 2006
Rubber D/A 2025

No. 12 PVC only
D/A 2007

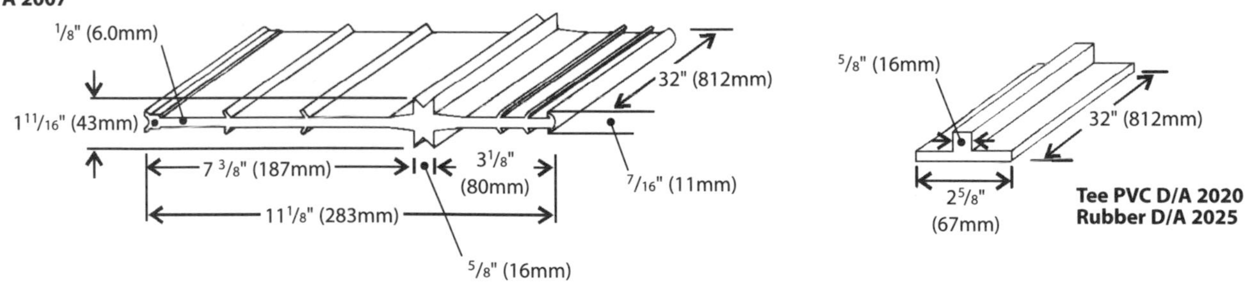

5/8" (16mm)
32" (812mm)
2 5/8"
(67mm)

Tee PVC D/A 2020
Rubber D/A 2025

Rapid® Soft-Joint/Expansion Joint

Keeps mortar and other foreign material outside horizontal or vertical expansion joints. Available in a variety of widths with adhesive on one side for convenient temporary attachment to steel shelf angles. 1/4" (6mm) thick for soft joint and 3/8" (10mm) thick for expansion joint. Closed cell neoprene material conforms to ASTM D1056 class RE41 or 2A1 compressibility exceeds 50%. Other widths and thicknesses available.

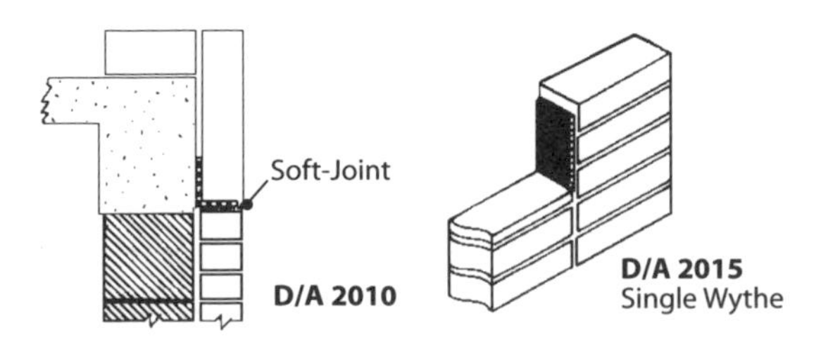

Figure 10-1
Movement joint materials

- ❖ Chases and recesses for piping, columns and fixtures
- ❖ Abutments of walls and columns
- ❖ Return angles in L-, T- and U-shaped structures
- ❖ One or both sides of wall openings

You usually need a control joint at one side of an opening less than 6 feet in width, and at both jambs of openings over 6 feet wide. If you don't use control joints at these points, add extra joint reinforcement above and below wall openings.

Where you use concrete masonry as a backup for clay brick or stone:

- ❖ Extend control joints through the facing if it's rigidly bonded (masonry bonded or with full-collar joints)
- ❖ Don't extend through the facing or veneer when the bond is flexibly tied (wire ties) to the backup and the collar joint isn't filled with mortar

Extend control joints through plaster that's applied directly to masonry units. If you apply plaster directly on lath that's furred out from masonry, you may not have to use vertical separation at every control joint. Unless it's specifically called for in the plans, joint reinforcement shouldn't extend through the control joints.

Don't space control joints more than four times the floor-to-floor height, or 100 feet on center, whichever is less.

Usually the best design to control cracking is a combination of joint reinforcement and control joints.

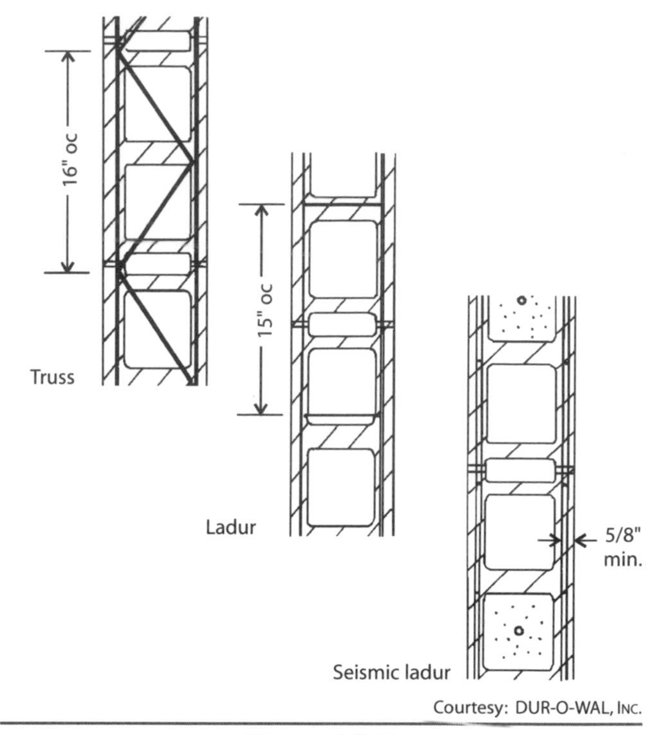

Figure 10-2
Truss, ladur and seismic ladur

Reinforcement for Masonry Walls

Reinforcement wire is the most common type of reinforcement used in masonry construction. This type of reinforcement is designed to:

- ❖ Control shrinkage cracking
- ❖ Bond different wythes of masonry together in composite and cavity walls
- ❖ Reinforce stack bond walls
- ❖ Bond intersecting walls

The most common type of wire reinforcement system is the truss wire. The truss wire system is used in conjunction with control joints, at a maximum vertical distance of 16 inches. There are three types of wire reinforcements: the truss, the ladur and the seismic ladur (Figure 10-2). They all come in 10-foot lengths and are easily cut with wire cutters. If you're caught without cutters, you can use a trowel or mason's hammer, bending the wire until it breaks. Installation methods for the truss and ladur systems are shown in Figure 10-3.

Place the ladur reinforcement across both the brick and block wall in a parallel and perpendicular direction. Use this type of wire in coursing that "lines up" — when the brick and block coursing meet in level coursing. This is usually the case in modular construction, when three courses of brick level out with one course of block.

When masonry walls aren't constructed simultaneously, or where bed joints don't line up at the same elevation, you need an adjustable system (Figure 10-4). This assembly provides a system of reinforcing

Use at least one longitudinal side rod for each bed joint. Out-to-out spacing of the side rods is approximately 2" (50mm) less than the nominal thickness of the wall or wythe in which the reinforcement is placed.

Splices

Side rods should be lapped 6" (150mm) at splices in order to provide adequate continuity of the reinforcement when subjected to normal shrinkage stresses.

Centering and Placement

Place joint reinforcement directly on masonry and place mortar over wire to form bed joint. This applies to both truss type (shown) and ladur type.

Block width t
t-2" (50mm) out to out
Centering — Dur-O-Wal is centered by "eye"
Mortar cover 5/8" minimum (16mm) on exterior face
Mortar cover 1/2" minimum (12mm) on interior face

A Installation methods for the truss and ladur systems

Prefabricated Corners and Tees

A complete line of prefabricated corner and tee sections are available in all design types and finishes. It is necessary to designate corners as inside or outside when using Trirod or Double side rod design types.

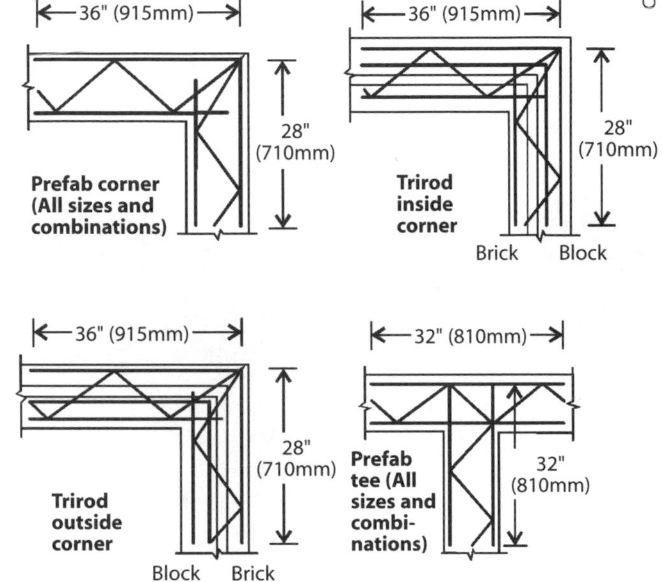

Courtesy: DUR-O-WAL, Inc.

Ladur corners and tees measure 30" x 30"

Truss and ladur are manufactured in accordance with ASTM A 951, Uniform Building Code Standard UBC 24-15 (1991) and UBC 21-10 (1994)

B Installation methods for corners and tees

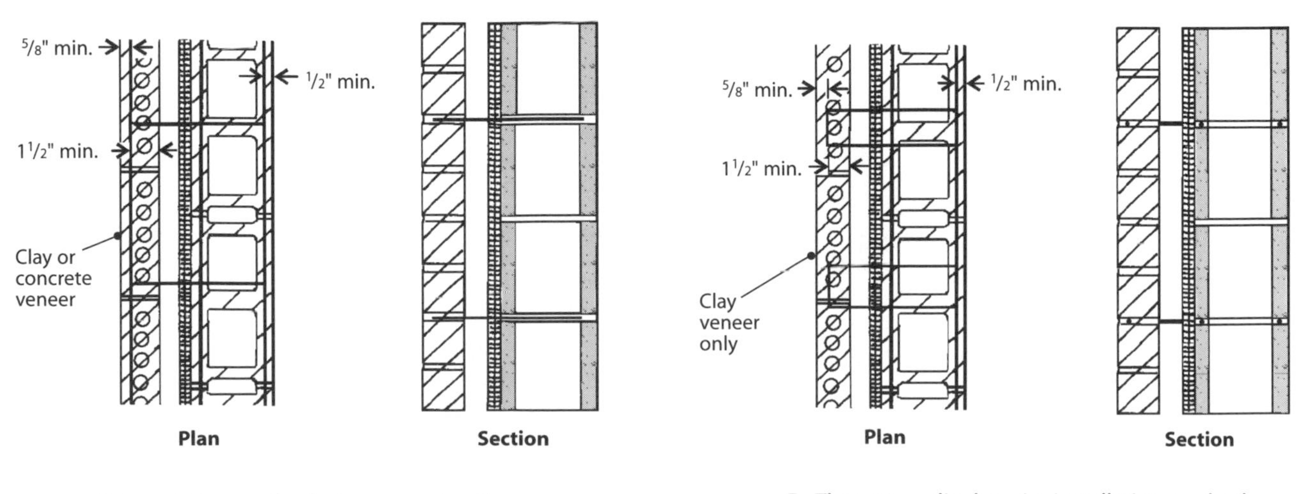

C Installation method where coursing lines up

D The perpendicular wire installation method

Figure 10-3
Installing the truss and ladur systems

wire laid up with the first wall that allows adjustable joint connections with later assemblies. See Figure 10-5 for typical spacing for truss and ladur joint reinforcement.

Anchors for Masonry Walls

Figure 10-6 shows a dovetail anchor that anchors a masonry wall to poured concrete. If you install the dovetail anchor slot in the concrete wall correctly, it's easy to put in the triangle anchor. Occasionally, the dovetail anchor will be out of plumb, causing problems in keeping the anchor flat. If that happens, the mason may have to cut the masonry unit or bend the anchor. Either method is time-consuming and produces a tie that's not as strong. Good workmanship in the concrete forming operation is essential.

Notched steel column anchors (Figure 10-7) come in various lengths to accommodate various thicknesses of steel flanges. Embed the D/A 601 anchor in both the mortar bed and into the core fill of a concrete block wall, for a stronger anchorage. For a brick wall system, use the corrugated anchor (D/A 604).

Figure 10-8 shows the channel slot anchor system. You can weld the channel slot to a steel column, or bolt it to a concrete column. The screw-on and

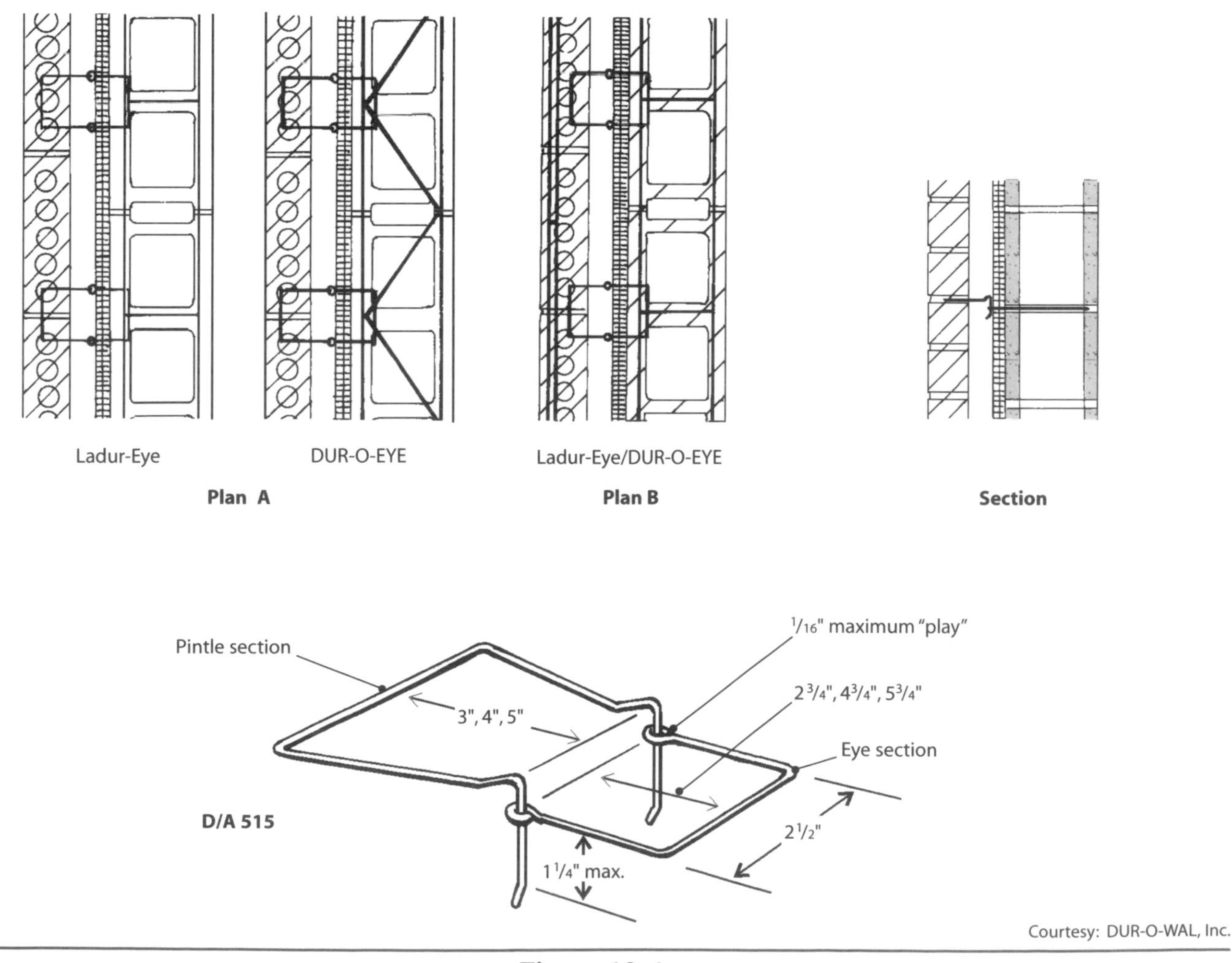

Figure 10-4
Ladur eye assemblies

Wall Openings — Unless otherwise noted, install Dur-O-Wal in the first and second bed joints, 8 inches apart immediately above lintels and below sills at openings and in bed joints at 16-inch vertical intervals elsewhere. Extend reinforcement in the second bed joint above or below openings two feet beyond the jambs. Use continuous reinforcement elsewhere but do not extend through vertical masonry control joints.

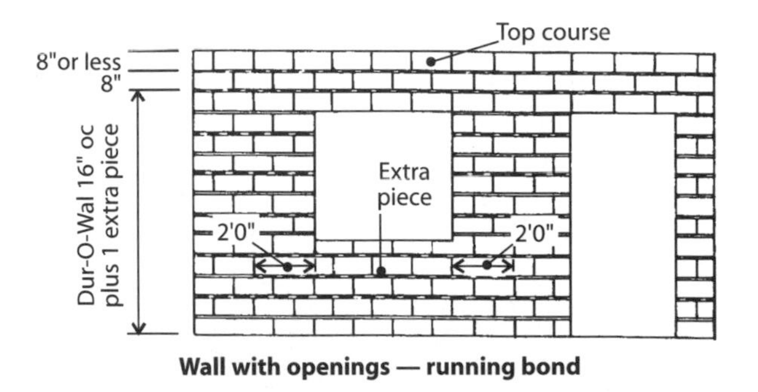

Wall with openings — running bond

Single Wythe Walls — Exterior and interior. Place Dur-O-Wal 16" oc and in bed joint of the top course.

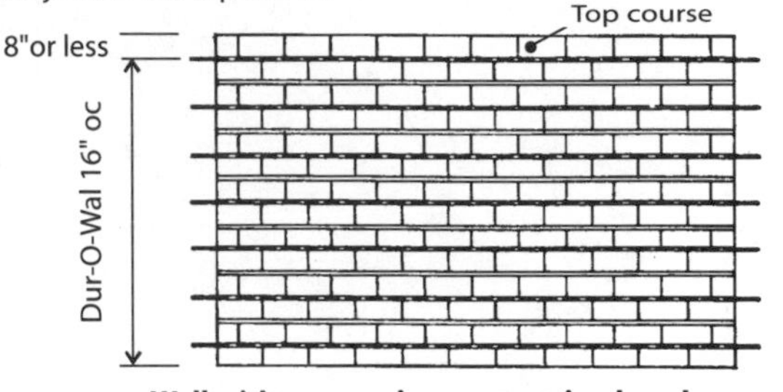

Wall with no openings — running bond

Foundation Walls — Place Dur-O-Wal 8" oc in upper half to two thirds of wall.

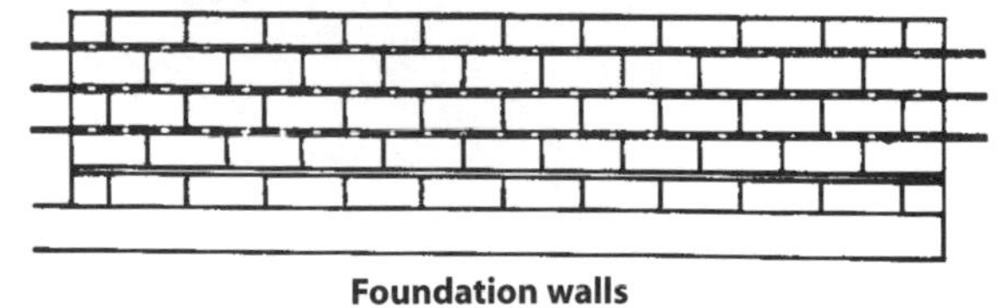

Foundation walls

Basement Walls — Place Dur-O-Wal in first joint below top of wall and 8" oc in the top 5 bed joints below openings.

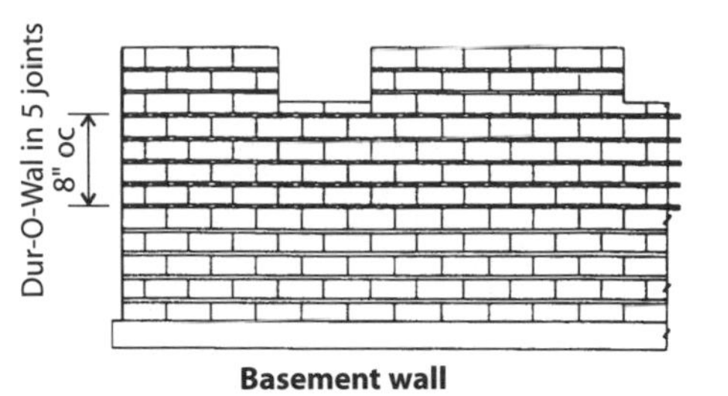

Basement wall

Courtesy: DUR-O-WAL, Inc.

Stack Bond — Place Dur-O-Wal 16" oc vertically in walls laid in stack bond. Place 8" oc for the top three courses in load bearing walls.

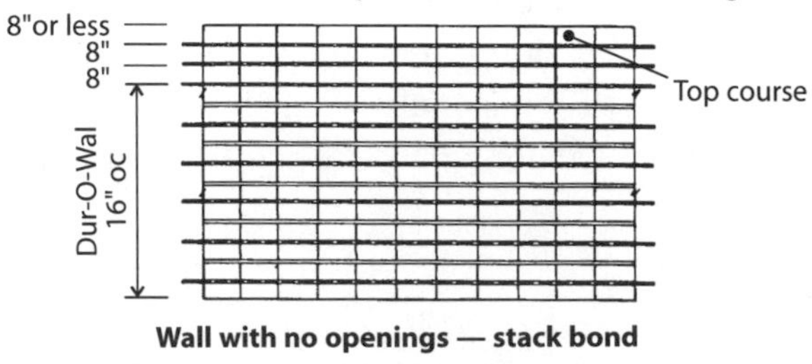

Wall with no openings — stack bond

Control Joints — Unless as otherwise noted, place all reinforcement continuously. Do not pass reinforcement through vertical masonry control joints.

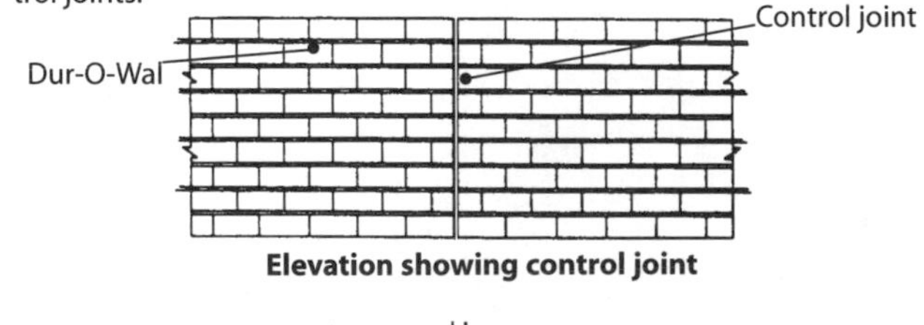

Elevation showing control joint

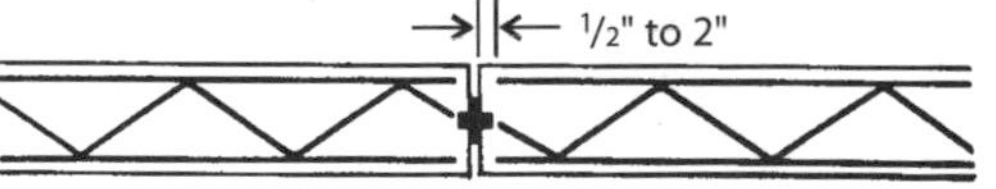

Horizontal section (different scale) Dur-O-Wal should not cross over control joint

Figure 10-5
Typical spacing of truss and ladur joint reinforcement

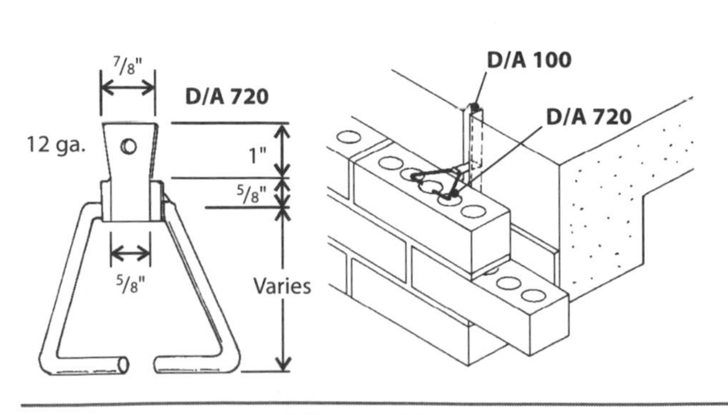

Figure 10-6
The dovetail anchor system

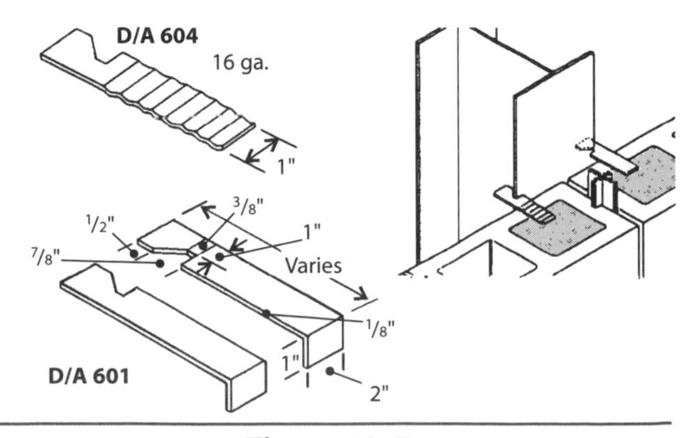

Figure 10-7
Notched steel column anchors

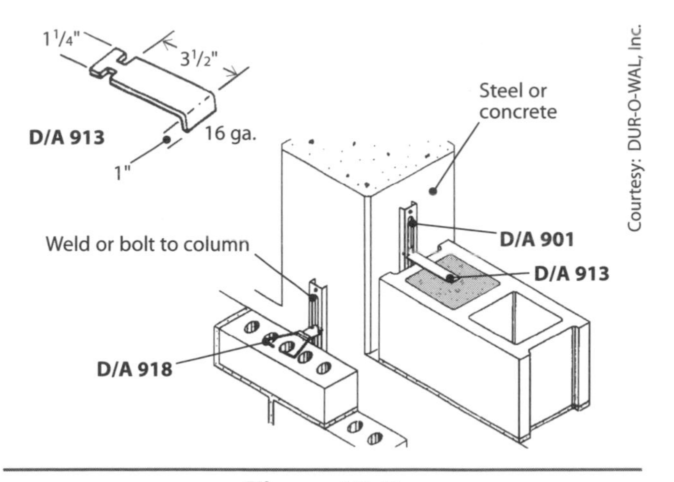

Figure 10-8
Channel slot anchor system

welded-on anchors in Figure 10-9 are designed for steel columns: Don't use them on steel studs or wood frames. When the wide flange of the column is perpendicular with the masonry coursing, the anchors have to be welded on. All of these anchors let you adjust the height of the anchor ties to the coursing of the masonry. This makes the anchoring operation easier for the mason.

Figure 10-10 shows vertical joints you can use for intersecting and connecting walls or connecting slabs and spandrel beams. These situations require interrupting the joint reinforcement and bond beam steel at control joints and expansion joints to allow movement within the plane of the wall. See Figure 10-11. There's no guidance in the building code for resisting

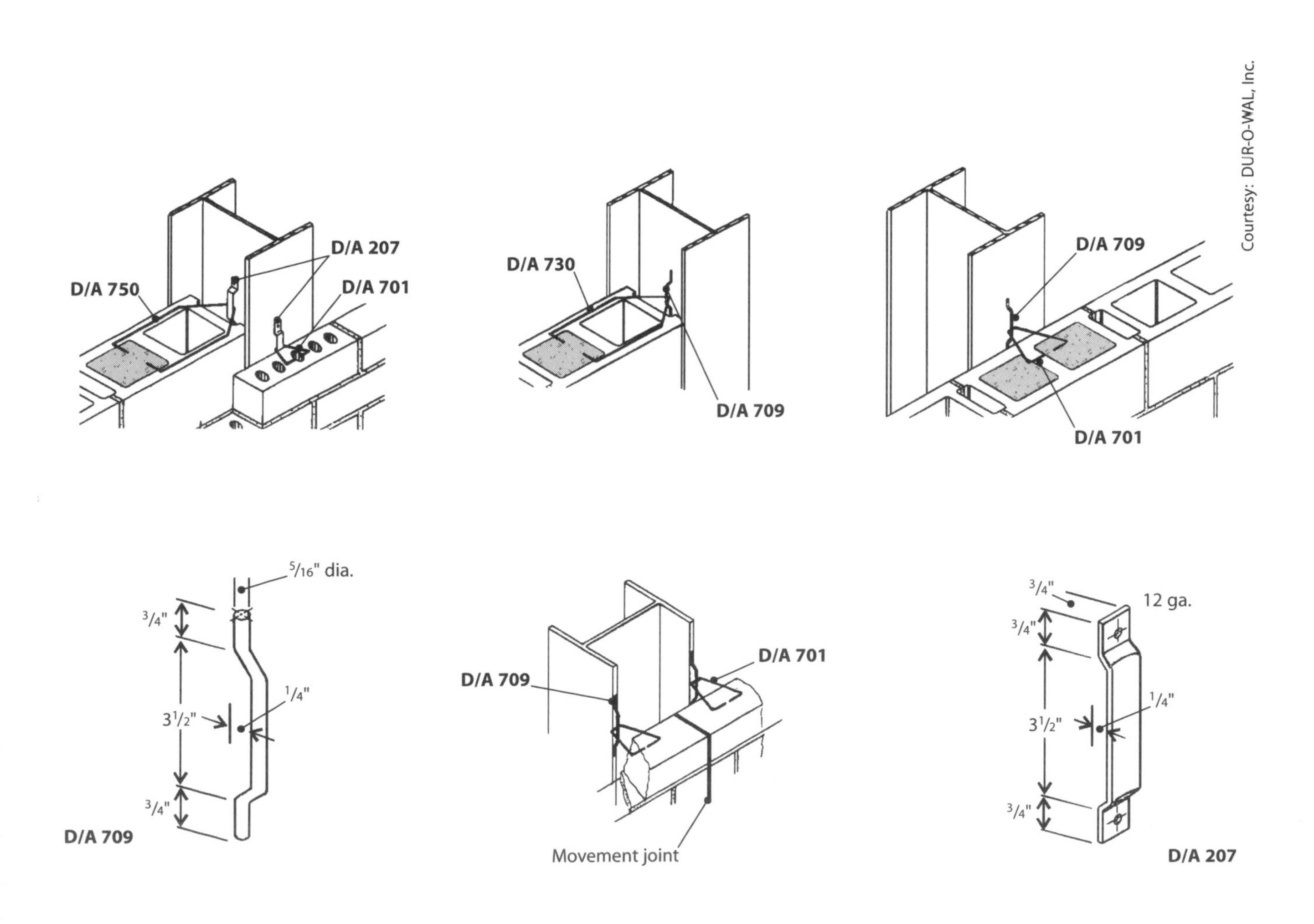

Figure 10-9
Screw-on, or welded-on column anchor

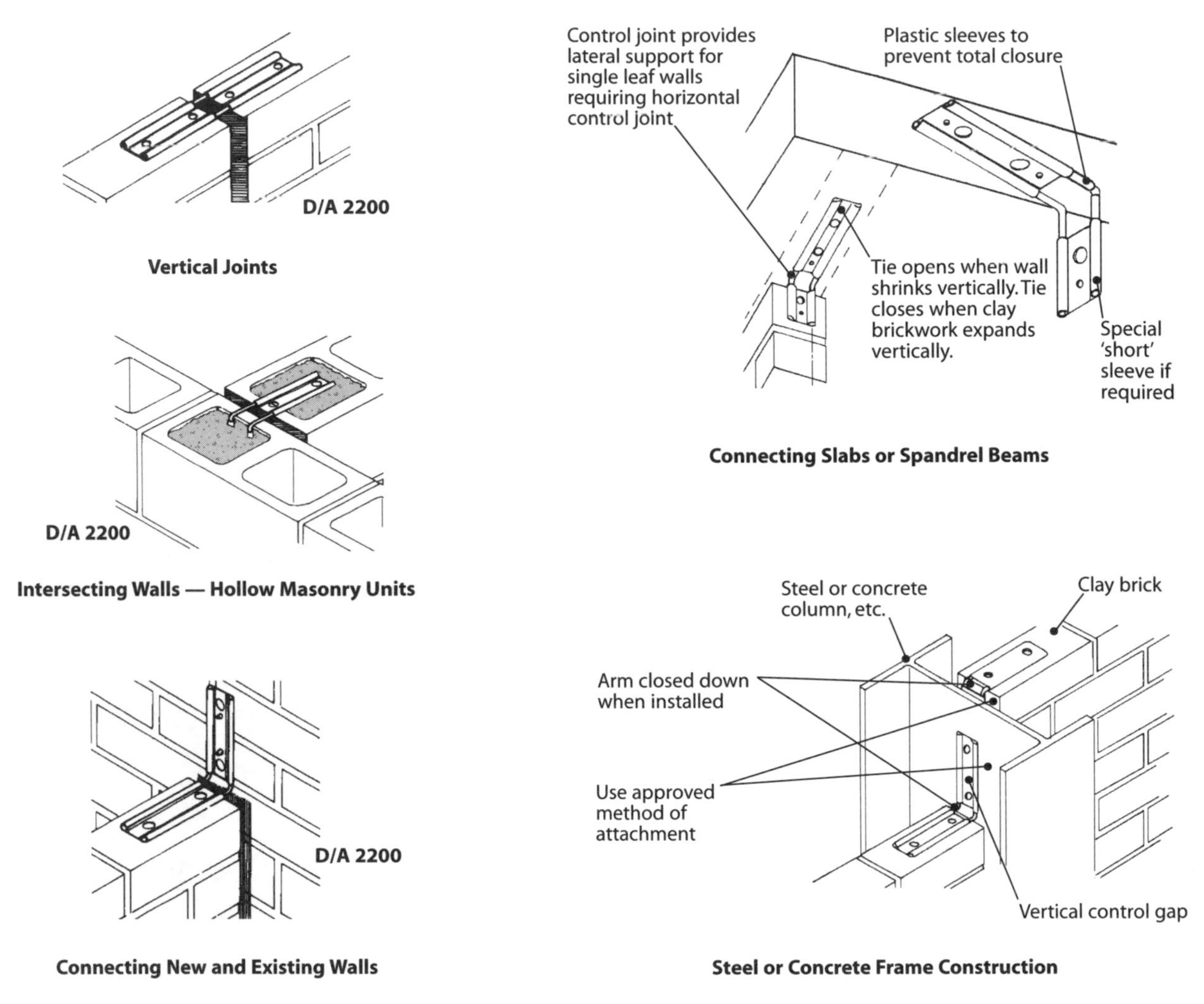

Figure 10-10
Vertical joints in intersecting and connecting walls or connecting slabs and spandrel beams

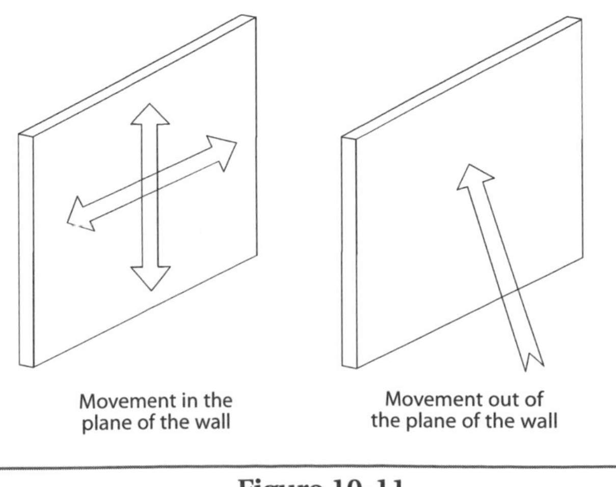

Figure 10-11
Wall movement

out-of-plane forces and movements of these joints. However, the UBC requires that at least two 9-gauge wires cross the head joints at 16 inches vertically when stack bond is used. The Masonry Standards Joint Committee Code requires an area of steel equal to 0.0003 times the wall area. Since control and expansion joints resemble stack bond head joints, the joint stabilizing anchors are made with two 8-gauge wires. Maximum spacing shouldn't exceed two courses of block (16 inches).

Use veneer anchor assemblies to anchor brick walls to block veneer assemblies when there's no cavity insulation. You can install these anchors as you build the wall, using the slot to adjust to the course levels.

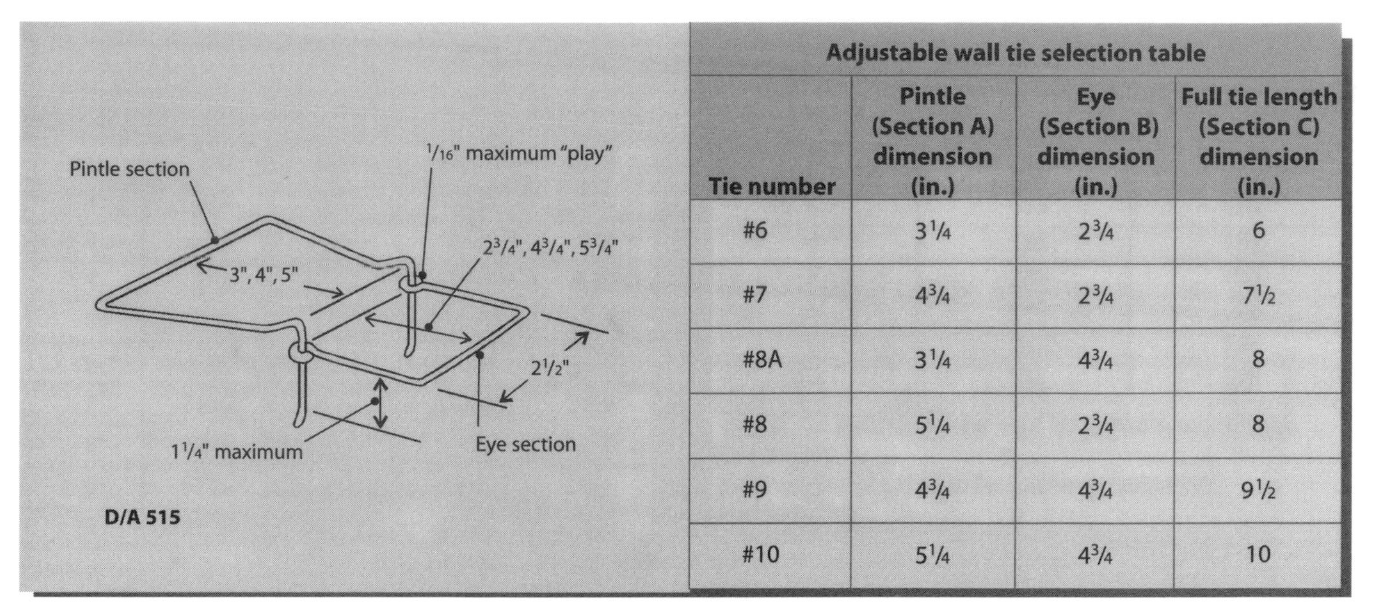

Adjustable wall tie selection table			
Tie number	Pintle (Section A) dimension (in.)	Eye (Section B) dimension (in.)	Full tie length (Section C) dimension (in.)
#6	$3\frac{1}{4}$	$2\frac{3}{4}$	6
#7	$4\frac{3}{4}$	$2\frac{3}{4}$	$7\frac{1}{2}$
#8A	$3\frac{1}{4}$	$4\frac{3}{4}$	8
#8	$5\frac{1}{4}$	$2\frac{3}{4}$	8
#9	$4\frac{3}{4}$	$4\frac{3}{4}$	$9\frac{1}{2}$
#10	$5\frac{1}{4}$	$4\frac{3}{4}$	10

A Types of adjustable wall ties

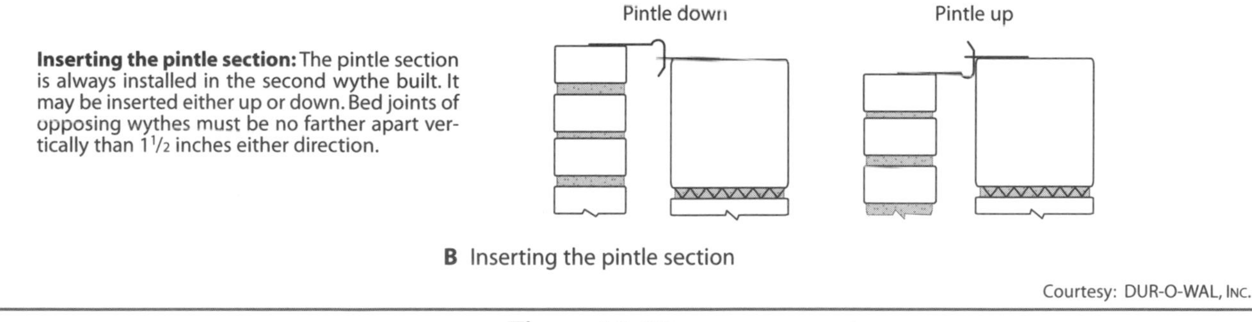

B Inserting the pintle section

Courtesy: DUR-O-WAL, INC.

Figure 10-12
Adjustable wall ties

There's also an anchor designed to be used with steel stud assemblies. Install it with two screws, either stainless steel or coated with a co-polymer.

Expansion Joints

You can use two types of expansion joints to prevent cracking in masonry walls. *Vertical joints* allow horizontal movement and *horizontal joints* allow vertical movement. However there's no single rule for positioning or spacing the joints that you can use on all structures. Each building design has to be analyzed by an engineer or architect.

Expansion joints are frequently placed at or near outside corners in both cavity and solid walls. This is particularly true where cavity walls rest on concrete foundations extending 2 or more feet above ground. If a vertical joint is interrupted by a lintel, design the joint to go around the end of the lintel and continue down the wall. Use horizontal joints in multistory walls where the lintels are a continuation of shelf angles that support masonry panels.

A vertical expansion joint with a pintle lets the masonry move horizontally (Figure 10-12). The joint filler material is hard rubber. There's a slot at the end of the blocks to hold the filler material. Add caulk to the joint later. At the bottom of the illustration, you

Advantage	Use
Reduce coursing problems	Use where the facing and backing don't course out at proper intervals or where you need to build out one wythe ahead of another
Increase productivity	Lets the mason concentrate on one wythe of the wall, to any given height, before changing to another wythe
May be used with rigid insulation	Provides the proper mechanical attachment of rigid insulation to the backing wythe
Make inspection easier	Easy to see if the ties are installed as specified
Eliminate bending ties	Mason doesn't have to bend or reshape conventional rigid ties when misalignment occurs
Anchor intersecting walls	Provides way to anchor intersecting walls when masonry bond at intersections isn't required
Improve waterproofing	Provides for better waterproofing of the backing wythe in cavity walls
Speed up and improve parging	Build wythe to be parged first with the eye section set out just far enough to accommodate the pintle section after parging (no projecting rigid ties)
Speed up construction	Inside wythe built to fully, or partially, enclose the building so that work can be started inside (exterior wythe built and tied in later)

Figure 10-13
Advantages and uses of adjustable wall ties

can see the continuous flashing that goes through the joint beneath the blocks. Install it to keep moisture from penetrating the joint.

Adjustable wall ties can save labor and help you overcome problems when you lay up walls that aren't conventionally designed. Figure 10-13 lists some advantages and uses of adjustable wall ties. See Figure 10-14 for various types of wall ties and how to apply them.

Sealants

The installation of sealants and caulking is often done incorrectly or incompletely. Almost all of the openings in masonry walls require a sealant or some kind of caulking. Caulking is sometimes used to hide imperfections in workmanship. But that's not its purpose. The real purpose of both the sealant and caulking is to prevent moisture penetration. In locations such as windows and doorways where there's no movement of the walls, caulking containing a solvent-based acrylic sealant or butyl caulk composition should be used. In large areas such as expansion joints it's best to use an elastomeric joint sealant. These include urethanes, silicones and polysulfides. Styrofoam rope should be used as a backer rod to complete the application.

Adjustable wall ties for faced or composite walls

Use adjustable wall tie No. 6 or No. 8 for faced or composite walls of all sizes.

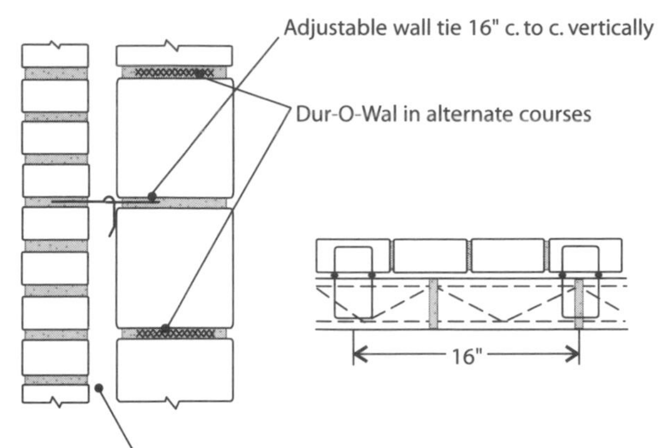

Adjustable wall ties for cavity walls with rigid insulation

Selection of tie (use selection table)

1. Select eye section long enough to extend approximately 2" into the masonry with center of eyelet at face of insulation.
2. Select pintle section long enough to extend approximately 2" into the masonry and reach across cavity to engage eyelet.

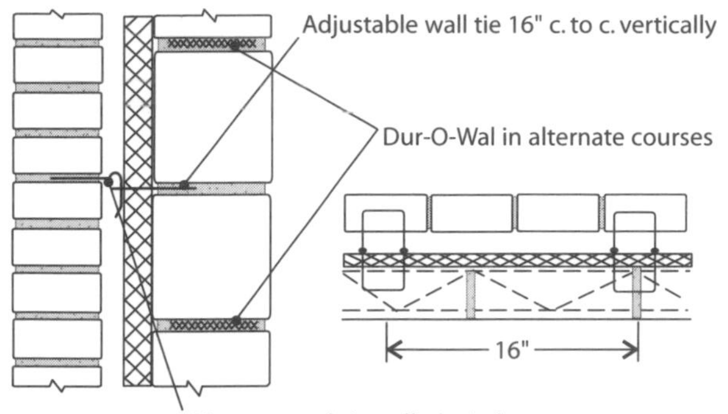

Adjustable wall ties for cavity walls with air space or loose insulation

Dur-O-Wal in alternate courses

Adjustable wall tie 16" c. to c. vertically

A B C

24"

Eye close to face of wythe

Overall wall width (in.)	Width of cavity and wythes			Tie number
	Exterior A (in.)	Cavity B (In.)	Backup C (in.)	
10	4	2	4	No. 7
11	4	3	4	No. 8
12	4	2	6	No. 7
13	4	3	6	No. 8
14	4	2	8	No. 7
15	4	3	8	No. 8

Courtesy: DUR-O-WAL, Inc.

Figure 10-14

Recommended applications for adjustable wall ties

11

Brick Wall Construction

For many years, the only brick generally available were standard, Roman and Norman. But now you can get brick in many sizes, ranging in thickness (bed depth) from a nominal 3 inches to 12 inches. In height, they range from a nominal 2 inches to 8 inches and lengths up to 16 inches. So you need to list brick dimensions carefully to avoid misunderstandings in both shipping and construction.

Figures 11-1 through 11-4 show the most typical brick currently being produced. However, few manufacturers produce all these sizes. Before you start a design or bid a job, talk with manufacturers or distributors in your area. Don't design a project with a brick size you'll have trouble finding. Even more important, don't bid a job until you're sure you won't have to pay a premium for the brick.

Although the sizes of brick are pretty standard, the names aren't — except for standard, Roman and Norman sizes. Manufacturers use their own names for certain sizes or they may use a common name for a brick with nonstandard dimensions. To avoid confusion, always identify a brick by size.

Except for nonmodular standard, oversize and 3-inch brick, most brick are made in modular sizes. The nominal size of modular brick is the manufactured dimensions, plus the thickness of the mortar joint the brick is designed for. In general, joint thickness is ⅜ inch or ½ inch.

The actual manufactured dimensions of brick may vary from the specified dimensions, as long as the variation is within the ASTM Specifications allowances. For standard specifications for facing brick, check ASTM Designation C216. For standard specifications for building brick, look at ASTM Designation C62. Standard specifications for hollow brick are in ASTM Designation C652.

The designated manufactured heights for all modular brick designed to be laid in three courses to 8 inches are the same: 2¼ inches. That includes standard brick and standard modular brick. Figures 11-5 and 11-6 show the vertical coursing dimensions for modular and nonmodular brick.

Unit designation	Manufactured size (in.)		
	Thickness	Height	Length
3-inch*	3	2⅝	9⅝
	3	2¾	9¾
Standard	3¾**	2¼	8
Oversize	3¾**	2¾	8

* In recent years, the "3-inch" has gained popularity in certain areas. The term 3-inch designates its thickness or bed depth. The sizes shown are the ones commonly produced under the name Kingsize. Other sizes of 3-inch brick are called Big John, Jumbo, Scotsman and Spartan. Originally developed as a veneer unit, they're also used for 8-inch cavity walls and 8-inch grouted walls.

** The manufactured thickness of standard or oversized nonmodular brick varies from 3½ to 3¾ inches. If you're not using a running bond, check with the brick manufacturer.

Figure 11-1
Sizes of nonmodular brick

Since labor is a major part of the cost of brick wall construction, many builders prefer to use larger units. The fewer units the mason has to lay, the lower the resulting masonry unit cost. For veneer or other non-structural brickwork, using thinner brick can save material costs. But make sure the face size of the brick goes with the size and architectural tone of the building it's on.

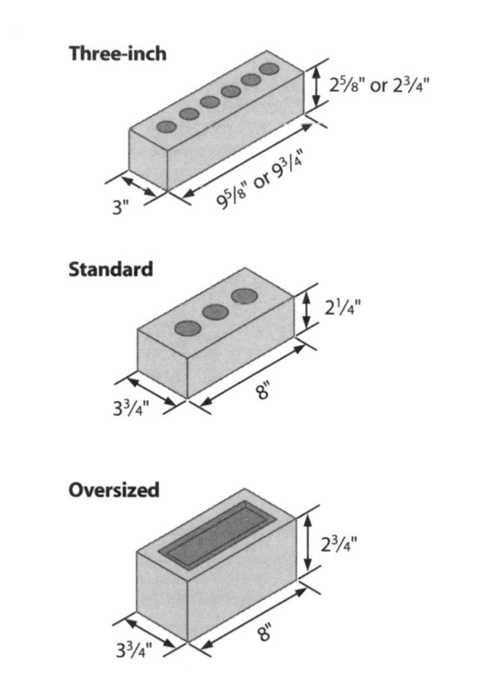

Figure 11-2
Actual dimensions of nonmodular brick

Bonds and Patterns in Brickwork

When you're talking about masonry, the word *bond* may mean:

- *Structural bond* — the method used to interlock or tie brick together so the entire assembly is a single structural unit
- *Pattern bond* — the pattern the brick and mortar make on the face of a wall
- *Mortar bond* — the adhesion of mortar to brick or reinforcing steel

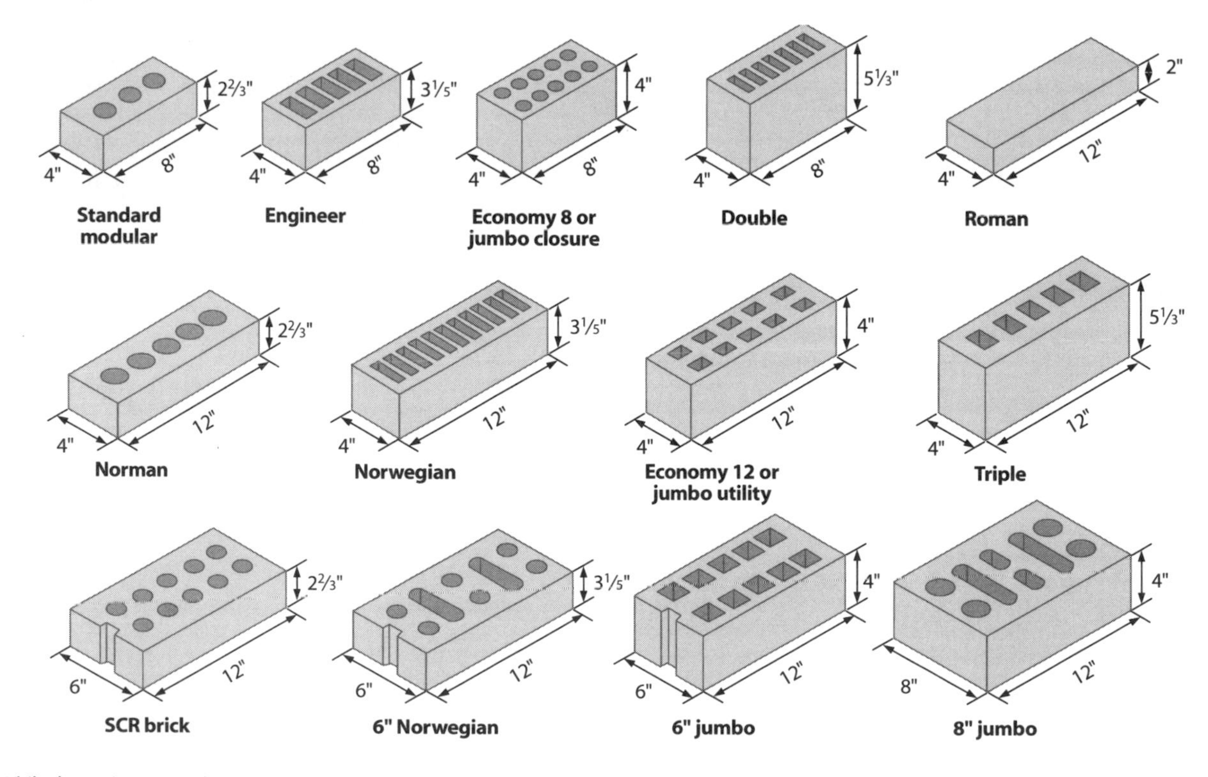

Note: While the coring types shown are typical for solid units, they do not necessarily apply to the specific types of units with which they are shown above. They will vary with the manufacturer.

Figure 11-3
Nominal dimensions of modular brick

Unit designation	Nominal dimensions (in.)			Joint thickness (in.)	Manufactured dimensions (in.)			Modular coursing (in.)
	t	h	l		t	h	l	
Standard modular	4	$2\frac{2}{3}$	8	$\frac{3}{8}$	$3\frac{5}{8}$	$2\frac{1}{4}$	$7\frac{5}{8}$	3C = 8
				$\frac{1}{2}$	$3\frac{1}{2}$	$2\frac{1}{4}$	$7\frac{1}{2}$	
Engineer	4	$3\frac{1}{5}$	8	$\frac{3}{8}$	$3\frac{3}{8}$	$2\frac{13}{16}$	$7\frac{5}{8}$	5C = 16
				$\frac{1}{2}$	$3\frac{1}{2}$	$2\frac{11}{16}$	$7\frac{1}{2}$	
Economy 8 or jumbo closure	4	4	8	$\frac{3}{8}$	$3\frac{5}{8}$	$3\frac{5}{8}$	$7\frac{5}{8}$	1C = 4
				$\frac{1}{2}$	$3\frac{1}{2}$	$3\frac{1}{2}$	$7\frac{1}{2}$	
Double	4	$5\frac{1}{3}$	8	$\frac{3}{8}$	$3\frac{5}{8}$	$4\frac{15}{16}$	$7\frac{5}{8}$	3C = 16
				$\frac{1}{2}$	$3\frac{1}{2}$	$4\frac{13}{16}$	$7\frac{1}{2}$	
Roman	4	2	12	$\frac{3}{8}$	$3\frac{5}{8}$	$1\frac{5}{8}$	$11\frac{5}{8}$	2C = 4
				$\frac{1}{2}$	$3\frac{1}{2}$	$1\frac{1}{2}$	$11\frac{1}{2}$	
Norman	4	$2\frac{2}{3}$	12	$\frac{3}{8}$	$3\frac{5}{8}$	$2\frac{1}{4}$	$11\frac{5}{8}$	3C = 8
				$\frac{1}{2}$	$3\frac{1}{2}$	$2\frac{1}{4}$	$11\frac{1}{2}$	
Norwegian	4	$3\frac{1}{5}$	12	$\frac{3}{8}$	$3\frac{5}{8}$	$2\frac{13}{16}$	$11\frac{5}{8}$	5C = 16
				$\frac{1}{2}$	$3\frac{1}{2}$	$2\frac{11}{16}$	$11\frac{1}{2}$	
Economy 12 or jumbo utility	4	4	12	$\frac{3}{8}$	$3\frac{5}{8}$	$3\frac{5}{8}$	$11\frac{5}{8}$	1C = 4
				$\frac{1}{2}$	$3\frac{1}{2}$	$3\frac{1}{2}$	$11\frac{1}{2}$	
Triple	4	$5\frac{1}{3}$	12	$\frac{3}{8}$	$3\frac{5}{8}$	$4\frac{15}{16}$	$11\frac{5}{8}$	3C = 16
				$\frac{1}{2}$	$3\frac{1}{2}$	$4\frac{13}{16}$	$11\frac{1}{2}$	
SCR brick[2]	6	$2\frac{2}{3}$	12	$\frac{3}{8}$	$5\frac{5}{8}$	$2\frac{1}{4}$	$11\frac{5}{8}$	3C = 8
				$\frac{1}{2}$	$5\frac{1}{2}$	$2\frac{1}{4}$	$11\frac{1}{2}$	
6-in. Norwegian	6	$3\frac{1}{5}$	12	$\frac{3}{8}$	$5\frac{5}{8}$	$2\frac{13}{16}$	$11\frac{5}{8}$	5C = 16
				$\frac{1}{2}$	$5\frac{1}{2}$	$2\frac{11}{16}$	$11\frac{1}{2}$	
6-in. jumbo	6	4	12	$\frac{3}{8}$	$5\frac{5}{8}$	$3\frac{5}{8}$	$11\frac{5}{8}$	1C = 4
				$\frac{1}{2}$	$5\frac{1}{2}$	$3\frac{1}{2}$	$11\frac{1}{2}$	
8-in. jumbo	8	4	12	$\frac{3}{8}$	$7\frac{5}{8}$	$3\frac{5}{8}$	$11\frac{5}{8}$	1C = 4
				$\frac{1}{2}$	$7\frac{1}{2}$	$3\frac{1}{2}$	$11\frac{1}{2}$	

[1]Available as solid units to ASTM C 216- or ASTM C 62-, or, in a number of cases, as hollow brick conforming to ASTM C 652-.

[2]Reg. U.S. Pat. Off., SCPI.

Figure 11-4

Sizes of modular brick[1]

Number of courses	2¼" high units		2⅝" high units		2⅜" high units	
	⅜" joints	½" joints	⅜" joints	½" joints	⅜" joints	½" joints
1	0'2⅝"	0'2¾"	0'3"	0'3⅛"	0'3⅛"	0'3¼"
2	0'5¼"	0'5½"	0'6"	0'6¼"	0'6¼"	0'6½"
3	0'7⅞"	0'8¼"	0'9"	0'9⅜"	0'9⅜"	0'9¾"
4	0'10½"	0'11"	1'0"	1'½"	1'½"	1'1"
5	1'1⅛"	1'1¾"	1'3"	1'3⅝"	1'3⅝"	1'4¼"
6	1'3¾"	1'4½"	1'6"	1'6¾"	1'6¾"	1'7½"
7	1'6⅜"	1'7¼"	1'9"	1'9⅞"	1'9⅞"	1'10¾"
8	1'9"	1'10"	2'0"	2'1"	2'1"	2'2"
9	1'11⅝"	2'¾"	2'3"	2'4⅛"	2'4⅛"	2'5¼"
10	2'2¼"	2'3½"	2'6"	2'7¼"	2'7¼"	2'8½"
11	2'4⅞"	2'6¼"	2'9"	2'10⅜"	2'10⅜"	2'11¾"
12	2'7½"	2'9"	3'0"	3'1½"	3'1½"	3'3"
13	2'10⅛"	2'11¾"	3'3"	3'4⅝"	3'4⅝"	3'6¼"
14	3'¾"	3'2½"	3'6"	3'7¾"	3'7¾"	3'9½"
15	3'3⅜"	3'5¼"	3'9"	3'10⅞"	3'10⅞"	4'¾"
16	3'6"	3'8"	4'0"	4'2"	4'2"	4'4"
17	3'8⅝"	3'10¾"	4'3"	4'5⅛"	4'5⅛"	4'7¼"
18	3'11¼"	4'1½"	4'6"	4'8¼"	4'8¼"	4'10½"
19	4'1⅞"	4'4¼"	4'9"	4'11⅜"	4'11⅜"	5'1¾"
20	4'4½"	4'7"	5'0"	5'2½"	5'2½"	5'5"
21	4'7⅛"	4'9¾"	5'3"	5'5⅝"	5'5⅝"	5'8¼"
22	4'9¾"	5'½"	5'6"	5'8¾"	5'8¾"	5'11½"
23	5'⅜"	5'3¼"	5'9"	5'11⅞"	5'11⅞"	6'2¾"
24	5'3"	5'6"	6'0"	6'3"	6'3"	6'6"
25	5'5⅝"	5'8¾"	6'3"	6'6⅛"	6'6⅛"	6'9¼"
26	5'8¼"	5'11½"	6'6"	6'9¼"	6'9¼"	7'½"
27	5'10⅞"	6'2¼"	6'9"	7'⅜"	7'⅜"	7'3¾"
28	6'1½"	6'5"	7'0"	7'3½"	7'3½"	7'7"
29	6'4⅛"	6'7¾"	7'3"	7'6⅝"	7'6⅝"	7'10¼"
30	6'6¾"	6'10½"	7'6"	7'9¾"	7'9¾"	8'1½"
31	6'9⅜"	7'1¼"	7'9"	8'⅞"	8'⅞"	8'4¾"
32	7'0"	7'4"	8'0"	8'4"	8'4"	8'8"
33	7'2⅝"	7'6¾"	8'3"	8'7⅛"	8'7⅛"	8'11¼"
34	7'5¼"	7'9½"	8'6"	8'10¼"	8'10¼"	9'2½"
35	7'7⅞"	8'¼"	8'9"	9'1⅜"	9'1⅜"	9'5¾"
36	7'10½"	8'3"	9'0"	9'4½"	9'4½"	9'9"
37	8'1⅛"	8'5¾"	9'3"	9'7⅝"	9'7⅝"	10'¼"
38	8'3¾"	8'8½"	9'6"	9'10¾"	9'10¾"	10'3½"
39	8'6⅜"	8'11¼"	9'9"	10'1⅞"	10'1⅞"	10'6¾"
40	8'9"	9'2"	10'0"	10'5"	10'5"	10'10"
41	8'11⅝"	9'4¾"	10'3"	10'8⅛"	10'8⅛"	11'1¼"
42	9'2¼"	9'7½"	10'6"	10'11¼"	10'11¼"	11'4½"
43	9'4⅞"	9'10¼"	10'9"	11'2⅜"	11'2⅜"	11'7¾"
44	9'7½"	10'1"	11'0"	11'5½"	11'5½"	11'11"
45	9'10⅛"	10'3¾"	11'3"	11'8⅝"	11'8⅝"	12'2¼"
46	10'¾"	10'6½"	11'6"	11'11¾"	11'11¾"	12'5½"
47	10'3⅜"	10'9¼"	11'9"	12'2⅞"	12'2⅞"	12'8¾"
48	10'6"	11'0"	12'0"	12'6"	12'6"	13'0"
49	10'8⅝"	11'2¾"	12'3"	12'9⅛"	12'9⅛"	13'3¼"
50	10'11¼"	11'5½"	12'6"	13'¼"	13'¼"	13'6½"
100	21'10½"	22'11"	25'0"	26'½"	26'½"	27'1"

[1]Brick positioned in wall as stretchers. Vertical dimensions are from bottom of mortar joint to bottom of mortar joint.

Figure 11-5

Vertical coursing dimensions of nonmodular brick[1]

Number of courses	Nominal height (h) of unit[2]				
	2"	2 2/3"	3 1/5"	4"	5 1/3"
1	0'2"	0'2 11/16"	0'3 3/16"	0'4"	0'5 5/16"
2	0'4"	0'5 5/16"	0'6 3/8"	0'8"	0'10 11/16"
3	0'6"	0'8"	0'9 5/8"	1'0"	1'4"
4	0'8"	1'10 11/16"	1' 13/16"	1'4"	1'9 5/16"
5	0'10"	1'1 5/16"	1'4"	1'8"	2'2 11/16"
6	1'0"	1'4"	1'7 3/16"	2'0"	2'8"
7	1'2"	1'6 11/16"	1'10 3/8"	2'4"	3'1 5/16"
8	1'4"	1'9 5/16"	2'1 5/8"	2'8"	3'6 11/16"
9	1'6"	2'0"	2'4 13/16"	3'0"	4'0"
10	1'8"	2'2 11/16"	2'8"	3'4"	4'5 5/16"
11	1'10"	2'5 5/16"	2'11 3/16"	3'8"	4'10 11/16"
12	2'0"	2'8"	3'2 3/8"	4'0"	5'4"
13	2'2"	2'10 11/16"	3'5 5/8"	4'4"	5'9 5/16"
14	2'4"	3'1 5/16"	3'8 13/16"	4'8"	6'2 11/16"
15	2'6"	3'4"	4'0"	5'0"	6'8"
16	2'8"	3'6 11/16"	4'3 3/16"	5'4"	7'1 5/16"
17	2'10"	3'9 5/16"	4'6 3/8"	5'8"	7'6 11/16"
18	3'0"	4'0"	4'9 5/8"	6'0"	8'0"
19	3'2"	4'2 11/16"	5' 13/16"	6'4"	8'5 5/16"
20	3'4"	4'5 5/16"	5'4"	6'8"	8'10 11/16"
21	3'6"	4'8"	5'7 3/16"	7'0"	9'4"
22	3'8"	4'10" 11/16"	5'10 3/8"	7'4"	9'9 5/16"
23	3'10"	5'1 5/16"	6'1 5/8"	7'8"	10'2 11/16"
24	4'0"	5'4"	6'4 13/16"	8'0"	10'8"
25	4'2"	5'6 11/16"	6'8"	8'4"	11'1 5/16"
26	4'4"	5'9 5/16"	6'11 3/16"	8'8"	11'6 11/16"
27	4'6"	6'0"	7'2 3/8"	9'0"	12'0"
28	4'8"	6'2 11/16"	7'5 5/8"	9'4"	12'5 5/16"
29	4'10"	6'5 5/16"	7'8 13/16"	9'8"	12'10 11/16"
30	5'0"	6'8"	8'0"	10'0"	13'4"
31	5'2"	6'10 11/16"	8'3 3/16"	10'4"	13'9 5/16"
32	5'4"	7'1 5/16"	8'6 3/8"	10'8"	14'2 11/16"
33	5'6"	7'4"	8'9 5/8"	11'0"	14'8"
34	5'8"	7'6 11/16"	9' 13/16"	11'4"	15'1 5/16"
35	5'10"	7'9 5/16"	9'4"	11'8"	15'6 11/16"
36	6'0"	8'0"	9'7 3/16"	12'0"	16'0"
37	6'2"	8'2 11/16"	9'10 3/8"	12'4"	16'5 5/16"
38	6'4"	8'5 5/16"	10'1 5/8"	12'8"	16'10 11/16"
39	6'6"	8'8"	10'4 13/16"	13'0"	17'4"
40	6'8"	8'10 11/16"	10'8"	13'4"	17'9 5/16"
41	6'10"	9'1 5/16"	10'11 3/16"	13'8"	18'2 11/16"
42	7'0"	9'4"	11'2 3/8"	14'0"	18'8"
43	7'2"	9'6 11/16"	11'5 5/8"	14'4"	19'1 5/16"
44	7'4"	9'9 5/16"	11'8 13/16"	14'8"	19'6 11/16"
45	7'6"	10'0"	12'0"	15'0"	20'0"
46	7'8"	10'2 11/16"	12'3 3/16"	15'4"	20'5 5/16"
47	7'10"	10'5 5/16"	12'6 3/8"	15'8"	20'10 11/16"
48	8'0"	10'8"	12'9 5/8"	16'0"	21'4"
49	8'2"	10'10 11/16"	13' 13/16"	16'4"	21'9 5/16"
50	8'4"	11'1 5/16"	13'4"	16'8"	22'2 11/16"
100	16'8"	22'2 11/16"	26'8"	33'4"	44'5 5/16"

[1]Brick positioned in wall as stretchers.

[2]For convenience in using table, nominal 1/3", 2/3" and 1/5" heights of units have been changed to nearest 1/16". Vertical dimensions are from bottom of mortar joint to bottom of mortar joint.

Figure 11-6

Vertical coursing dimensions for modular brick[1]

Structural Bonds

There are three ways to make a structural bond in a masonry wall:

- ❖ Overlapping (interlocking) the masonry units
- ❖ Using metal ties embedded in the connecting joints
- ❖ Using grout to adhere adjacent wythes of masonry

Overlapping bond is based on variations of two traditional methods of bonding — English bond (Figure 11-7) and Flemish bond (Figure 11-8). Modern building codes require that headers make up at least 4 percent of the wall surface. The headers can't be more than 24 inches on center, vertically or horizontally.

Metal ties are common for structural bonding in both solid-wall and cavity-wall construction. See Figure 11-9. Most building codes let you use rigid steel bonding ties in solid walls. Use at least one metal tie for each 4½ square feet of wall surface and stagger the ties in alternate courses. The distance between adjacent ties shouldn't be more than 24 inches vertically and 36 inches horizontally. Provide additional bonding ties at each opening, spaced within 12 inches of the opening. If you use ties smaller than 3/16 inch in diameter, reduce the tie spacing so the tie area per square foot of wall isn't less than specified previously.

Sometimes you can make a structural bond of solid and reinforced brick masonry walls by pouring grout into the cavity or collar joint between wythes of masonry.

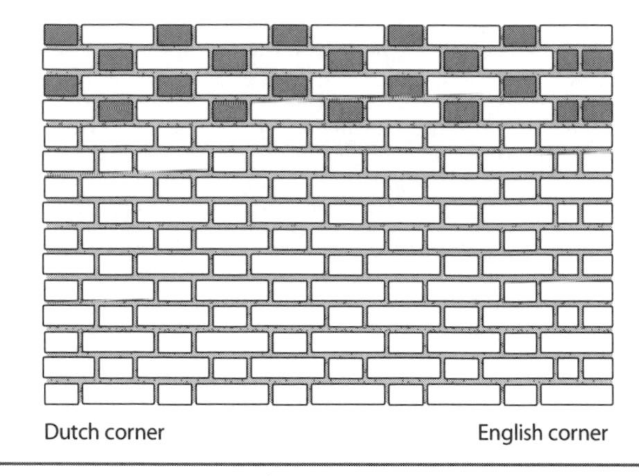

Figure 11-8
Flemish bond — structural

The metal-tie method is generally recommended for exterior walls. It's easier to do and it resists rain well. Using metal ties also lets the facing and backing move a bit, which may prevent cracking. But the method of bonding you use will depend on the use requirements, wall type and other factors.

Pattern Bonds

You can use structural bonds such as English or Flemish to create patterns in the face of a wall. But in the strict sense of the term, *pattern* refers to the arrangement of brick texture or color used in the face. It's possible to create many patterns using the same

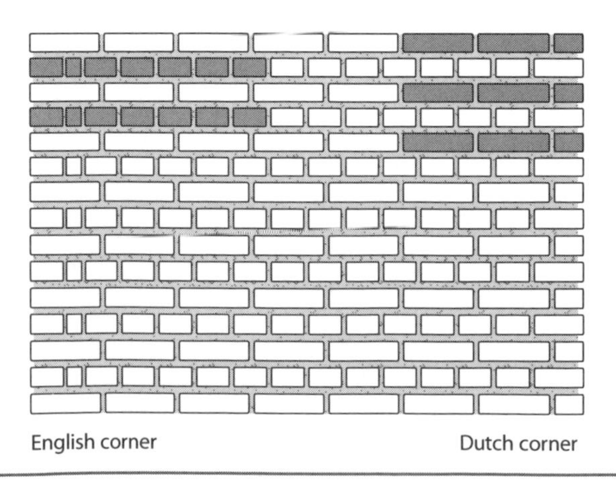

Figure 11-7
English bond — structural

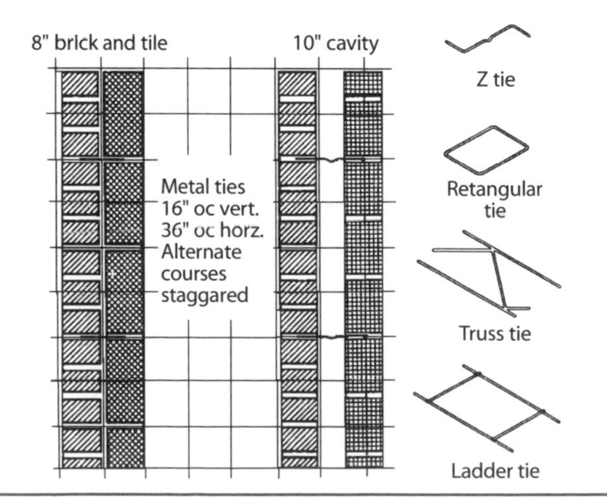

Figure 11-9
Metal-tied masonry walls

structural bond. There are five basic structural bonds commonly used today which create typical patterns:

- Running bond
- Common or American bond
- Flemish bond
- English bond
- Block or stack bond

Using these bonds and variations of brick color and texture and joint types, you can make an almost unlimited number of patterns. That makes a distinctive texture that's not solely dependent on the texture of the individual brick. Figure 11-10 shows a Flemish bond at the header course and a stack bond for the projecting brick.

Figure 11-10
Flemish bond at the header course with brick projecting in stack bond

Now let's learn more about the five basic bonds.

- *Running bond* — The simplest of the basic pattern bonds, you make a running bond using only stretchers. Since there are no headers in this bond, you also usually use metal ties. Use running bond for cavity wall construction and veneered walls of brick, and for facing tile walls using extra-wide stretcher tile. The wall in Figure 11-11 A is finished with raked joints.
- *Common or American bond* — This is a variation of running bond with a course of full-length headers at regular intervals. See Figure 11-11 B. These headers provide structural bonding as well as pattern. Usually you use header courses at each fifth, sixth or seventh course. In laying out any bond pattern, it's important to start the corners correctly. For common bond, start a three-quarter brick each way from the corner at the header course. You can vary the common bond by using a Flemish header course.
- *Flemish bond* — For this bond, alternate headers and stretchers in every course. That makes the headers and stretchers in every other course appear in vertical stacks. See Figure 11-11 C. You'll lay the stretchers longitudinally with the length of the wall, and the headers transversely across the length of the wall.

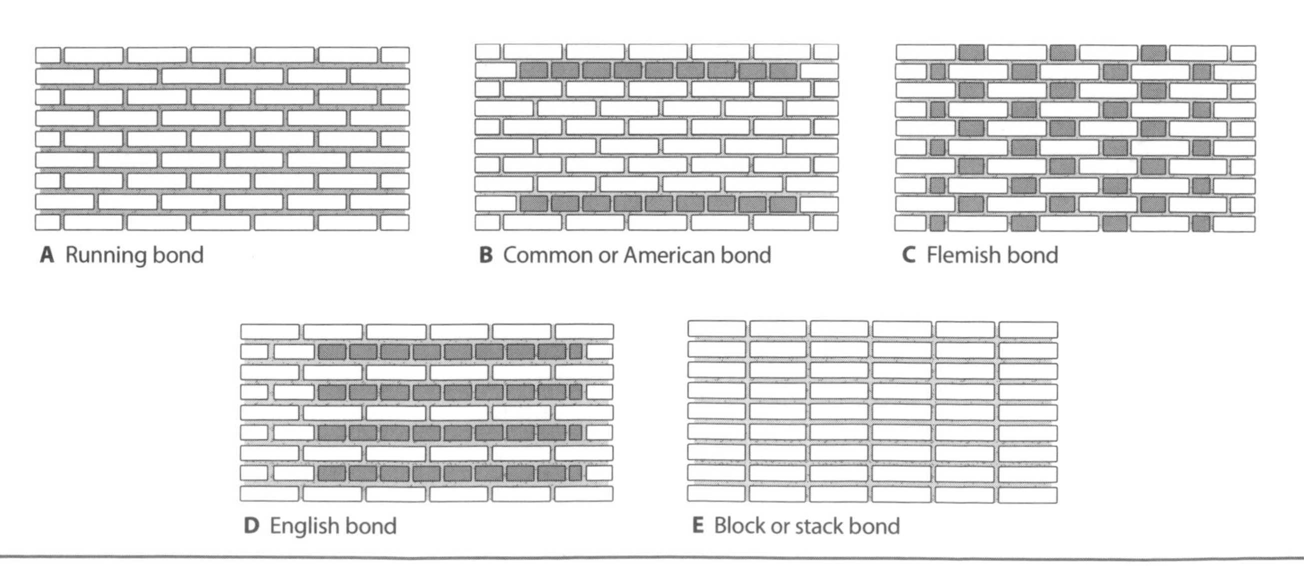

Figure 11-11
Five basic pattern bonds for brick

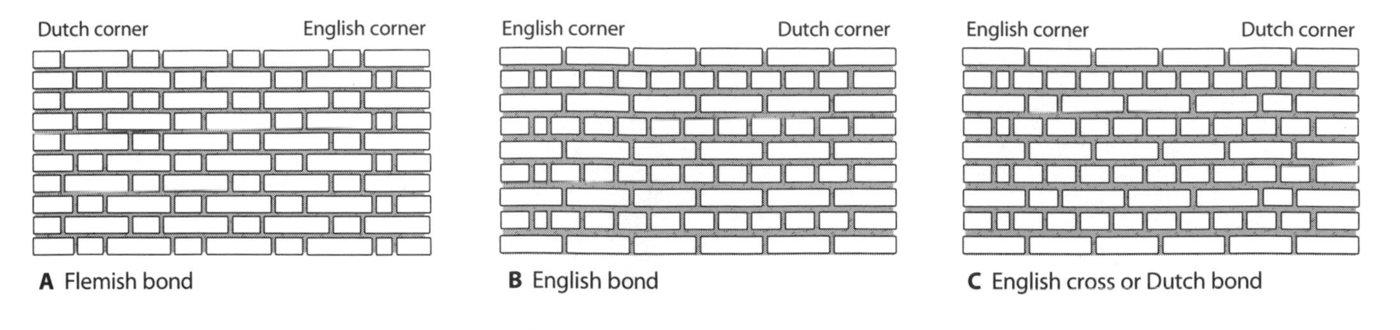

Figure 11-12
Dutch and English corners

- *English bond*—You make this bond by alternating courses of headers and stretchers. Center the headers on the stretchers, and align the joints between stretchers in all courses vertically. See Figure 11-11 D. Use snap headers in courses which are not structural-bonding courses. There are variations of English bond, called English Cross or Dutch bond, where you don't align the vertical joints between the stretchers in alternate courses vertically. These joints center on the stretchers themselves in the course above and below.

A Double stretcher with units in diagonal lines

B Units set in dovetail fashion

Figure 11-13
Garden wall bond

- *Block or stack bond*—This is just a pattern bond. The brick don't overlap since all the vertical joints are aligned. See Figure 11-11 E. You can bond this pattern to a backing with rigid ties, or use 8-inch bonded brick when you can get them. In large wall areas and in loadbearing construction, reinforce the wall with steel reinforcement in the horizontal mortar joints. You must use prematched or dimensionally accurate masonry units to keep the head joints aligned.

There are two methods to start the corners in Flemish and English bonds. Figure 11-12 shows the so-called *Dutch corner*, which uses a three-quarter brick closure, and the *English corner*, which uses a 2-inch *queen closure* or quarter-brick closure. Always place the 2-inch closure 4 inches in from the corner, never at the corner.

Figure 11-13 A shows a double-stretcher garden-wall bond with the pattern units in diagonal lines. Figure 11-13 B shows the garden-wall bond with the pattern set in dovetail fashion.

Mortar Joints

Mortar is what bonds brick together and seals the spaces between them. It also bonds brick to reinforcing steel and can be used to make up for the dimensional variations in brick. You can also use mortar decoratively to make shadow lines. Use colored mortar to enhance a pattern by coloring entire joints. Or you can use the tuck point method by going over an entire wall with a 1-inch deep raked joint and carefully filling in the colored mortar later.

Mortar Joint Finishes

There are two types of mortar joint finishes — *tooled* and *troweled* (struck). For a tooled joint, you use a special tool to compress and shape the mortar in the joint. Figure 11-14 shows how to use a concave jointing tool to make a strong water-repellent joint.

For a troweled joint, you cut off the excess mortar with a trowel and then finish with the same trowel. See Figure 11-15.

Figure 11-16 shows the five common types of mortar joints:

- *Struck or troweled joint* — This is a common joint in ordinary brickwork, and it's easy to do. Usually the mason reaches over from the opposite side of the wall to press the trowel into the bottom of the joint. Although the trowel causes some compaction in the concrete, the small ledge doesn't shed water readily. It's not a completely watertight joint.
- *Weathered joint* — The weathered joint is similar except the mason faces the wall and presses the trowel in at the top of the joint. This type of joint sheds water more readily than the struck joint.
- *Rough-cut or flush joint* — This is the simplest joint since you just hold the edge of the trowel flat against the brick and cut in any direction. This makes an uncompacted joint with a small hairline crack where the mortar is pulled away from the brick by the cutting action.
- *Raked joint* — You make this joint by removing the surface of the mortar while it's still soft. While the joint may be compacted, it's not easy to make this joint weathertight. I can't recommend it where

Figure 11-14
Making a tooled joint with a concave jointing tool

Figure 11-15
Struck or troweled joint

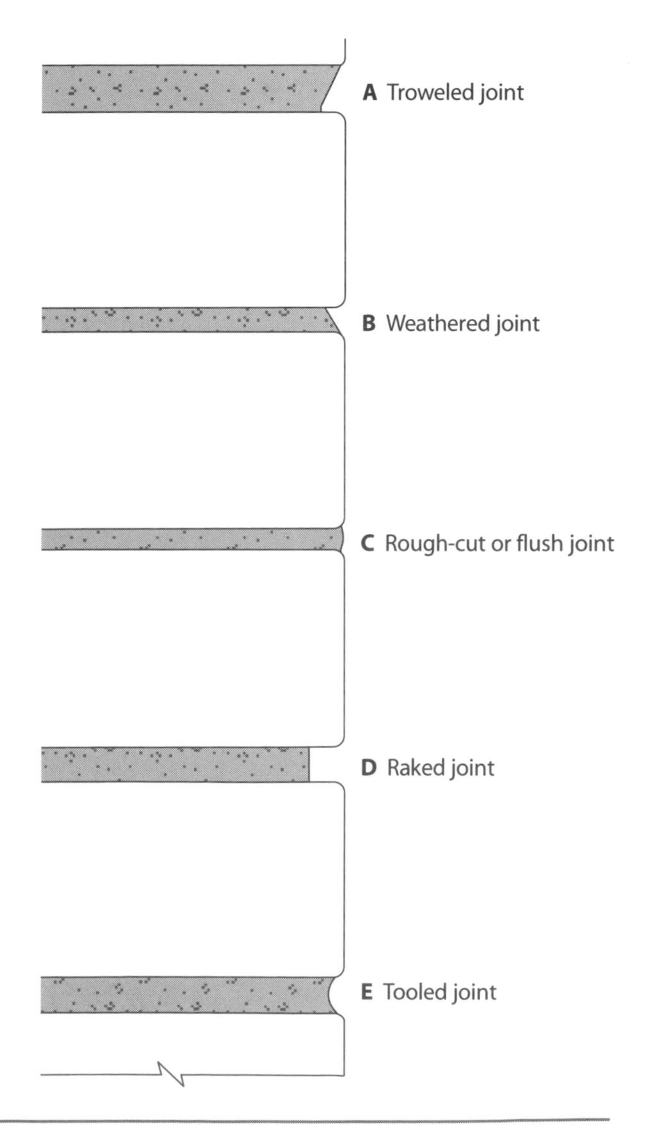

Figure 11-16
Common mortar joints

heavy rain, high wind, or freezing is likely to occur. This joint produces marked shadows and tends to darken the overall appearance of a wall. Figure 11-17 A shows how to use a tool called a joint raker or skate wheel to rake the mortar to an even depth. After you rake a joint, smooth it out to compress the joint and make a neater appearance. Use a slicker or caulking trowel for smoothing, as shown in Figure 11-17 B.

- *Tooled joints* — The tooled joint is made with a convex or round tool to compress the mortar and make it shed moisture. This is the strongest joint as it helps bond the mortar to the brick.

There's a less common joint, called the *weeping joint*, that's not appropriate for cold climates because water will seep into the joints. It's a difficult joint to construct anyway because you have to leave the mortar hanging from the wall to form the weeping effect.

A Making a joint with a special tool

B Smoothing out the joint with a slicker

Figure 11-17
Raked joints

Salvaged Brick

Some of your clients will request salvaged brick for its rugged appearance or low initial cost. But salvaged brick is generally weaker and less durable than new brick. Most salvaged brick comes from demolished buildings which have stood 40 to 50 years or more. It's next to impossible to salvage brick from a modern structure. Brick today are laid in portland cement-based mortar and the cement sticks almost permanently to the brick.

Strength and Durability

Fifty years ago manufacturing methods and conditions produced brick that varied widely in absorption properties and color. The kilns used then didn't produce even temperatures, so the finished brick varied according to its placement in the kiln. Brick in areas with higher firing temperatures had greater shrinkage and a darker color. In cooler areas, underburned brick were more porous, slightly larger and lighter colored than the harder-burned brick. These are called salmon brick, named for their pinkish-orange color. Generally, salmon brick aren't durable when exposed to weathering.

Most water penetrates through flaws in joints rather than directly through the materials. Comparative tests have shown that mortar doesn't bond as well to salvaged brick as to similar new brick. So a wall built of salvaged brick probably has an inferior mortar bond. That makes it more likely to let water in and weaker under lateral loading than a wall with new brick. The durability of masonry depends on the quality of the materials and the mortar bond.

There is a use for used brick, however. With today's thinner masonry walls, consider used brick primarily as a facing material. But you'll still probably have some salmons exposed to weathering because it's impossible to accurately sort and grade salvaged brick. It's likely that masonry of salvaged brick will spall, flake, pit and crack due to freezing if it's exposed to excessive moisture. With modern brick, you can depend on a reasonable degree of uniformity in size, strength and color. That's not true with salvaged brick. You'll find it difficult to determine whether salvaged brick meets current material specifications or building code requirements.

Mortar joint thickness (in)	Bricks per 100 SF of wall	SF of wall per 1000 bricks	Number of wall ties per 100 SF of wall	Mortar required (CF) per 100 SF of wall
1/4	698	143	100	4.48
3/8	655	153	93	6.56
1/2	616	162	88	8.34
5/8	581	172	83	10.52
3/4	549	182	78	12.6

Figure 11-18
Brick and mortar required for 4-inch-thick wall

Appearance

The architectural demand for a variety in colors has led manufacturers to use raw clays which turn other colors than dark red when fired to maturity. Today you can find pink brick that are hard-burned and durable. Many pink brick conform to grades SW or MW (severe or moderate weathering) under ASTM C 216 or C 62.

Many manufacturers blend different colored brick to provide a rustic appearance similar to salvaged brick. Using this brick can give the desired aesthetic effect without sacrificing durability or strength, a feat which is nearly impossible when you use salvaged brick.

Estimating Brick Quantities

It isn't hard to estimate brick and mortar quantities. But simple arithmetic mistakes can be expensive. If you overestimate, you'll have to pay someone to return the materials (and usually pay a restocking charge). Underestimate and you may have to stop work and wait for a delivery.

The most common way to estimate is by the square feet of wall area. Just multiply the length times the height of the wall, then deduct the wall openings. Once you know the total area, multiply by 7. Since there are 6.75 standard brick per square foot, multiplying by 7 gives you an allowance for broken or unusable brick. For modular brick, the number of brick per square foot is the same regardless of the bed joint thickness.

There's more than one way to estimate mortar. You can calculate the number of cubic feet per 100 square feet of wall, as shown in Figure 11-18. Or you can use a base figure of 8 bags of mortar for each 1000 brick (or 125 brick per bag of mortar).

Sand is the least expensive of the materials, so you can overestimate slightly. Use a base figure of 1 ton for each 1000 brick. This allows for a small amount of waste. But don't waste it on purpose. Store the sand with a waterproof tarp both under and on top of it. That keeps it under control and protects it from possible salt contamination.

12

Brick Veneer Construction

❖❖❖

Brick veneer uses a nominal 3- or 4-inch-thick exterior brick wythe tied to a backup system with a 1-inch clear space between the veneer and the backup system. The backup system may be wood frame, metal stud, concrete or masonry. If the backup system is masonry, adding brick veneer creates a cavity wall. Figure 12-1 shows a typical brick veneer wall.

Brick veneer is popular because it makes existing walls look and wear better. It's often used to refinish dwellings and commercial buildings, even high-rise ones. There's more on brick veneer in high-rise construction later. Here I'll talk about brick veneer in buildings three stories or less in height. Figure 12-2 shows how high you can make brick veneer.

The foundation of a building supports the brick veneer on the building. The brick itself carries no vertical loads other than its own weight. It should carry only its proportional share of the lateral load, but it often carries more than its share because the backup isn't stiff enough.

For a good brick veneer job, you need to:

- Make an adequate foundation
- Attach the veneer properly to a strong, rigid, well-braced backup system
- Use good flashing and weep holes
- Use proper materials
- Use proper ties and space them correctly
- Use good workmanship

Good workmanship is as essential in constructing brick veneer as it is in other types of masonry construction.

Properties of Brick Veneer

Brick veneer is very resistant to rain penetration. But be sure you keep the 1-inch clear space between the brick veneer and the backup to ensure proper drainage. Don't use clear coatings, such as silicones, on brick masonry walls. They can trap moisture or salts inside the walls and cause the brick to spall.

Brick veneer walls have several advantages:

- They usually have fire ratings of up to 2 hours. The typical brick veneer shown in Figure 12-1 has a 2-hour fire rating.
- They lower peak heating and cooling loads, and give you the opportunity to add insulation.
- Finally, they reduce sound transmission. The brick veneer wall in Figure 12-1 has an estimated STC of 40 to 44.

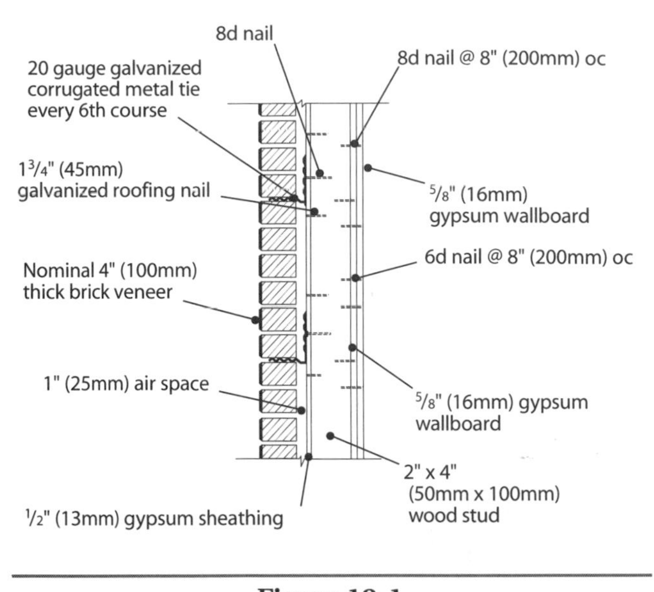

Figure 12-1
Typical brick veneer wall assembly

When you apply brick veneer over existing construction, you'll also insulate the existing exterior walls. Just install the insulation materials directly over the existing finish before adding the new brick veneer. But be sure to keep a 1-inch air space between the brick veneer and the rigid insulation. If the existing wood-frame or metal-stud walls have little or no insulation, remove the existing siding and install insulation within the wall. Then reapply what you took off before adding the brick veneer.

Nominal thickness of the brick veneer, (in) (mm)	Empirical height limitations		
	Stories	Height at plate, (ft)(m)	Height at gable, (ft)(m)
3 (75)	2	20 (6.10)	28 (8.53)
4 (100)	3	30 (9.14)	38 (11.58)

Figure 12-2
Empirical height limitations for brick veneer

Foundations for Brick Veneer

Figure 12-3 shows three typical foundation details for brick veneer. Make the foundation that supports brick veneer at least as thick as the total thickness of the brick veneer wall and foundation together. Many building codes will let you use a nominal 8-inch foundation wall under a single-family dwelling constructed of brick veneer, as long as you corbel the top of the foundation wall, as shown in Figure 12-3 C. But don't let the corbel project more than 2 inches. Individual corbels shouldn't project more than one-third the thickness of the masonry unit and one-half its height. Make the top corbel course a full header course that's not higher than the bottom of the floor joist.

When you add veneer to existing construction, extend the brickwork down to the existing foundation where possible, as shown in Figure 12-4 A. But before you put any masonry on an existing foundation, make sure the foundation is clean and free of loose soil and debris.

If the existing foundation isn't wide enough to support the entire thickness of the brick wythe, add a new foundation, as shown in Figure 12-4 B, at the same depth as the existing foundation. Install bond breaks between the old and new foundation.

Selecting Materials

For a veneer wall, use brick that conforms to ASTM C 62 or C 216. Because the brick wythe is isolated from the remainder of the wall, you'll need Grade SW when you're building in a freezing climate. Don't use salvaged brick. As explained earlier, it's weaker and less durable than new brick.

Mortars

Mortar has an important effect on the strength of a brick veneer wall. The bond between mortar and brick is the single most important factor affecting wall strength. Type N mortar is suitable for most brick veneer, but your building code may require Type S or

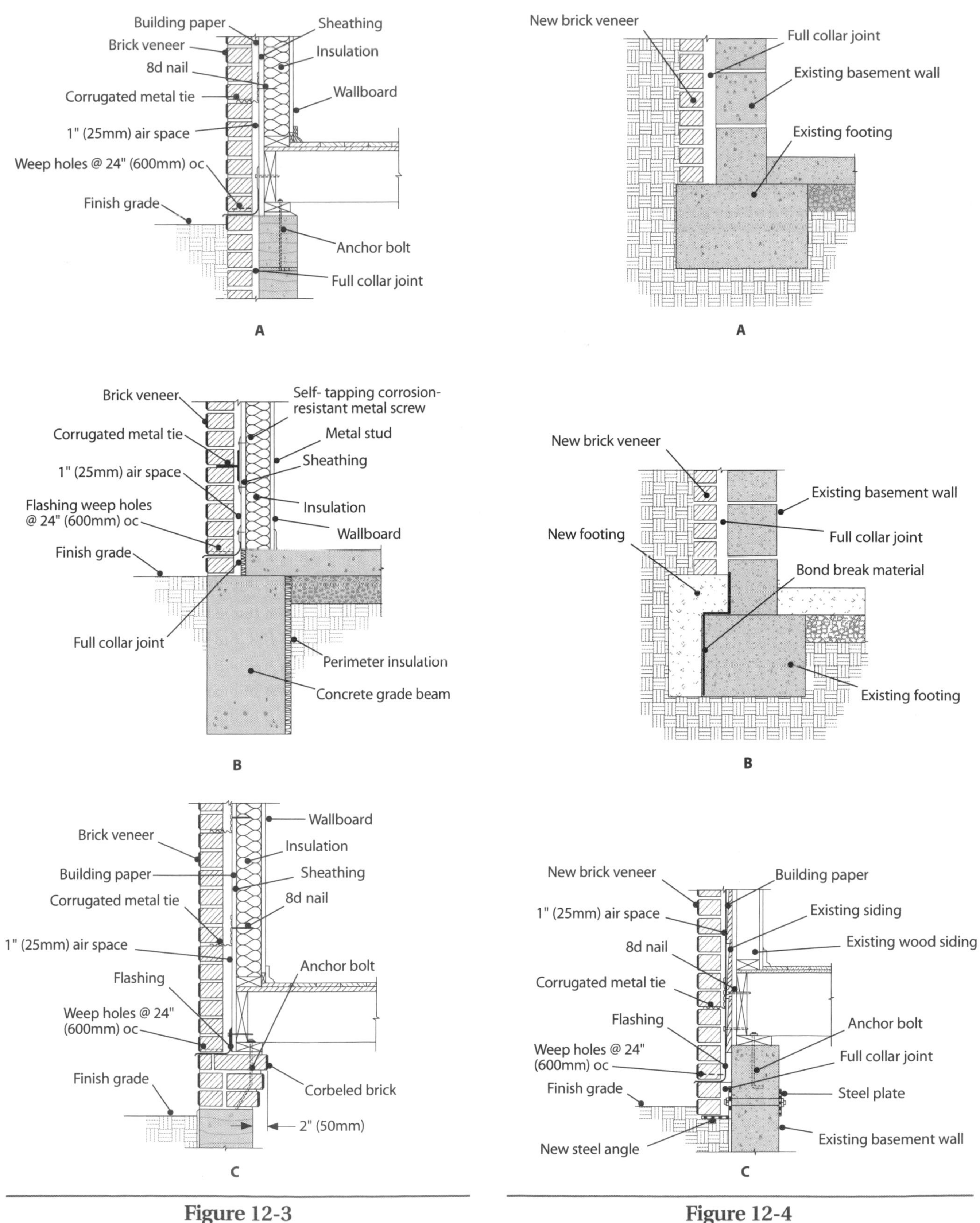

Figure 12-3
Typical foundation details for brick veneer in new construction

Figure 12-4
Typical foundation details in existing construction

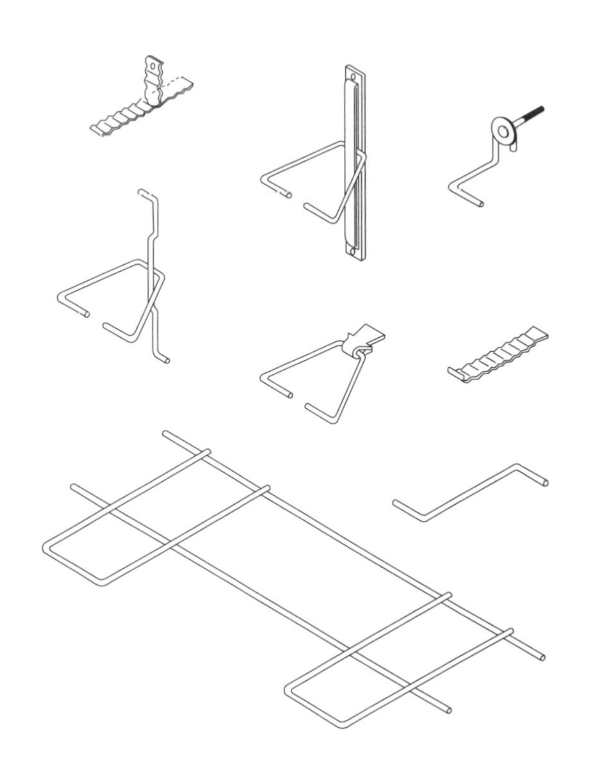

Figure 12-5
Typical ties for brick veneer

Type M. To make the best bond between masonry and mortar, use portland cement-lime mortars that meet the standards in ASTM C 270 or BIA M1-72, Type S.

Many other masonry cements contain air-entraining agents or other ingredients which sacrifice good bond for workability, color, or other qualities. Use Type S mortar in brick-veneer walls at locations where winds can be expected. Use Type M for brick veneer below grade where the brickwork is in contact with the earth. Use only portland cement lime-based mortars for brick masonry with veneer panel and curtain walls above three stories.

Joints

You need to keep all joints that will receive mortar clean and free of droppings, so the wall assembly performs as a drainage wall. If mortar blocks the air space, it may make a bridge for water to travel to the interior. Tool the joints with a jointer as soon as the mortar has become thumbprint hard. This firmly compacts the mortar against the edges of the adjoining brick. The types of joints I recommend for use with brick veneer are Concave, V, and Grapevine. Don't use other types of joints because they don't resist moisture well enough.

Ties

Figure 12-5 shows some typical ties. Here are some facts about ties:

- Corrugated-metal ties — Should be corrosion resistant. They should be at least 22 gauge, 7/8 inch wide and 6 inches long.
- Metal wire ties — Should be at least 9 gauge and corrosion resistant. Metal wire ties should comply with ASTM A 28 or A 185.
- Corrosion resistance — Should conform to Zinc-Coating of Flat Metal — ASTM A153, Class B-1, B-2, B-3; Zinc-Coating of Wire — ASTM A116, Class 3; Copper-coated Wire — ASTM B 227, Grade 30 HS; Stainless Steel — ASTM A 167, Type 304.

The tie you use will depend on the backup system.

Ties for Wood-Frame Backup

Use corrosion-resistant corrugated metal ties, at least 22 gauge, 7/8 inch wide, 6 inches long, to attach brick veneer to wood-frame backup. See Figure 12-5 A. Fasten the ties to the wood frame with corrosion-resistant nails that penetrate the sheathing, and drive them a minimum of 1½ inches into the studs.

Ties for Metal-Stud Backup

Use corrosion-resistant wire ties that are at least 9 gauge to attach brick veneer to metal studs. Figures 12-5 B and C show typical wire ties for connecting brick veneer to metal studs. Use wire ties with a minimum diameter of 3/16 inch to attach brick veneer to structural steel (Figure 12-5 D). Figure 12-6 shows one way to securely attach these ties to a steel frame.

Ties for Concrete Backup

If you're attaching brick veneer to concrete, use corrosion-resistant wire or flat-bar dovetail anchors. Wire anchors should be at least 6 gauge and 4 inches wide, with the wire looped and closed. Flat-bar dovetail anchors should be at least 16 gauge with a minimum width of 7/8 inch. Fabricate flat-bar dovetail anchors so that the end embedded in masonry is turned up ¼ inch. Embed dovetail anchors at least 2 inches into the bed joint of the veneer, and anchor them to dovetail slots placed in the concrete. Dovetail anchors are shown in sections E and F in Figure 12-5. Figure 12-6 shows one method for attaching brick veneer to a concrete structural member.

Ties for Masonry Backup

Figures 12-5 G and H show rectangular ties, U ties and Z ties you can use to attach brick-veneer masonry to masonry backup systems. Don't use Z ties with hollow units. Use continuous horizontal joint reinforcement with tab ties for concrete masonry backup. All ties should be of at least 9-gauge corrosion-resistant wire.

Tie Fasteners

The type of fastener you use to attach ties to an existing wall also depends on the construction of the existing wall. For wood-frame construction, use corrosion-resistant nails to attach the corrugated-metal ties. Hammer the nails at least 1¼ inches into the wood studs. For metal construction, use corrosion-resistant, self-tapping metal screws to attach metal wire ties to metal construction. The screws should penetrate at least ½ inch into the metal. For concrete or masonry construction, you can attach metal wire ties with lag bolts and expansion shields or masonry nails. Use corrosion-resistant fasteners and anchors.

Flashing and Weep Holes

Moisture can enter masonry at vulnerable spots. To prevent damage, locate flashing and weep holes at the bottom of the wall and at all openings. Place them above and as near to grade as possible so the wall will drain properly. Install continuous flashing at the bottom of the air space to divert moisture out through the weep holes. Securely fasten the flashing to the backup system and extend the flashing through the face of the brick veneer to form a drip. Where flashing isn't continuous, such as at heads and sills, turn up the ends approximately 1 inch.

Install flashing carefully to prevent punctures or tears. Where several pieces of flashing are required to flash a section of the veneer, lap the ends of the flashing and properly seal the joints. If veneer continues below the flashing at the base of the wall, grout the space between the veneer and the backup to the height of the flashing.

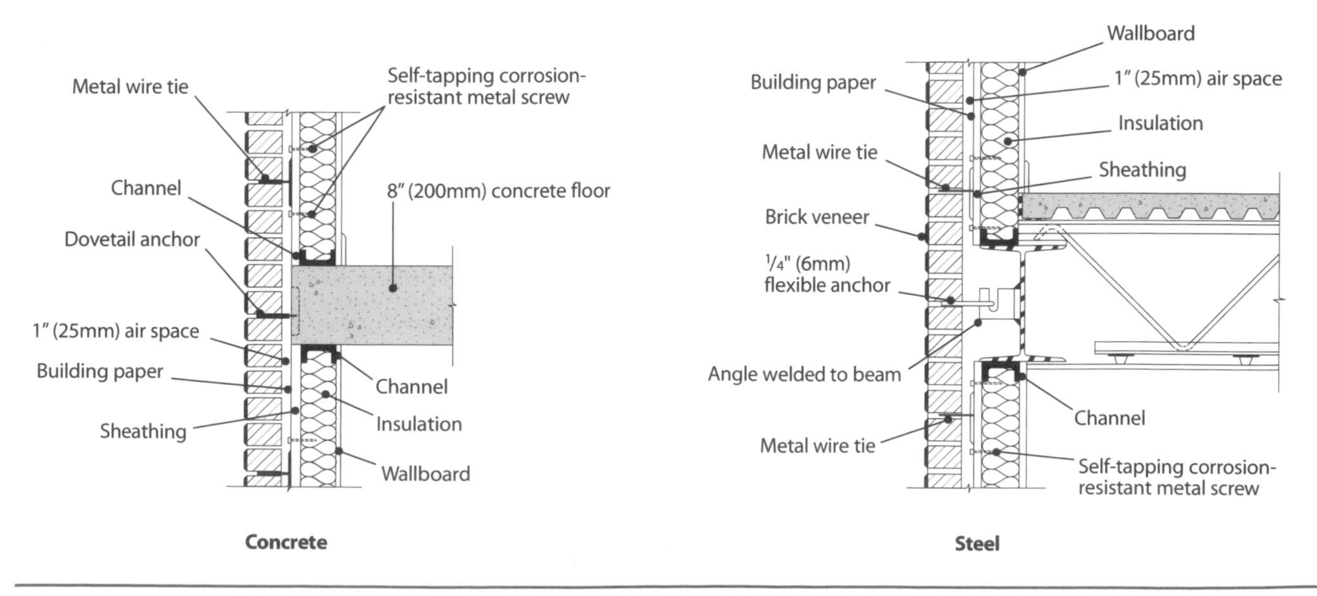

Figure 12-6
Attachment of ties to steel and concrete frames

There are many types of flashing you can use in brick veneer walls. Choose from sheet metals, bituminous membranes, plastics or combinations of these. Asphalt-impregnated felt isn't an acceptable bituminous membrane. Of course, cost is an important factor in the selection of flashing material. But I recommend that you only use top-quality materials because replacement in the event of failure is expensive, if not impossible. Most people use sheet metal. Copper is the best, but it's expensive. Never mix copper with other sheet metals. Mixing dissimilar metals causes corrosion.

Form weep holes by omitting mortar from all or part of the head joint when constructing the veneer. Or you can use a removable material, like a well-oiled rod, that forms unobstructed openings. Other options are plastic tubing, rope wicks or other materials which you leave in place. Space weep holes no more than 24 inches on center and locate them in the head joints immediately above all flashings. Space weep holes formed with wick material a maximum of 16 inches on center.

Sometimes metal screening or other materials are placed in open weep holes, but don't do this indiscriminately. Metal screening can corrode and stain the masonry. Make sure the building owners understand the importance of the weep holes, so they don't cover them with earth during landscaping.

Supporting Veneer

If possible, design the brick veneer to support its own dead weight on the foundations. You can do this unless the wall has so many openings in the veneer that it needs vertical support by the structural frame. Then the veneer can be a continuous vertical member spanning several supports. That reduces the deflection and the tensile stresses from flexural stresses in the wall.

Metal Studs

To provide lateral support for the brick veneer, securely attach sheathing to both sides of the metal studs. This sheathing must be rigid, properly detailed, and correctly attached for it to be effective. Brace the studs horizontally at mid-height for added strength, stiffness and fire resistance.

Steel Angles

When you use a continuous steel angle to support brick veneer at a foundation wall, it should be made of steel conforming to ASTM A 36, and treated or coated to resist corrosion. Also make sure the bolts or other fasteners are corrosion resistant. Size the angle and size and space the bolts according to the specifications or as determined by structural analysis. Use steel angles at least ¼ inch thick with at least 3-inch legs for lintels.

When constructing brick veneer on continuous corrosion-resistant steel angles, lay the first course of brick in a mortar-setting bed. That way you can compensate for any variations and misalignment of the steel angles.

Figure 12-4 C shows another way to support brick veneer. With this method, you attach a continuous corrosion-resistant steel angle to the existing foundation wall. Install the angle at, or slightly below, grade. Installing the angle below the frost line cuts down on the possibility of damage from freeze-thaw actions. Attach the angles to existing basement or foundation walls constructed of concrete or masonry. Never attach or anchor angles to wood plates or framing members.

Use this method of support with caution. First, analyze the loads on the angles. Carefully compute the sizing and spacing of bolts, taking into account not only the loads, but also the strength of the foundation wall itself. In general, use this method on one-story structures where the total height of the wall is less than 14 feet.

Shelf Angles

If building codes or other factors don't permit the brick veneer to be self-supporting for its full height, support it at each floor, or at least every other floor, by shelf angles, as shown in Figure 12-7. Anchor and shim

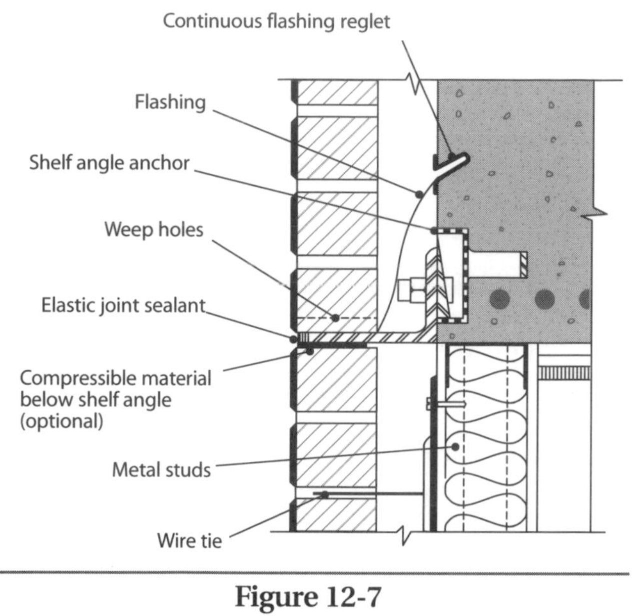

Figure 12-7
Steel shelf angles

these angles to keep them from moving more than 1/16 inch. Use structural steel shelf angles that are the correct size. For severe climates and exposures, consider using galvanized or stainless-steel shelf angles and install continuous flashing to cover the angles. Whatever type shelf angles you use, make a space at intervals to allow for thermal expansion and contraction.

Anchorage

When you use brick veneer to enclose a skeleton-frame structure, be careful to anchor the masonry veneer to the skeleton frame so both the brick veneer and the frame can move. A skeleton frame is more flexible than brick veneer, so it moves more during temperature and moisture changes. Use flexible anchors to tie the veneer to the structural frame so the veneer won't crack. See Figure 12-8.

Allowing for Movement

Small brick-veneer buildings usually don't need provisions for movement. But for structures larger than single-family houses, you may need bond breaks, expansion joints, flexible anchorage, joint reinforcement and sealants.

Bond Breaks

In general, if a masonry wall moves ¼ inch in 15 feet, it will crack. I've seen walls that moved more than ½ inch and had no cracks, but this is rare. Bond breaks let a wall move without cracking. Figure 12-9 shows a typi-

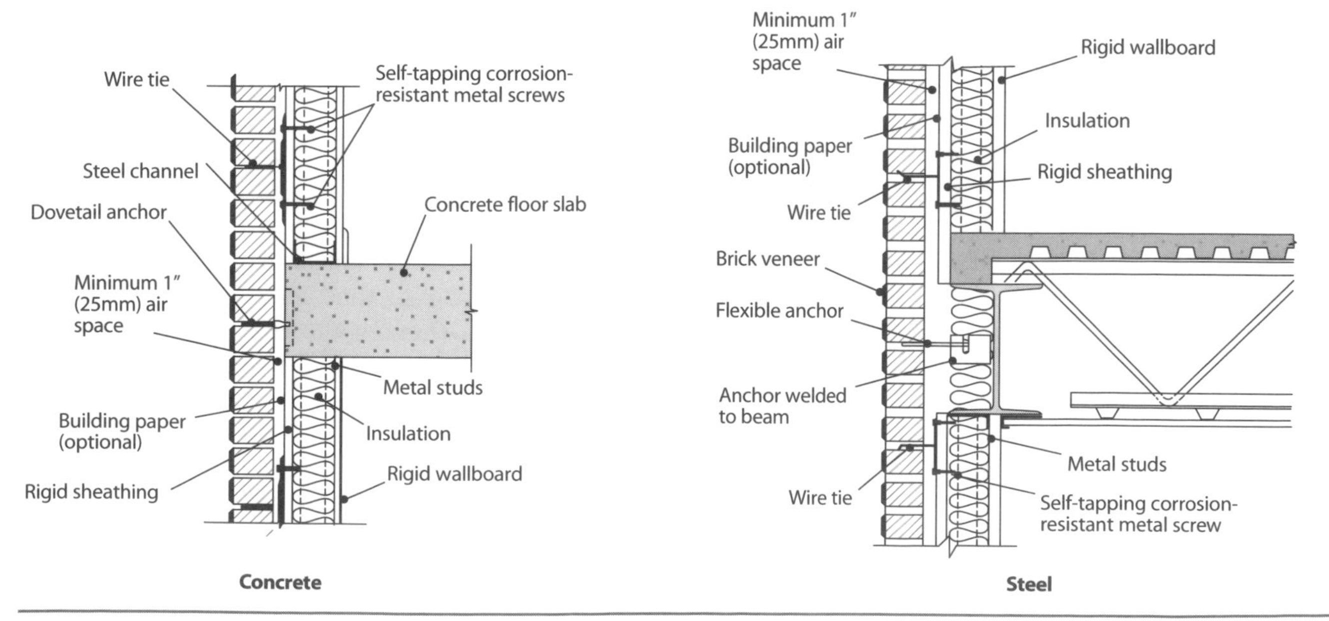

Figure 12-8
Attachment of brick veneer to structural frame

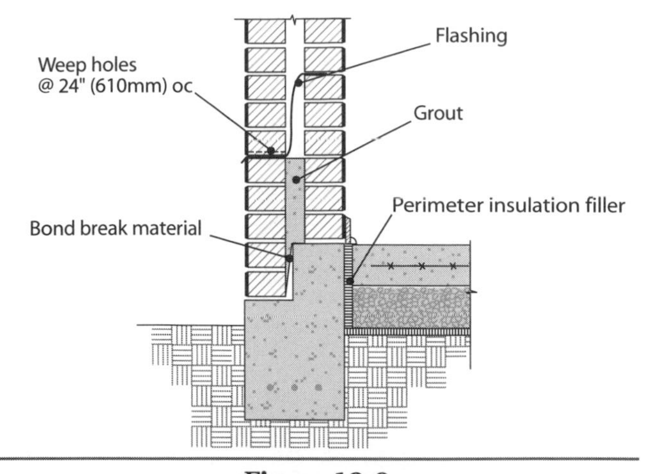

Figure 12-9
Typical foundation detail

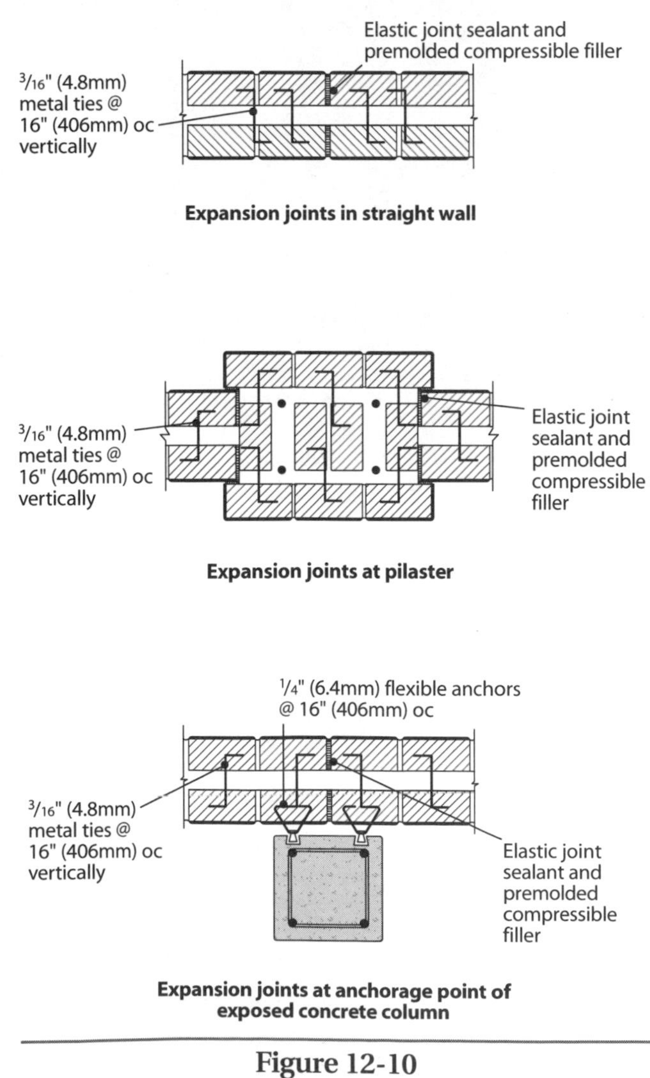

Figure 12-10
Expansion joint locations

cal foundation detail with a bond break made of building paper between the base of the cavity wall and the top of the concrete beam.

You can make a bond beam or tie beam at the bottom of a wall by placing reinforcing bars and filling the cavity with grout. This ties the inner and outer wythes of masonry together and distributes any strain over a longer length of wall. This can also be accomplished by a closer spacing of the horizontal joint reinforcement at the bottom of the wall.

Expansion Joints

Use expansion joints to accommodate the movement of brick masonry walls due to changes in temperature and moisture. Typical details of expansion joints and their locations are shown in Figures 12-10 and 12-11.

There's no single recommendation for positioning and spacing expansion joints that will apply on all structures. You just have to figure it out for each building. Brick veneer walls may move more during temperature and moisture changes than solid or composite walls exposed to the same environment. So brick veneer may need more expansion joints. Expansion joints are rarely required in residential construction. In commercial buildings, however, they may be required. Check your building code.

Vertical Expansion Joints

Vertical expansion joints accommodate potential horizontal movements in masonry. See Figure 12-12. Design and locate expansion joints so they won't weaken the wall. For parapet walls, carry all vertical expansion joints through the wall. Space additional expansion joints through the parapet approximately halfway between those running full height, unless the parapet is reinforced.

Horizontal Expansion Joints

If you have to use shelf angles to support brick veneer, place horizontal expansion joints immediately beneath each angle, as shown in Figure 12-7. This is particularly important in reinforced concrete-frame

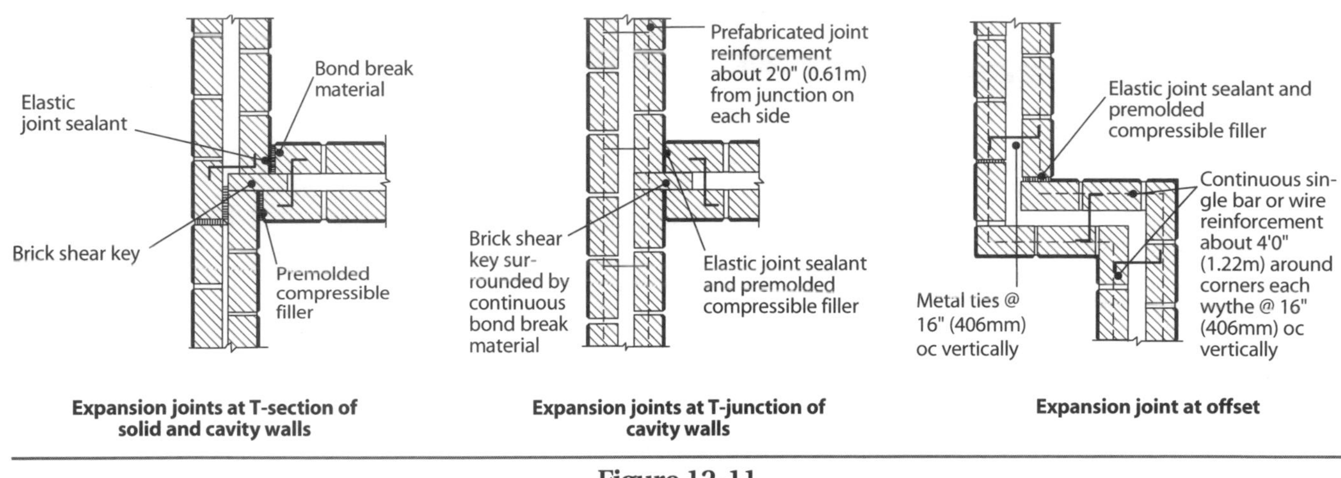

Figure 12-11
Expansion joint locations

buildings. Construct pressure-relieving joints by leaving an airspace or placing a fully compressible material under the shelf angle. In either case, seal the joint with a permanent elastic sealant.

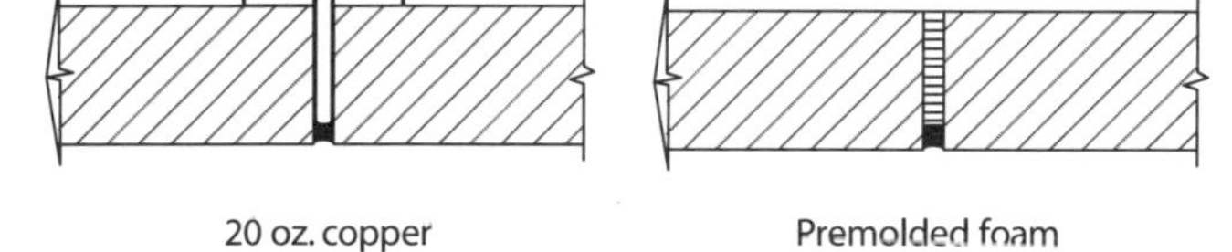

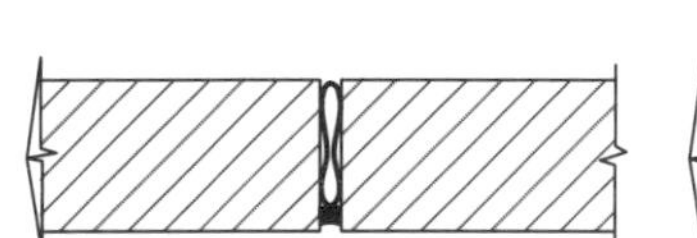

Figure 12-12
Vertical expansion joint fillers

Horizontal Joint Reinforcement

Brick veneer doesn't require horizontal joint reinforcement. But you may want to use some horizontal reinforcement so you can increase the spacing between expansion joints. For example, you can use single-wire horizontal joint reinforcement at corners, offsets or intersecting walls to make the veneer stronger there. Don't use horizontal joint reinforcement at expansion joints. Fabricate horizontal joint reinforcement from wire meeting ASTM A82 or ASTM A185 requirements. It should have a corrosion-resistant coating which complies with ASTM A116, Class 3, or ASTM A153 B-2.

Concrete Slabs

A concrete slab that curls can pick up the brick bonded to it. Unfortunately, this behavior of concrete is frequently overlooked by designers. Figure 12-13 shows a typical concrete slab detail that can help prevent this problem. In this design, building paper breaks the bond between the concrete slab and the brick wall. This lets the slab move with respect to the wall. The slab is thickened onto a beam over the interior wythe to help stiffen the slab and minimize curling.

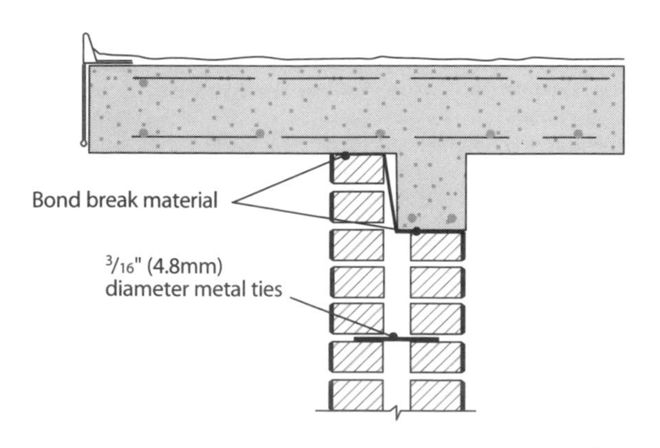

Figure 12-13
Concrete roof slab detail

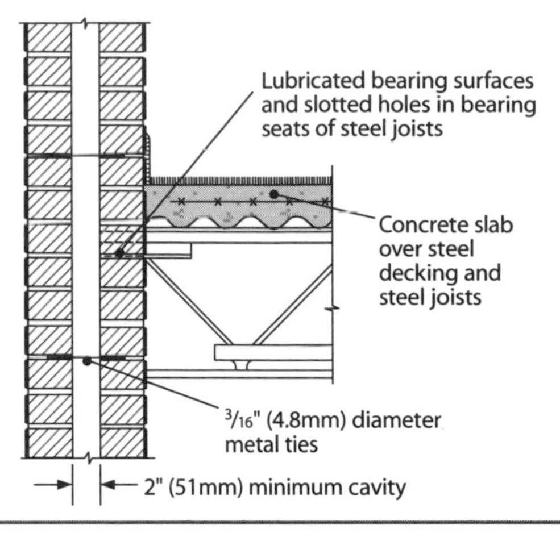

Figure 12-14
Steel joist structural floor assembly

Structural Steel

Steel expands more from temperature changes than brick masonry does. To allow for this, lubricate the bearing surfaces and provide slotted holes in the seats of the steel members. Tighten the anchor bolts by hand only. Figure 12-14 shows a structural system using steel joists bearing on a masonry wall.

Lintels, Sills and Jambs

Figure 12-15 shows typical details for lintels, sills, and jambs for new construction. The advantages of using reinforced brick lintels are more efficient use of

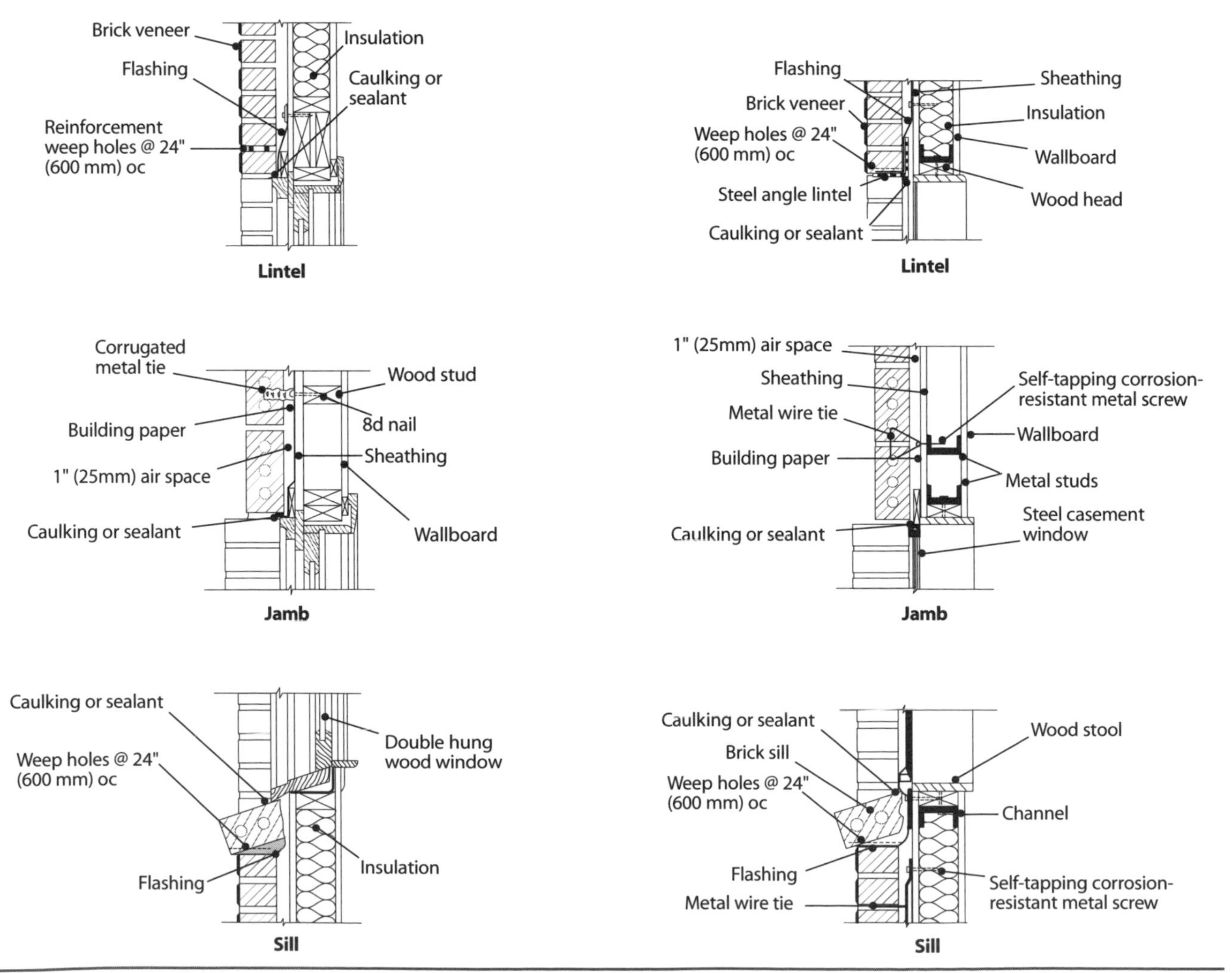

Figure 12-15
Typical lintel, jamb, and sill details

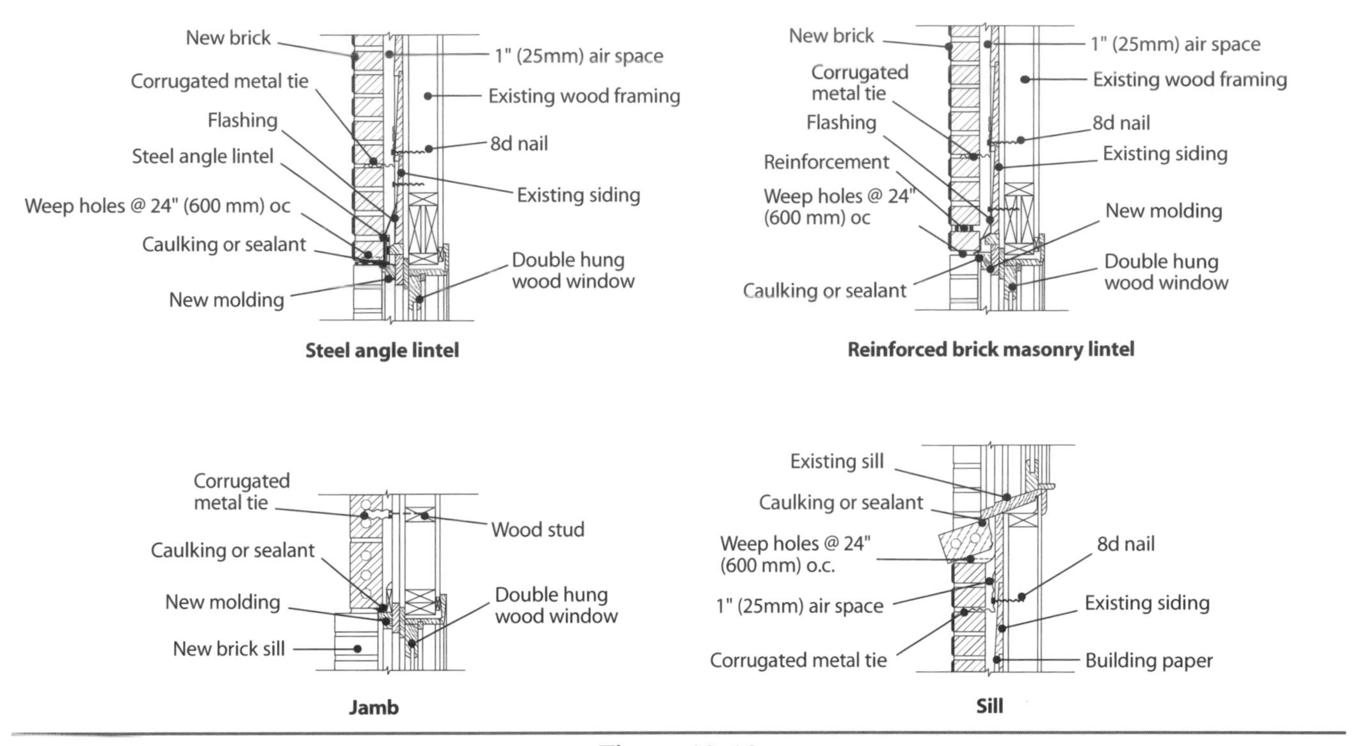

Figure 12-16
Typical lintel, jamb, and sill details for veneer on an existing building

materials, built-in fireproofing, elimination of differential movement, and no required painting or other maintenance.

To reinforce brick lintels, use steel bars manufactured in accordance with ASTM A615, Grades 40 or 60, that conform to ASTM A36. Use steel angle lintels at least ¼ inch thick with a horizontal leg of at least 3½ inches with nominal 4-inch-thick brick veneer, and 2½ inches with nominal 3-inch-thick brick veneer. The minimum required bearing length for steel angle lintels is 4 inches. But local building codes may change the spans and sizes of steel angle lintels to meet fireproofing requirements.

Figure 12-16 shows typical lintel, jamb, and sill details for construction on an existing building. Usually you can make new brick sills so that the existing sill overlaps the new brick sill. Install new molding at existing jambs and heads of openings so the framing extends enough to form an air space between the brick veneer and the existing construction. You can use lintels made of reinforced brick masonry, steel angles, or precast concrete. Reinforced brick masonry and steel angle lintels are the most common in brick veneer construction.

Figure 12-17 shows typical lintel, jamb and sill details for panel or curtain walls with metal studs and gypsum drywall. Attach windows to either brick veneer or the backup system, but don't attach it rigidly to both.

Attaching Brick Veneer to an Existing Structure

To attach brick veneer here, you'll need one tie for each 2⅔ SF of wall area. Figure 12-18 shows how to space the ties. Don't space ties, either horizontally or vertically, more than 24 inches on center. This tie spacing applies both above and below grade.

Use at least 9-gauge corrosion-resistant wire ties of the type shown in Figure 12-19. Don't use corrugated metal veneer ties. Embed all ties at least 2 inches into the bed joints of the brick veneer. Securely attach them to the metal studs through the sheathing, and not just to the sheathing alone. Place ties so they are completely surrounded by the mortar.

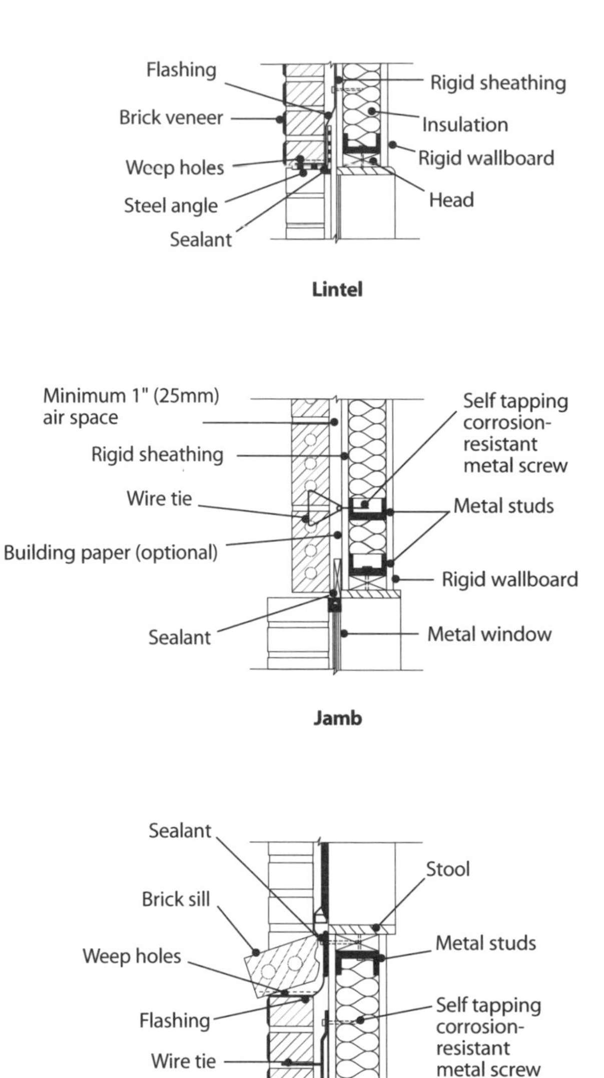

Figure 12-17
Typical lintel, jamb, and sill details for panel or curtain walls

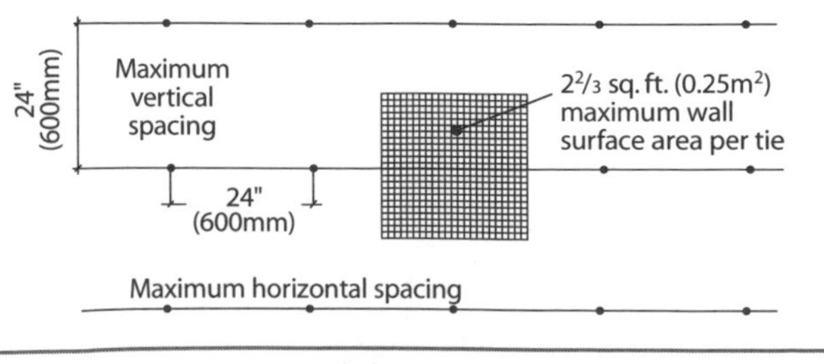

Figure 12-18
Spacing of metal ties

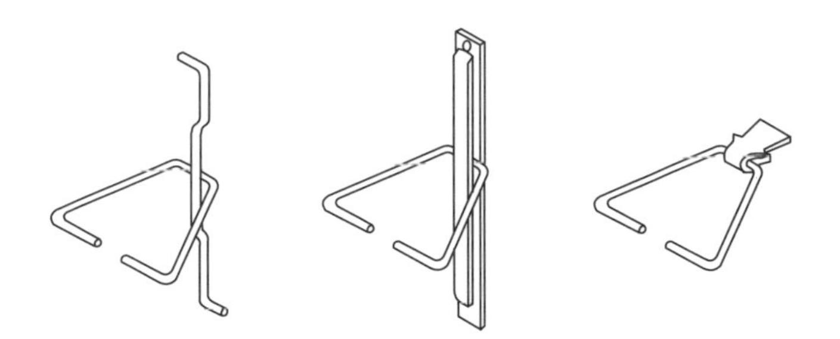

Figure 12-19
Typical ties

Finishing the Top of Veneer

Figure 12-20 shows a typical detail for the top of brick veneer at an existing eave. Leave at least a ⅛-inch clear space between the top of the last course and the bottom of the soffit. Cover the space with a new molding strip and sealant or caulking. If the eave doesn't properly cover the top of the veneer, extend the eave.

Brick Veneer Over Metal Studs

Many contractors like to use brick veneer over a metal stud backup system with gypsum sheathing. See Figure 12-21. To provide for full lateral support of the metal studs, you have to be sure to properly attach sheathing on both sides of the studs. For example, sometimes loading the scaffold from the inside of a building means the exterior sheathing gets left out in various locations. Make sure that doesn't happen. And you must attach the backup system properly at the top and bottom.

Here are the advantages of this wall system when it's properly designed, detailed, and constructed:

- Provides a space for insulation
- Lets the building stay dry so the brickwork can continue at a convenient time when weather permits
- Provides the appearance and attributes of a brick building

Figure 12-22 shows brick veneer construction on a high-rise building. Note the type of masonry tie used. This brick veneer is attached to a metal-stud system with insulating boards.

Attach the window to either the brick veneer or to the backup system, but not rigidly to both. If you use this system to add brick veneer to an existing building, be sure the new assembly is stiff enough to take its share of the load. If not, here are some methods you can use to add stiffness:

- ❖ Use thicker brick veneer.
- ❖ Use hollow brick units with reinforced and grouted cells. This type of wall is most effective in severe earthquake zones.
- ❖ Change the backup system from metal studs to masonry. It may be better to use a cavity wall or insulated cavity wall design instead of a brick veneer design.

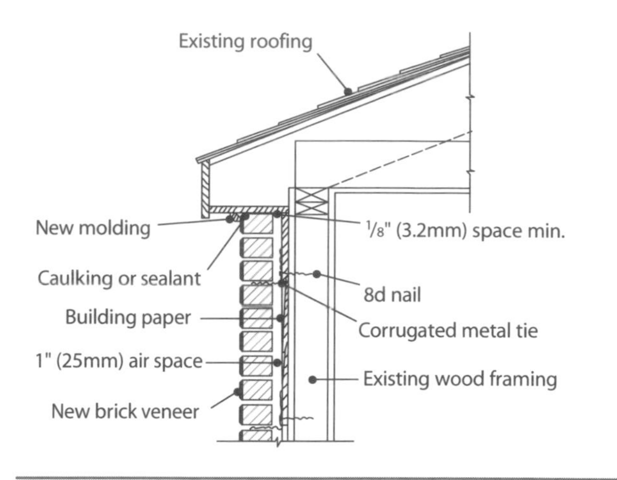

Figure 12-20
Typical eave detail at the top of brick veneer

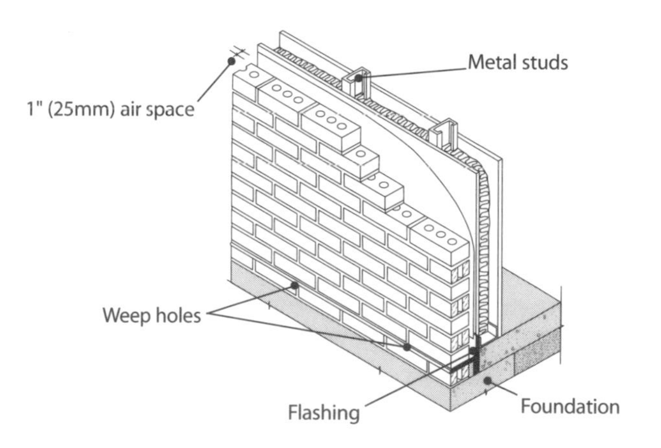

Figure 12-21
Typical brick veneer and metal stud wall

Figure 12-22
Brick veneer construction on a high-rise building

Cavity Walls

You can use cavity walls as curtain walls in concrete- and steel-frame buildings. When you do, support the inner masonry wythe by the frame at each floor level and lay it to the column faces. Tie the outer wythe to the main structure with metal ties. Also tie the outer wythe, supported by shelf angles, to the inner masonry wythe with metal ties. There are several ways to secure the shelf angle, which supports the outer wythe of masonry at each floor, to the spandrel. Be careful to anchor and shim the angle. Design angles so they don't move more than $^{1}/_{16}$ inch. Even if you use galvanized shelf angles, install continuous flashing as one piece. Provide a space at intervals for expansion and contraction in the wall.

Wood Floor Joists

Wood floor joists normally have a 3-inch fire-cut end and will bear only on the interior wythe of a cavity wall. If the ends project into the cavity, they can form a ledge which may create a moisture bridge across the cavity. All building codes require you to anchor joists to the masonry walls at specified inter-

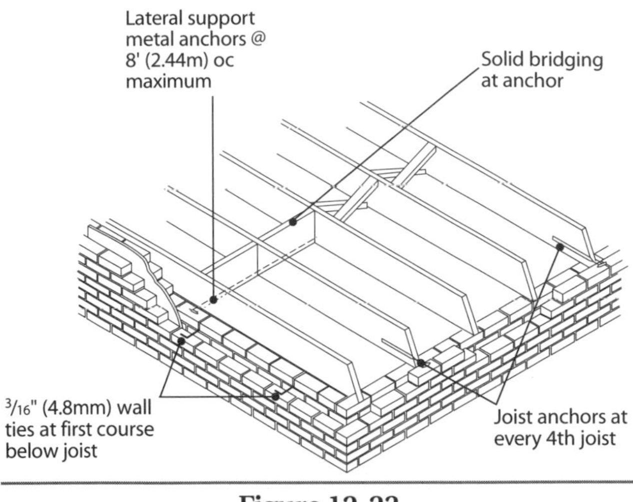

Figure 12-23
Anchorage of wood floor to cavity wall

vals in a prescribed manner. Codes generally require an anchor at the end of every fourth joist. Where the joists are parallel to a wall, have anchors meet three joists at less than 8-foot intervals. Cavity wall ties are usually required within 8 inches of the joist bearing level. See Figure 12-23.

Wood Rafter Plates

Figure 12-24 shows two ways to anchor a wood roof to a cavity wall. The detail on the left shows a method using solid units in both wythes. You can use the detail on the right with vertical cell backup units. Grout anchor bolts into the hollow cells to anchor them securely. Regardless of the method, extend anchor bolts holding roof plates into the masonry at least 16 inches, which is normally about six standard-size brick courses. After the wood plate is installed, tighten the nut by hand.

Doors and Windows

Avoid solid masonry jambs at windows and doors in cavity walls. However, for steel windows, the jamb must be partially solid to accept most standard jamb anchors. Use wood or steel surrounds to adapt non-modular steel casement windows to modular cavity walls. Place cavity wall ties spaced at 3 feet or less around all openings, and not more than 12 inches from each opening.

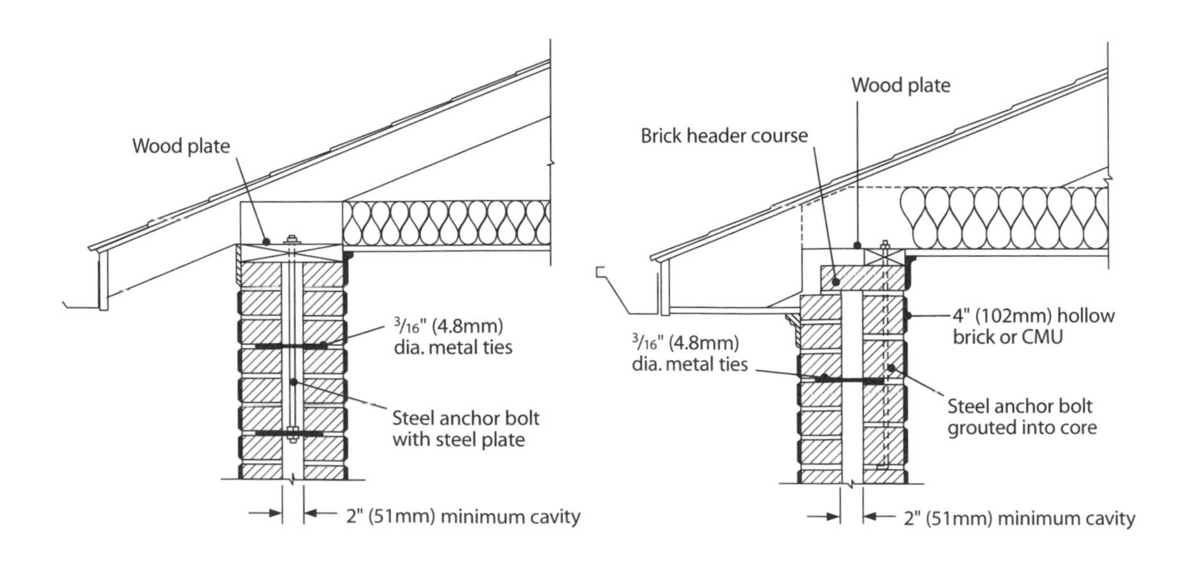

Figure 12-24
Anchorage of wood framing to cavity walls

Caulking and Sealants

Too frequently, caulking is used just as a means of correcting or hiding poor workmanship rather than as an integral part of construction. Detail and install it with the same care as the other elements of the structure. Use a good-grade polysulfide, butyl or silicone rubber sealant. Don't use oil-based caulking.

Properly prime all joints before placing caulking compounds or sealants. Use compressible backer rope materials for joints deeper than ¾ inch or wider than ³/₈ inch. Be sure to carefully caulk joints at masonry openings for door and window frames, expansion joints, and other locations susceptible to rain penetration. See Figures 12-25 through 12-27. Make caulking joints at the perimeter of exterior door and window frames between ¼ inch and ³/₈ inch thick. Fill these joints solid with an elastic caulking compound or sealant forced into place with a pressure gun.

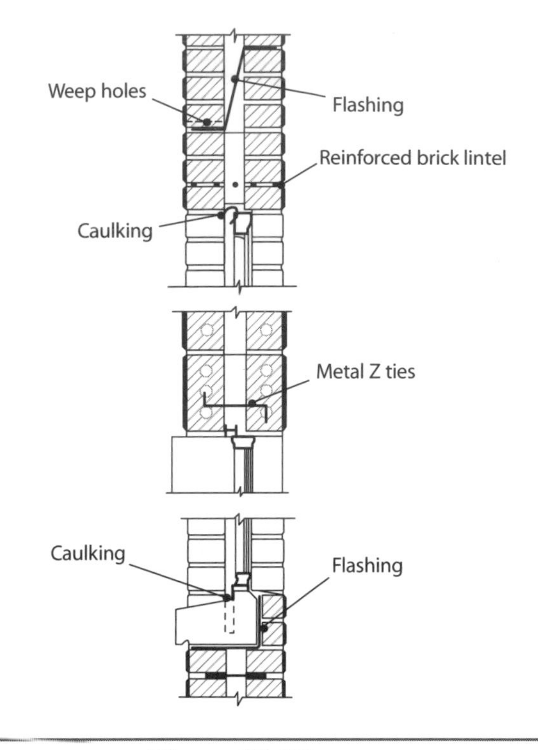

Figure 12-26
Metal casement window

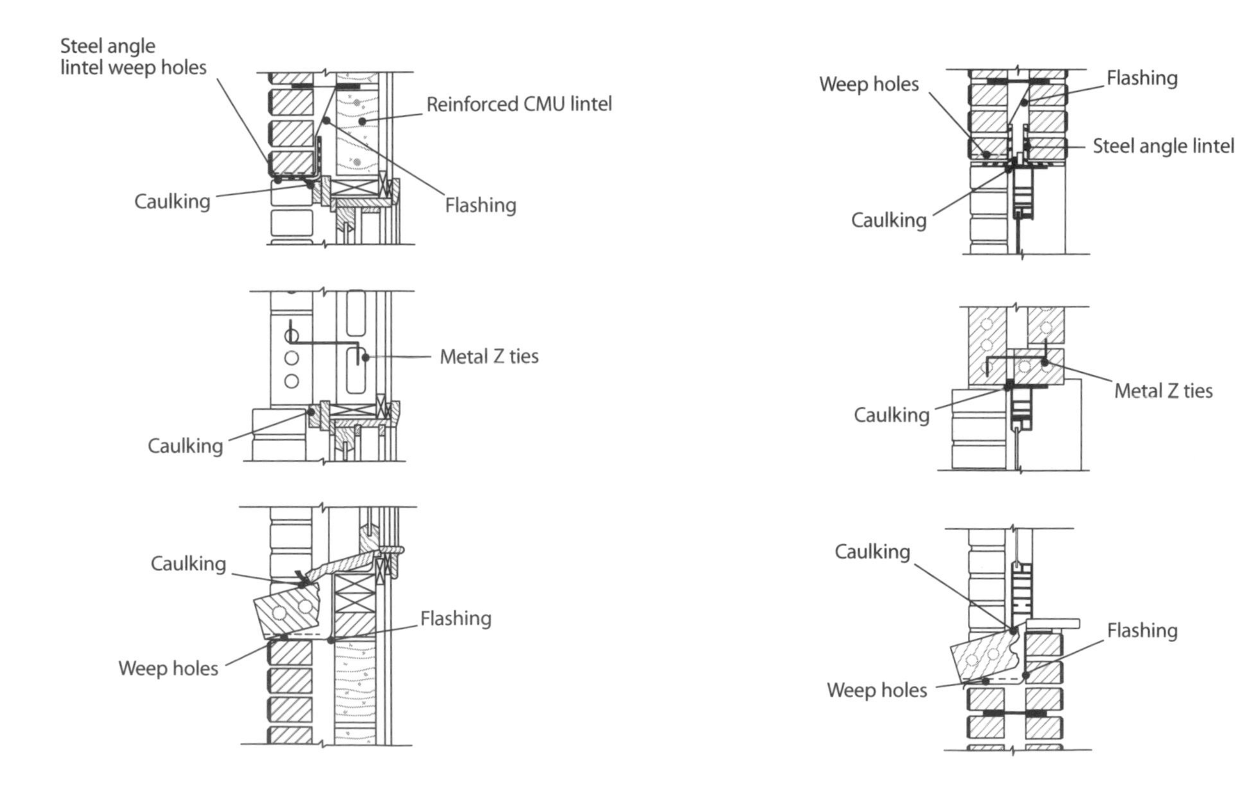

Figure 12-25
Double hung wood window

Figure 12-27
Commercial metal window

13

Limestone

Masons who work on commercial construction are occasionally required to apply limestone to the work. Limestone is suitable for caps, lintels, copings and moldings as well as a veneer over other masonry.

You'll find several different finishes in limestone block:

- *Smooth finish* is generally recognized as the standard finish. It looks and feels smooth, and can be produced by a variety of machines.
- *Diamond gang sawed finish* is a smooth finish which may contain some parallel markings and scratches. The direction of these markings can be either horizontal or vertical. You can specify which way you want the marks to run.
- *Chat-sawed finish* (gang sawed) is a medium rough finish. It's produced by sawing with a coarse abrasive containing some metallic minerals which may add permanent brown tones to the natural color variations. This finish may contain parallel saw marks.
- *Shot-sawed finish* (also gang sawed) is a coarse uneven finish ranging from a pebbled surface to one ripped with irregular parallel horizontal or vertical groves. Steel shot used in the sawing process produces the random markings. The shot markings are uncontrolled and the joint lines may show deviations in the sawn face. Rust stains from the steel shot may cause additional color tones.

Besides these standard finishes, you can order a number of custom finishes. Architects often request these custom designed, textural finishes.

When you plan a job, bear in mind that holes and sinkages for attaching stone to a building will appear in the shop drawings, but you must arrange for any sinkages you need for your own job handling. If you don't, the supplier will cut only those sinkages he needs for his equipment. Get in touch with the limestone suppliers as soon as the job is awarded. Since most limestone is cut to order, there may be a waiting time for quarrying and delivery. Blueprints and job specifications will direct the type, finish and dimensions. Occasionally job specifications may state where limestone should be purchased. Some projects will use standard size copings, sills and lintels that are designed to fit the modular building system. There have been times when jobs have been held up because of the availability of limestone.

How to Store Limestone

Before limestone (or any material) is delivered to your job site, prepare a safe, well-organized storage area. Be sure there's good access for the equipment you'll use to handle the material. Your supplier will probably expect you to return pallets and other transportation aids promptly, so have appropriate storage

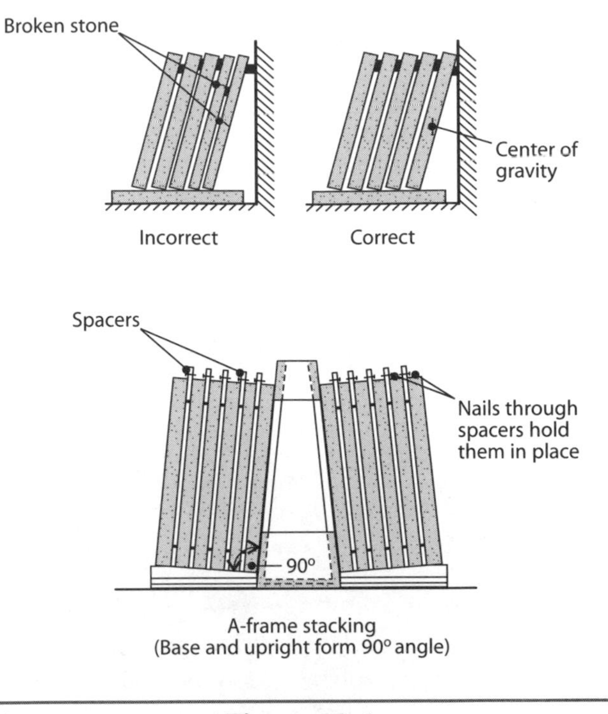

Figure 13-1
Stacking limestone

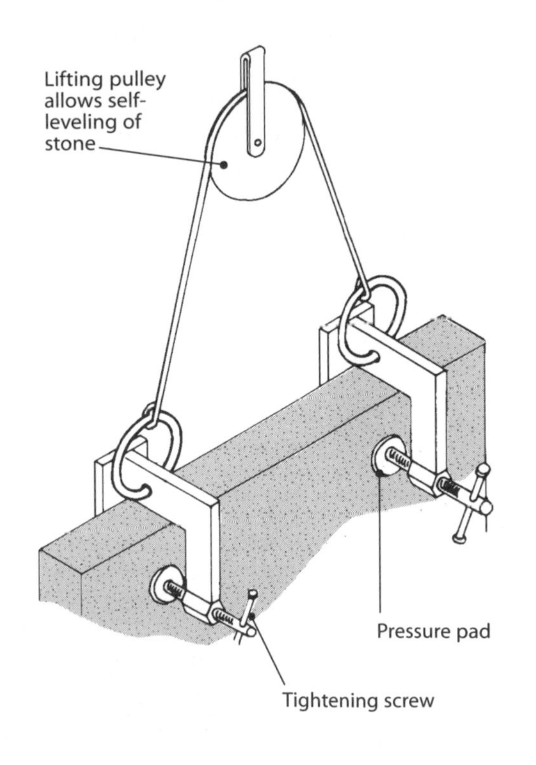

Figure 13-2
Lifting limestone with a lifting pulley

materials ready. Dealers will occasionally charge a deposit on pallets used for the delivery of the stone. These are hardwood pallets and are expensive. The deposit charge is in the 8 to 10 dollar range per pallet. It's unlikely that you'll be able to keep the supplier's shipping materials for the duration of the job.

When the stone arrives, check immediately for cracked, chipped, or broken pieces. You may be able to used chipped stones if the chips are on the back or if you'll be trimming stones to fill small areas. Check the color, finish and grade to make sure they comply with the specs and plans.

Store the stone on sturdy pallets, skids, timbers or A-frames in a well-drained space. Cover the ground with chips or gravel to protect the stone from mud splatters.

Stack or lean stone panels with spacers (pine boards will do) between them to keep them from getting scratched. Don't cover up the lifting holes as you store the panels. Alternate the panels, face to face, then back to back. See Figure 13-1. Discard the packing material delivered with the limestone and cover the stone with plastic or waterproof paper. Leave room for air to circulate around and between stone blocks or panels.

How to Handle Limestone

Avoid sliding one face surface over another. Don't bump the stone or let it strike support angles, other structures or parts of the building.

Never hoist or suspend stones over workers or bystanders. If you must hoist stones over scaffolding, make sure workers are clear until the stones are secure.

You can use a lifting pulley to raise limestone, as shown in Figure 13-2. But always use a safety sling under stones held or lifted by devices attached to their tops or top edges. Keep the sling in place until stones are at their final installation position. And make sure the slings are long and wide enough. If they're not, they'll pinch the stone as it rises and that may damage the edges (Figure 13-3). The load will also be unstable. When you move stones with slings, stack them back to back, or insert padding between them to avoid chips and scratches. See Figure 13-4.

Be sure any lifting device you use is strong enough for the heaviest stone you're going to lift. Be especially careful when handling large panels. Use tag lines if it's windy. Always use safety slings, and keep them in place until you've got the stones within a few inches of where you want them.

To avoid breaking the stone or injuring workers, don't lift stone panels from horizontal to vertical using a clamp or other top-only attachment.

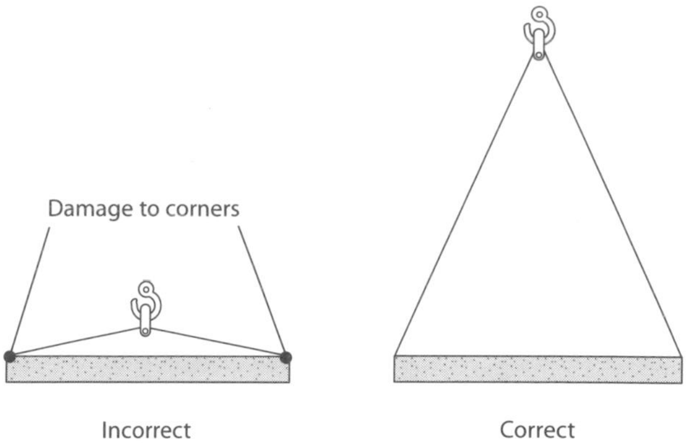

Damage to corners

Incorrect

Correct

Figure 13-3
The correct way to use a sling

Supporting Limestone

When you arrive on the job, check the framing and support systems. Steel that supports limestone has to bear the weight of the stone. Usually it's a series of clips or continuous angles, with or without bars or blades on the nose for lateral restraint. Where support steel is installed in epoxy-filled slots, as in preassemblies, the parts of the support which are in the stone must be stainless steel.

I recommend either stainless steel or galvanized angle irons. A36 or other hot formed angles will work, but even if they're primed and painted they may rust and cause problems. Stainless steel is expensive so you may want to use galvanized instead. If galvanized steel is scratched, has to be cut to a shorter length, or if welded, coat it with galvanized paint on the job before setting the stones.

Figure 13-5 shows some types of angle iron supports and how they're attached to the superstructure.

Anchors

An anchor is any device which attaches stone to a structure or stone to stone. An anchor doesn't carry any load except wind or slight movement of the different materials during temperature changes. All building materials expand and contract at different rates. Figure 13-6 shows commonly-used anchors. Sometimes the specs provide you with the names of anchor manufacturers or local suppliers.

When a supplier shows a particular anchor on his shop drawing, it doesn't necessarily mean he'll supply that anchor with the stone. One exception is the steel required in an assembly of multiple-stone units which are preassembled and shipped as a unit to the job. This is specified by the architect and will appear on the shop drawings.

Any anchor or other metal you insert in slots or sinkages must be stainless steel. You can use multipart anchors that contain other metals as long as any part which touches the stone is stainless steel.

Figure 13-4
Using a sling and padding to protect the stones

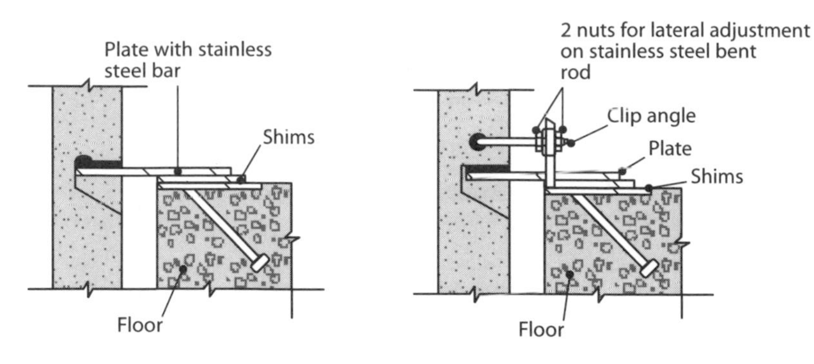

A Supports above the floor

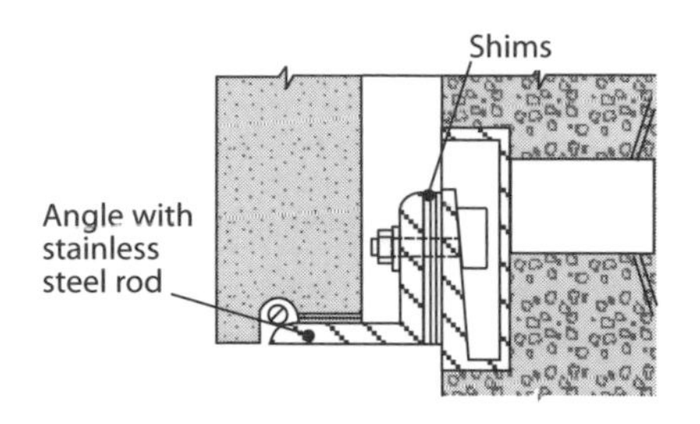

D Metal anchor embedded in concrete structure

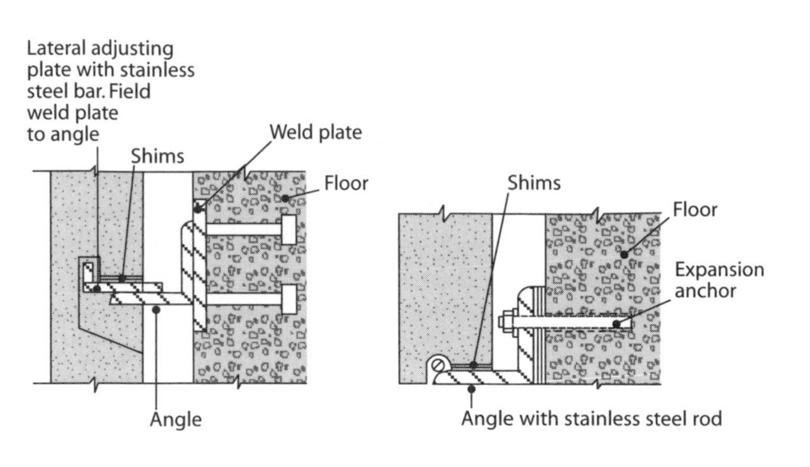

B Supports in front of the floor

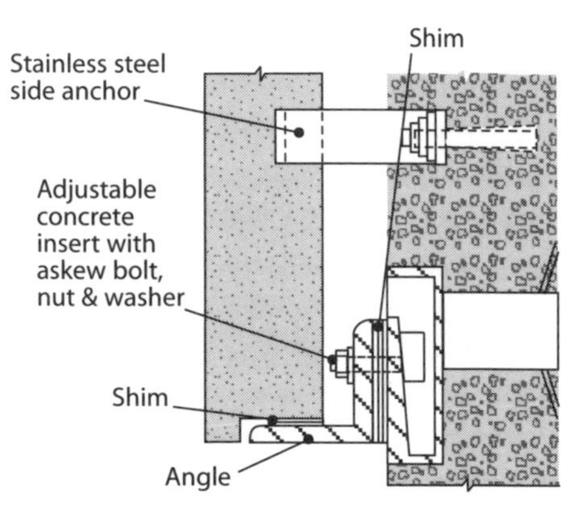

E Similar system with stainless steel side anchor

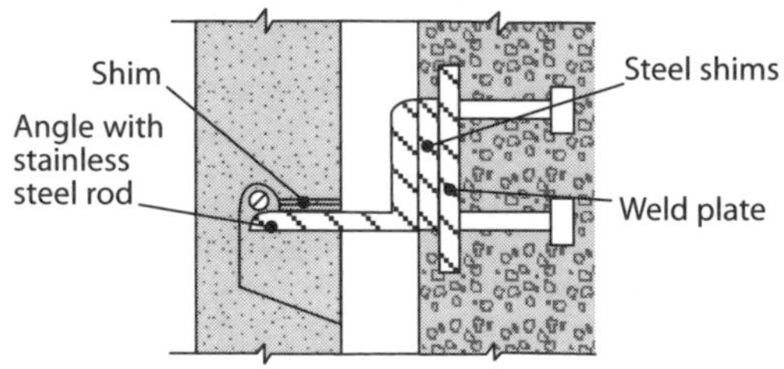

F Stainless steel welding insert embedded in concrete

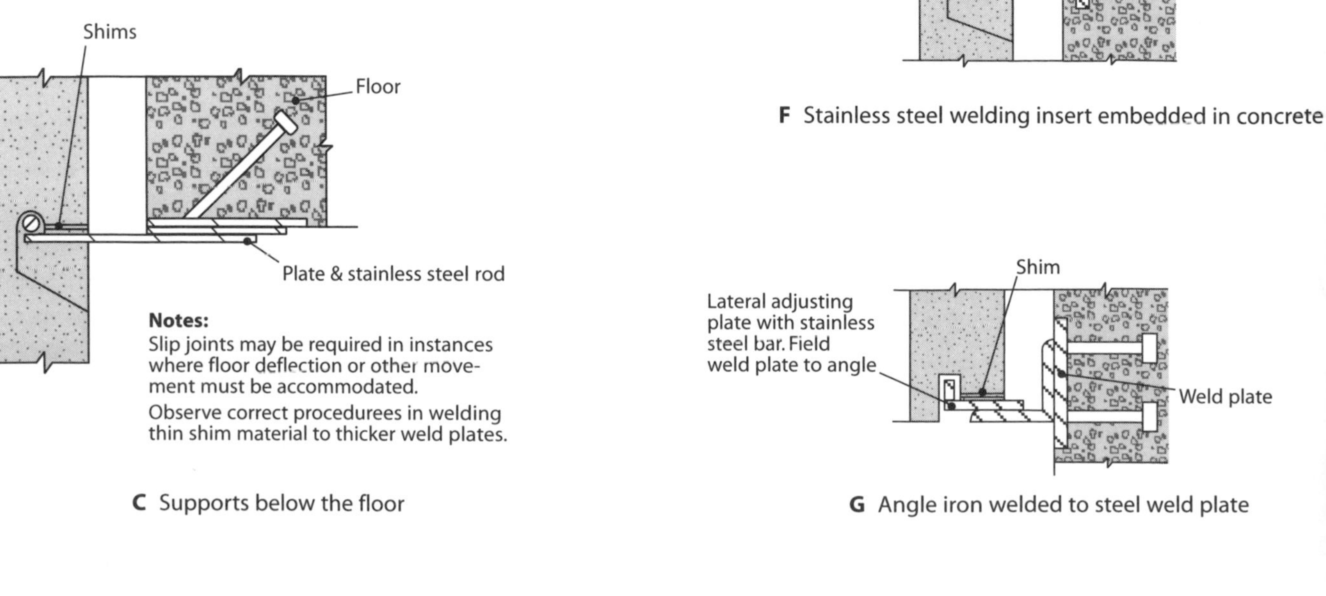

Notes:
Slip joints may be required in instances where floor deflection or other movement must be accommodated.
Observe correct proceduree in welding thin shim material to thicker weld plates.

C Supports below the floor

G Angle iron welded to steel weld plate

Figure 13-5
Limestone support systems

Figure 13-7 shows how you would typically place an anchor. You slip a Z-anchor into anchor slots and secure it with mortar. Notice the C clamp at the joint. It keeps the stone in place while the mortar sets.

Making Joints

If you use portland cement to make cement/lime mortar for limestone, make sure it conforms to ASTM C150. If you use masonry cement, it should conform to ASTM C 91. Either material will make suitable mortar. Any lime you use should conform to ASTM C270. Lime helps make mortar easier to work with and helps reduce shrinkage.

Make portland cement/lime mortar with one part cement, one part lime and six parts sand, all by volume. This 1:1:6 ratio follows the requirements of ASTM C270, Type N. This mortar has a compressive strength of about 750 psi when it's cured. It bonds well and resists bad weather, so it's good mortar to use for most limestone construction. You can change the components of mortar for specific applications. See

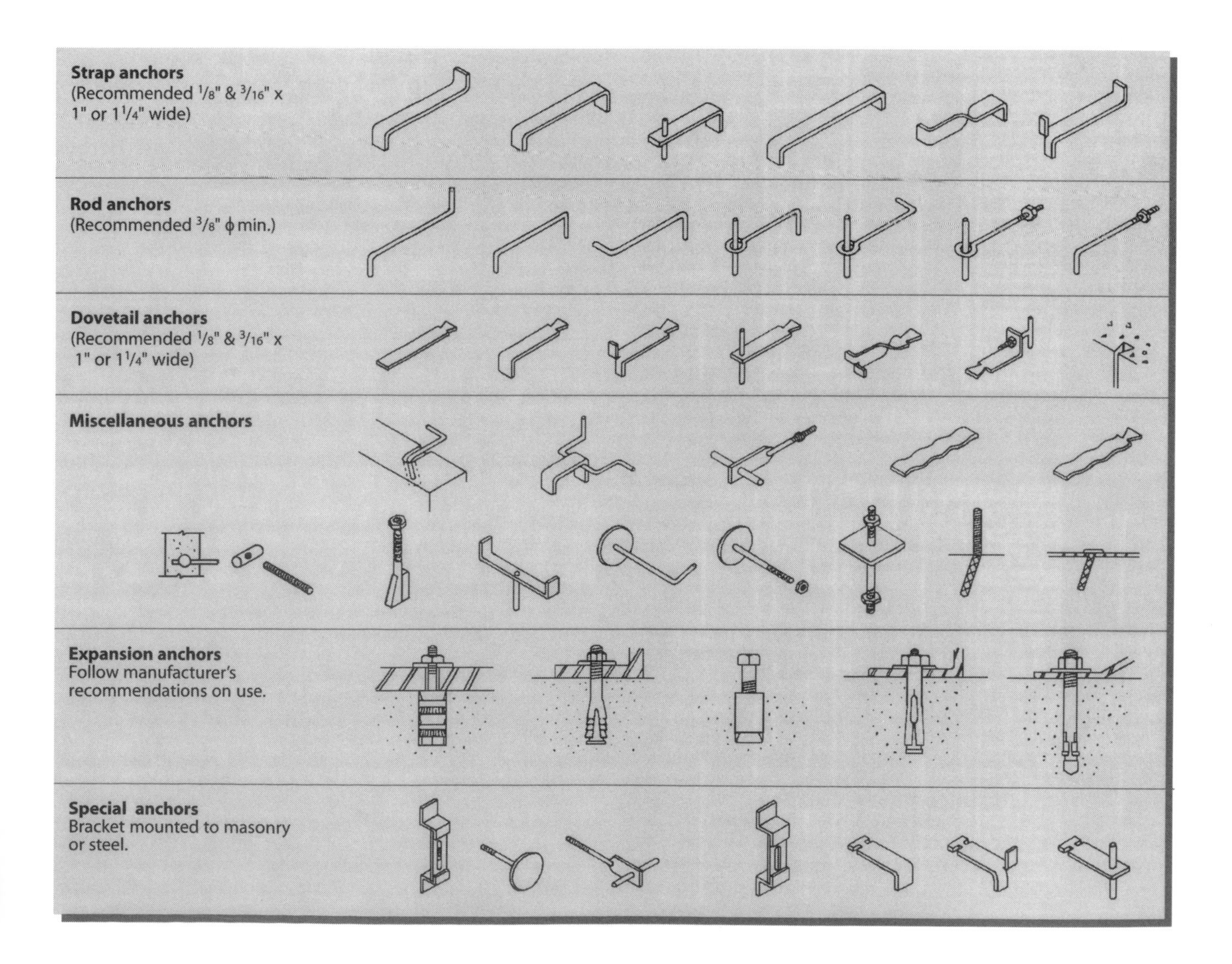

Figure 13-6
Recommended anchors

Figure 13-7
How to place an anchor

the architect's specifications for guidance on situations outside the ordinary. Refer to the tables in Figures 13-8 and 13-9.

Mortar

Usually it's best to use mortar to fill anchor holes. However, in certain circumstances, the architect may want you to use another material. Lead wool is one alternative. You can use noncorrosive shim stock (lead, nylon, aluminum, stainless steel or plastic) as long as it isn't packed into the joint too tight. But you'll probably use building joint sealers because you can gun them into place directly from the tube. Don't use nonshrink, high strength or expansive grouts to fill anchor slots or holes.

Pointing

If you're going to stick or point mortar to form the finished joint, leave extra mortar until it reaches its initial set. Then brush the joint. Don't bag or wet wipe the joints because it'll make smears. Use fiber brushes and detergents or soap powder for the final cleaning. Never use acids for cleaning limestone. Protect limestone when you use acid to clean any areas adjacent to it with acids. There's more information about this later in the section on cleaning limestone.

It's best to set limestone, rake the mortar out to a depth of $1/2$ to $1^1/2$ inches and point with mortar or apply sealant later. Pointing seals shrinkage cracks in the setting mortar and the concave joint will make the best protection against leakage. Use a grout bag or mortar gun to help you place pointed mortar without smears. When tooling joints, leave excess mortar in place until it sets, then remove it with a brush or trowel edge.

Setting Limestone in Cold Weather

Don't use admixtures to lower the freezing point of mortar for limestone. These compounds work as accelerators because of the calcium chloride in them.

Mortar type	Parts by volume of portland cement	Parts by volume of hydrated lime	Aggregate ratio (measured in damp, loose conditions)
M	1	$1/4$	Not less than $2^1/4$ and not more than 3 times the sum of the volumes of cement and lime used.
S	1	over $1/4$ to $1/2$	
N	1	over $1/2$ to $1^1/4$	
O	1	over $1^1/4$ to $2^1/2$	

Figure 13-8
Mortar proportions by volume

Mortar type	Average compressive strength at 28 days, min. psi	Water retention, min %	Air content, max %
M	2500	75	12
S	1800	75	12
N	750	75	14
O	350	75	14

Figure 13-9
Property specification requirements

You can't use calcium chloride on limestone because salts cause efflorescence and may cause spalling or flaking through recrystallization (crystal growth).

Cover newly-set stone with plastic or waterproof canvas and keep it clean and dry. The cover will also prevent smoke stains caused by using salamanders or other combustion heaters during cold weather construction. Refer to the chapter on mortar for more information on masonry work during cold weather.

Dry Setting

Today, much stonework is set dry — that is, without mortar. As a general rule, installations made up of large panels or preassembled stone units are set dry. So are stones individually supported by attachment to a structure rather than by mortar joints to other stones. Position the limestone accurately above the supports, then lower it slowly into place. By working slowly, you'll avoid damaging support surfaces or angle pockets.

Close the joints with sealants. Use noncorrosive shims to level the stone. The shim material you use is less important than the quantity. Be sure you use enough shim surface area to avoid concentrating the loads.

Where it's possible for anchors which cross joints to bottom out in anchor slots or holes, place a compressible material at the bottom of the slot or hole. Be sure the mortar you use in anchor holes doesn't pile up and fill the space behind the sealant and the backer rod. That would defeat the purpose of the soft joints.

Sealants and Joint Movement

Sealants are similar to mortar — they both close joints. But a mortar joint typically carries the weight of the stone above it, while a sealant doesn't. You can finish mortar joints using a sealant bead instead of pointing if you use a joint tape or other sealant stop. It's better to use a sealant to close joints where stone abuts different material such as aluminum, glass, plastic or other man-made material. It's also preferred for long vertical joints such as on column covers or 30- and 45-degree miters.

To make a good sealant joint, you need a backer rod, joint tape or other "caulk stop" which will flex with the sealant (Figure 13-10). In a typical sealant joint, use a backer rod to establish the depth of the sealant bead. Install the backer rod into the joint with a caulking trowel. Don't make any joint deeper than it is wide. Most of the problems with joints are due to joints that are too deep.

In some cases, you may have to transfer weight from stone to stone in an otherwise nonloading system. If you haven't used any mortar on the project, you can use shims. Be sure the shims you use are of noncorrosive material such as lead pads, aluminum sheet, plastic or nylon. Use enough shim material to avoid squeezing, which narrows the joint, and to avoid point loading, since the goal is to spread out the load. Point loading occurs when there's a high point in the shims causing stress on one point along the base of the stone. This could result in a fracture.

Sealant joints require a backer rod or caulk stop. The placement of the backer rod determines how deep (thick) the sealant bead will be. And to a large extent, thickness determines how the joint performs. Generally the beads in joints that fail are too thick because the backer rod is placed too deep in the joint. Backer rod is sold in rolls of different thickness.

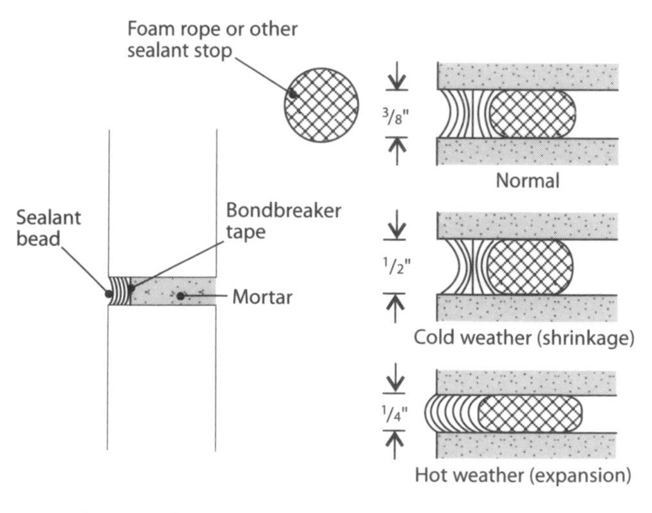

Note Shims may be substituted in sufficient area to support the load.

Figure 13-10
Joint sealant design

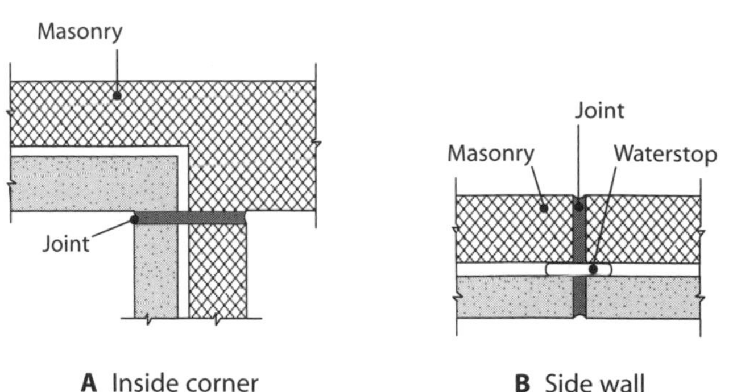

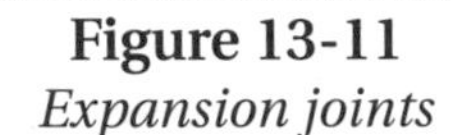
Figure 13-11
Expansion joints

Choosing a Sealant

There is no best sealant for limestone, or for any other material. The specific conditions determine the proper sealant. Major factors are the width of the gap between the materials and how much you expect the joints to move due to changes in temperature. In general construction, the expansion joints for limestone are 3/8 inch wide. Expansion joint backer material must be flexible. The blueprints will illustrate the type of fastener.

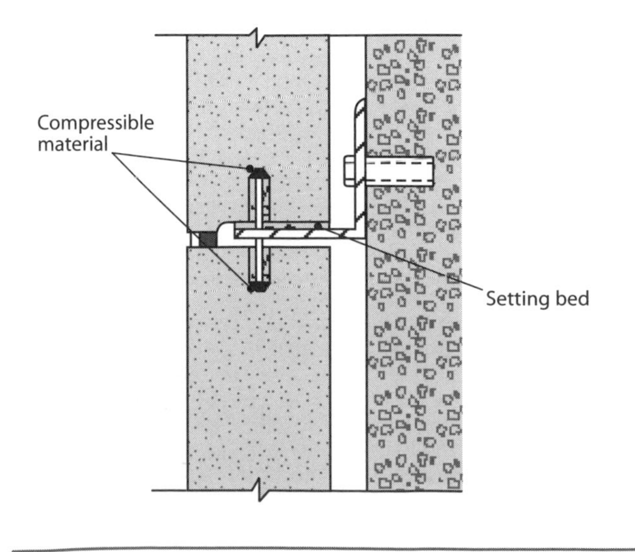

Figure 13-12
Pressure-relieving joints

Limestone will move less than other building materials as the temperature changes. But you have to allow for even a slight movement. When limestone is supported by or set adjacent to materials that are sensitive to temperature changes, you have to be especially careful about the joints. Mortar joints accommodate very little movement. You may need joint sealants to keep any wall watertight. Some of the best sealants are urethane/polyurethane, acrylic polymeric, silicones and acrylic latex.

Expansion Joints

You can install expansion joints to relieve the pressure between the skin and frame in a masonry building. The rule of thumb is an expansion joint every 50 feet for a limestone building on a steel frame in an area with wide temperature variations. The project engineer usually determines the placement of these joints.

Figure 13-11 A shows an expansion joint at an inside corner of a limestone wall backed up by masonry. The engineers or architect will design expansion joints in strategic locations to prevent cracks in the structure. Figure 13-11 B is an expansion joint in a side wall. Waterstops in connection with expansion joints prevent the penetration of water into the cavity.

In support systems where large panels span floor to floor or to column, or where smaller stones are installed with a combination of mortar and sealant joints, leave room for expansion between support clips or continuous angles. The easiest way to do this is to leave an empty space (usually about 1/4 inch) under the angle or clip. This space is enough to let the lower stone panel or assembly expand. Another solution is to fill the space with a compressible backer rod.

Where a lateral restraint anchor crosses an expansion joint, you must build flexibility into the anchor system as well. A simple and effective way to do this is to place a compressible material in the bottom of dowel holes, disc anchor slots, or other metal anchor chases or slots. This is particularly important where a single anchor restrains two stones. Figure 13-12 shows this.

Always use a sealant to close pressure-relieving joints. Never use pointing or mortar for these joints.

Dampproofing

Keep moisture away from the backs, bed and joint surfaces to prevent alkali water stains on stone work. You can use dampproofing material to isolate stone from surface and ground water containing alkali, chlorides and other stain-producing or crystal-forming materials.

Be sure you apply dampproofing wherever limestone is at or below grade. Where stone backs or beds may get wet periodically (for example where there are waterstops in cavity walls), it's a good idea to use dampproofing. Always apply suitable dampproofing wherever limestone can absorb moisture from grade or from within the walls.

Figure 13-13 A shows a masonry backup system where limestone is set on a concrete footing level with the masonry backup. There's a bituminous coating on the exterior side of the masonry backup and the inside and beneath the limestone. In section B, there's a ledge for the limestone on the concrete footing. The ledge and the back and bottom of the limestone are coated with a bituminous coating. The blueprints will indicate the height of the coating. Section C shows an independent limestone wall on a concrete footing ledge coated with dampproofing.

You also need to protect stone from surface water by grading for proper drainage. Figure 13-14 illustrates methods of dampproofing using flashing and weep holes. This is preferable to the simple method of applying just bituminous coating. Figure 13-14 A shows a masonry backup with a flashing applied from the top of the first course of masonry to the concrete coating for the stone. There's a bituminous coating under the flashing at the concrete. There is also the addition of a weep hole indicated by the dotted lines. There are several ways you can create a weep hole. You can install plastic or metal pieces to carry the moisture out or use a piece of wick-type material such as clothesline. Figure 13-14 B shows a concrete brick or stone starter course under the limestone. There's waterproof material between the masonry backup and the starter course. A solid mortar collar joint between the walls is slanted to carry off moisture through the weep hole. The starter course will help keep the limestone away from ground water.

Although it's helpful to dampproof the face of the backup or structural concrete, you also need to paint the back of the stone with a suitable material. Carry the dampproofing up the partially-exposed face at least to grade level when soil or paving at grade will cover the limestone and when the stones will present an evaporation surface above grade.

Dampproofing Materials

Any of the bituminous preparations made for dampproofing are acceptable, including asphalt emulsion paint, spray-grade mastic materials and cementitious job-mixed powders. Generally, the same materials you'd use below grade on structures are acceptable for dampproofing stone.

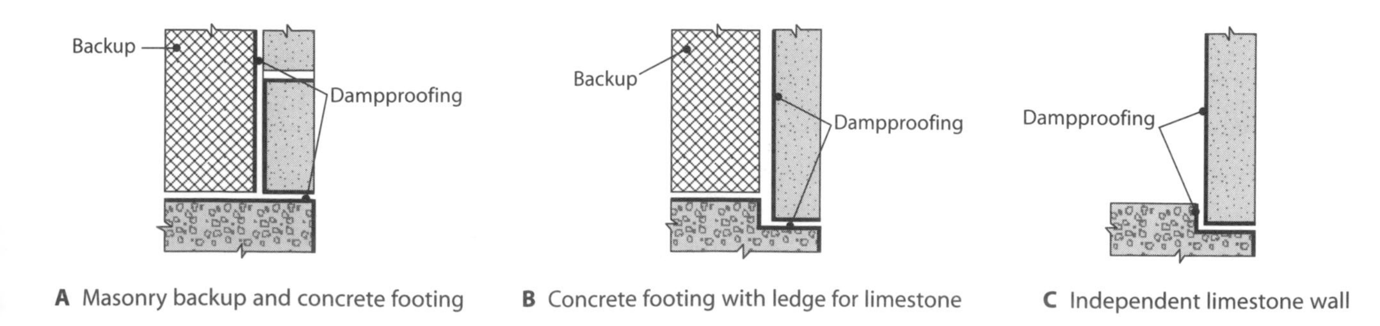

A Masonry backup and concrete footing **B** Concrete footing with ledge for limestone **C** Independent limestone wall

Figure 13-13
Dampproofing limestone

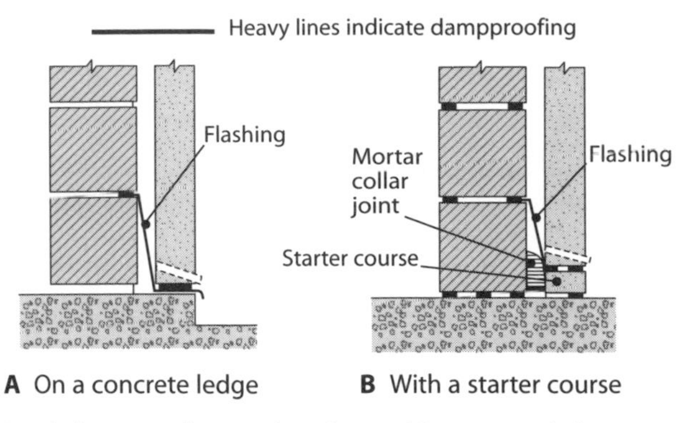

Figure 13-14
Dampproofing with flashing and weep holes

Specifications sometimes call for parging — applying a coat of white mortar to the backs of stones. Their intent is to dampproof the stone. But the Indiana Limestone Institute believes that the materials and procedures shown in this section, if done correctly, will produce superior results.

There are also ways to protect limestone from moisture at ground level. Figure 13-15 A shows gravel in a trench in front of the limestone wall. Because the limestone doesn't touch the dirt, it's protected from moisture and stains. If you must place limestone below grade, take precautions to direct water away from the wall (Figure 13-15 B). These precautions may include a gravel-filled trench with a drain tile to carry water away, waterproofing material on the outside slightly above grade, and on the inside according to the specs. Figure 13-16 illustrates the right and wrong way to grade outside the exterior of limestone walls.

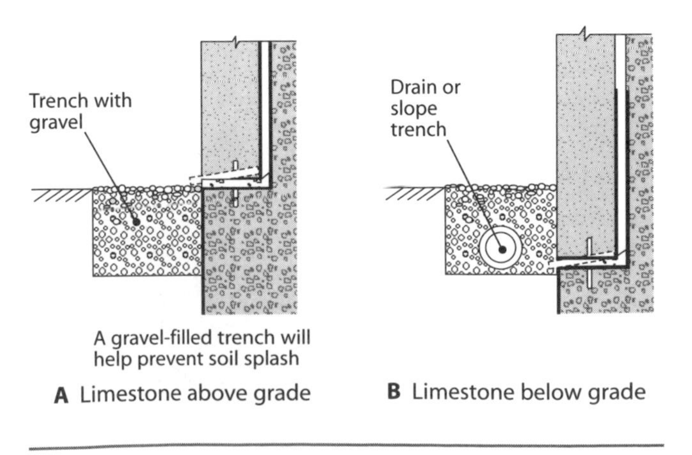

Figure 13-15
Protecting limestone from ground-level moisture

When you build stacked limestone walls, I recommend providing weep holes at each level. Figure 13-17 shows two ways to protect the wall. In section A, you can see the weep holes at the base of the limestone, sloping downward for good drainage. The air space between the limestone and the masonry back in section B prevents the transfer of moisture via capillary action from the backup to the limestone.

Cleaning Limestone

Limestone is wet when it's quarried. This moisture, called quarry sap, contains varying amounts of organic and chemical matter. Gray stone usually contains more moisture than buff. As the material dries and stabilizes, the stone seasons. The time it takes for seasoning varies with the amount of moisture in the ground and the shape of the stone mass. Slabs season faster than blocks.

Your contract may require you to buy the stone in time for it to season and stabilize before you install it. But sometimes a quarry may have to ship unseasoned stone. This stone may vary in color for several months or even a year. During that time, there's nothing you can do to make the stone look better or season faster. Eventually the stone will get its characteristic light, neutral, even color. Don't apply water repellents or surface sealants during the time the limestone is seasoning (drying).

Limestone delivered to the job site may be polluted by raw slush, planer dust and road dirt. It's best to wash the stones when you unload the truck and shake out the load. If that's not practical, you can remove the debris during the final washing after installation. But you'll have to thoroughly clean the joint surfaces before setting to make sure the mortar and sealant will stick properly.

Sometimes you'll get brown (alkali) stain on new construction. This problem is second only to that caused by mortar smears. Brown stain occurs when highly alkaline water contacts the stone backs and beds. To minimize brown stain, avoid these situations:

- ❖ Unglazed window openings which let rain onto the concrete block backup or concrete floors. Cover openings with heavy plastic or tarps until the windows are glazed.
- ❖ Wash from a concrete pour leaking down into wall cavities or otherwise getting behind installed limestone
- ❖ Limestone used as a form panel for poured concrete
- ❖ Sidewalks poured against installed limestone

Brown stain can be especially troublesome on paving, curbing, steps or platforms if they're not dampproofed. Stone here gets a constant supply of moisture from the ground, causing wet concrete pads and mortar beds.

Mortar used above grade usually doesn't contain enough water to make an intense stain. And the typical ratio of mortar bed area to stone area is very low. When mortar is the culprit, the stain is usually quite light-colored and limited to areas next to the joints.

Walls which are slushed or grouted are candidates for alkali stain. Because grout usually has a pourable consistency, it contains much more moisture than ordinary pointing or setting mortar.

Removing Stains from Limestone

When you can't prevent stains, you'll have to try to remove them.

Brown (Alkali) Stain

Nothing you can do will prevent recurring brown stain caused by alkali leached from cement when a wall is wet. Any treatment is wasted effort unless you can stop all leaks in the wall first. The good news is that as the wall dries, natural weathering will usually remove the stain without a trace.

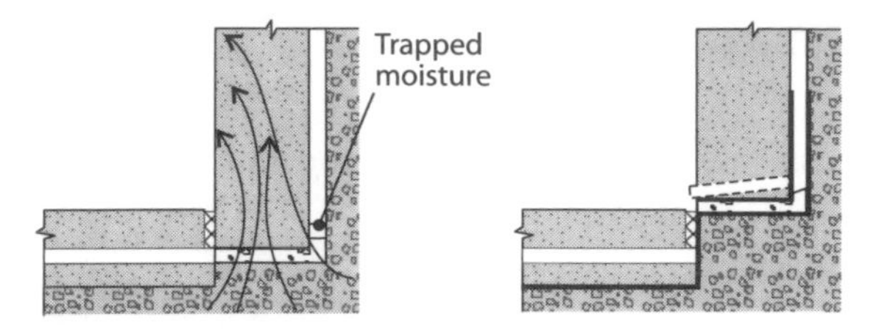

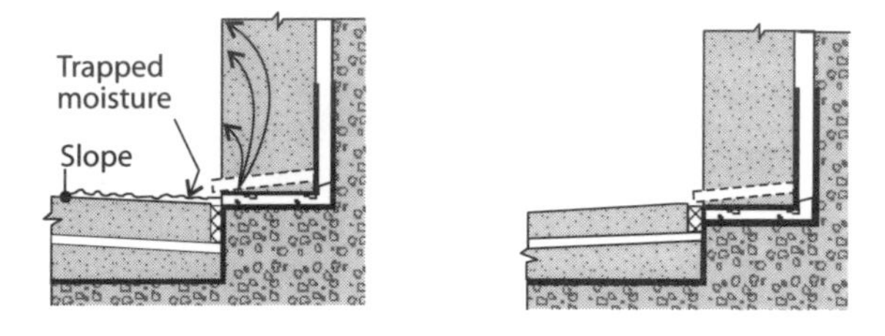

Figure 13-16
Grading outside of the exterior of limestone walls

But if you must remove the stain more quickly, wait for the wall to dry then wash affected areas with trisodium phosphate or other detergent. Scrub the solution on with fiber brushes, then rinse with water. If you use a high-pressure spray to rinse, keep the pressure below 1200 psi and use a fan-shaped nozzle no closer than 6 inches from the face of the stone. Steam cleaners are also effective. Any method may require more than one pass. But don't let any of these solutions contact the stone beds or backs.

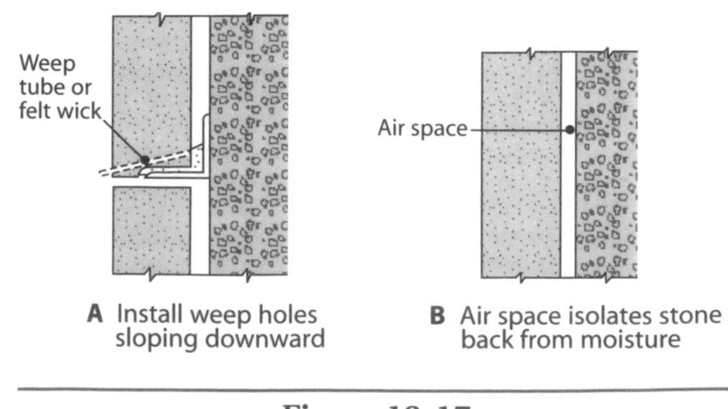

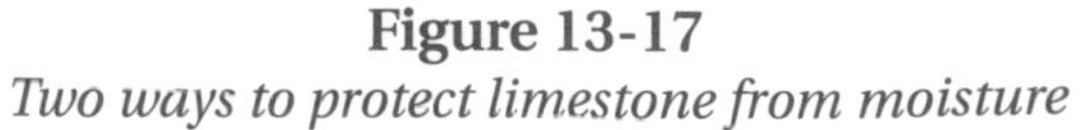

Figure 13-17
Two ways to protect limestone from moisture

Green Stain from Copper or Bronze

Scrub the stone with a solution of potassium cyanide diluted at the rate of 1 to 2 ounces of cyanide per quart of water. You can also use a sodium cyanide solution of the same strength. Follow this treatment by thoroughly drenching the surface with water. Here's a note of caution: These compounds are poisons and must be stored, handled and disposed of properly.

Oil Stains

If the stains are fresh, you can sometimes draw them out with a lime poultice. However, if more drastic treatment is necessary, saturate a blotter with either amyl acetate or benzene and place it over the stain. Make sure the blotter extends well beyond the edges of the stain. Place a hot iron over the blotter. The heat speeds the evaporation of the solvent and removal of the stain. In severe cases, you might have to repeat the process. You can remove any ghosting which remains by washing with trisodium phosphate or other detergent and rinsing with clear water.

Tar Stains

Slice off all the tar possible with a razor blade, being careful not to smear the tar onto adjoining surfaces. It helps to chill the tar with ice before you try to cut it away. Next, follow the directions above for removing oil stains.

Blood Stains

Wet the stain with water, then dust with a layer of sodium peroxide. Sprinkle with water and let the peroxide stay for a few minutes. Then scrub vigorously using lots of water. Once the sodium peroxide is thoroughly rinsed away, scrub again with a 10 percent solution of formic acid.

Rust Stains

Scrub with a hot concentrated solution of oxalic acid. Drench the surface thoroughly with clean water.

For more information on removing stains from limestone, get the booklet *Design & Procedure Aids* published by the Indiana Limestone Institute.

Repairing Damaged Stone

Some damage, chips or cracks are inevitable in limestone construction. The way you handle that damage, and its repair, defines the difference between a good job and a marginal one.

Limestone fabricators have developed handling methods which reduce the amount of damage at the mill. When damage occurs that's judged repairable, it's industry practice to repair and ship the affected stone. Damage that occurs in transit, unloading and stacking can usually be repaired on the job, often with materials furnished by the supplier. You can prevent damage by handling stone properly. For example, pad stone edges under slings as shown Figure 13-18 to avoid compression chips.

You can usually leave small chips and snips alone if they don't detract from the finished appearance, especially if they're not at eye level. You can repair larger

Figure 13-18
Pad stone edges under slings to avoid compression chips

chips with cementitious materials made for this purpose. Skillfully placed, these repairs are nearly invisible. Repaired stone is suitable for use on a building's upper reaches, or any place where it's not visible at close range.

If you chip a block or panel and can salvage the chip, re-attach it with either a thermosetting resin adhesive or cyanoacrylate super glue. If the chip is large or in a critical location, you can usually install a *dutchman* to reclaim the stone's usefulness. A dutchman is a separate piece of stone cut to fit tightly in the squared-up void and glued in place.

You can usually repair broken corners satisfactorily with thermosetting adhesives. You may have to use dowels, plates or angles in addition to the adhesive, depending on the location of support surfaces in relation to the damage. The resulting hairline joints are usually not objectionable, and such repairs are structurally sound if you do them properly.

Sometimes you can repair cracked stone. But when a stone has been so severely shocked that it develops a crack, it may not be strong enough to use. I don't recommend using fractured stone. The stone could come loose, or you could be liable if water penetrates through the crack. Order a new stone.

14

Cleaning and Painting Brick Masonry

❖❖

The final appearance of a brick masonry wall depends primarily on the attention you give the masonry surfaces during construction and cleaning. Many of the problems of brick masonry walls result from improper cleaning methods. Some walls have been irreparably damaged from lack of attention to cleaning procedures.

Cleaning Brick Masonry

Here are some important points to remember when you clean brick masonry:

- ❖ Saturate the wall surface before you apply any cleaning solution. A dry wall will absorb cleaning solution and you may get mortar smear, white scum, efflorescence or green stain.
- ❖ Mix chemical solutions carefully. Improperly mixed or overly concentrated acid solutions can etch or wash out cementitious materials from the mortar joints. Or they can discolor brick, particularly lighter shades. This is called acid burn. Chemical cleaning solutions are generally more effective when the outdoor temperature is 50 degrees F or above.
- ❖ Protect windows, doors and trim. Many cleaning agents, particularly acid solutions, have a corrosive effect on metal. They may also pit and stain masonry surfaces and trim materials such as limestone and cast stone.

Before actually cleaning a surface, test and evaluate all the cleaning procedures and solutions on an area about 20 square feet in size. Wait at least a week to judge whether the cleaning was effective or not.

Keep all walls as free as possible from mortar smears. Here are some general precautions you can take to get a cleaner wall:

- ❖ Protect the base of a wall from rain-splashed mud and mortar spatter. Use straw, sand, sawdust or plastic sheeting spread on the ground 3 to 4 feet out from the wall surface and 2 to 3 feet up the wall.
- ❖ Turn scaffold boards near the wall on edge at the end of the day to keep rain from spattering mortar and dirt directly on the wall.
- ❖ Cover walls at the end of the workday to keep water off the mortar joints.
- ❖ Protect stored brick from mud. Cover and store brick off the ground.
- ❖ Be careful about making too many mortar droppings. Tool the joints when they're thumbprint hard. After tooling, cut off mortar trailing with a trowel and brush extra mortar and dust from the surface. Use a bricklayer's brush with medium-soft bristles. Avoid any motion that rubs or presses mortar into the brick faces.

Cleaning New Masonry

Figure 14-1 is a general cleaning guide for new masonry. As a general precaution, have your workmen use protective clothing and accessories. There are three basic cleaning methods for new masonry:

- ❖ Hand cleaning with bucket and brush
- ❖ High-pressure water
- ❖ Sandblasting

Bucket and Brush Hand Cleaning

This is the most popular but most misunderstood of all the methods used for cleaning brick masonry. It's popular because it's easy to do and muriatic acid and proprietary cleaning compounds are easy to get. Here are some recommended general procedures for using detergent, acid or commercial compound solutions:

- ❖ Mix a 10 percent solution of muriatic acid in a nonmetallic container. Pour the acid into the water. Don't let your tools come in contact with acid solutions.
- ❖ For commercial compounds, make sure you pick one that's suitable for brick. Follow the manufacturer's recommended dilution instructions. Test each product on a panel or inconspicuous wall area.
- ❖ Use detergent or soap solutions to remove mud, dirt and soil. A suggested solution is 1/2 cup dry measure of trisodium phosphate (Calgon or equal) and 1/2 cup dry measure of laundry detergent dissolved in 1 gallon of clean water.
- ❖ For very difficult mortar stains on glazed brick and tile, use no more than 1 part high-grade hydrochloric acid (chemically pure) to 25 parts clean water.

Schedule cleaning last, if possible. Make sure the mortar is thoroughly set and cured. But don't wait too long to start cleaning. Mortar smears and spatters left over a long period of time (six months to a year) can cure on a wall surface and be very difficult to remove.

Protect metal, glass, wood, limestone and caststone surfaces. Mask or otherwise protect windows, doors and ornamental trim from acid solutions.

Starting at the top of a wall, apply the cleaning solution. Use a long-handled stiff-fiber brush or other type recommended by the cleaning solution manufacturer. Let the solution stay on the wall five to ten minutes. For proprietary compounds follow the manufacturer's instructions for application and scrubbing. Use wooden paddles or other nonmetallic tools to remove stubborn particles. Don't use metal scrapers or chisels. Metal marks will oxidize and stain. When cleaning glazed brick or tile, don't use metal cleaning tools or brushes.

Clean a small area, preferably about 20 square feet. Heat, direct sunlight, warm masonry and drying winds affect the drying time and reaction of the acid solutions. Work on the shady side of a building to avoid rapid evaporation.

Rinse thoroughly. Flush walls with large amounts of clean water from top to bottom before they dry. Acid solutions generally lose their strength after 5 to 10 minutes of contact with mortar particles. Flush the wall of cleaning solution and dissolved matter completely from top to bottom to keep white scum from forming.

High-Pressure Water Cleaning

To cut labor costs, many cleaning contractors use high-pressure water. Some high-pressure systems feature a high-pressure gun and nozzle equipped with a control switch. This setup lets the operator apply solutions to a wall from the base unit. Other systems have two separate hoses, one with plain water and one with a cleaning solution. Nozzle pressures generally range between 400 and 700 psi at a flow rate of 3 to 8 gallons per minute.

Make sure the cleaning compounds you use with this method are compatible with the equipment. Some cleaning manufacturers are careful to recommend that only specific cleaning compounds be pumped through their equipment. Others build pumps that will resist hydrochloric acid solutions for reasonable lengths of time.

Equipment should be as portable as possible. Units may be on wheels, skids, trailers or pickup-truck beds. More elaborate systems include pumps, engines, acid

Brick category	Cleaning method	Remarks
Red and red flashed	Bucket and brush hand cleaning High pressure water Sandblasting	Hydrochloric acid solutions, proprietary compounds, and emulsifying agents may be used. **Smooth Texture** Mortar stains and smears are generally easier to to remove; less surface area exposed; easier to presoak and rinse; unbroken surface, thus more likely to display poor rinsing, acid staining, poor removal of mortar smears. **Rough Texture** Mortar and dirt tend to penetrate deep into textures; additional area for water and acid absorption; essential to use pressurized water during rinsing.
Red, heavy sand finish	Bucket and brush hand cleaning High pressure water	Clean with plain water and scrub brush, or *lightly* applied high pressure and plain water. Excessive mortar stains may require use of cleaning solutions. *Sandblasting is not recommend.*
Light colored units, white, tan, gray, specks, pink, brown and black	Bucket and brush hand cleaning High pressure water	*Do not use muriatic acid!!* Clean with plain water, detergents, emulsifying agents, or suitable proprietary compounds. Manganese colored brick units tend to react to muriatic acid solutions and stain. Light colored brick are more susceptible to "acid burn" and stains, compared to darker units.
Same as light colored units, plus sand finish	Bucket and brush hand cleaning High pressure water	Lightly apply either method. (See notes for light colored units, etc.). *Sandblasting is not recommend.*
Glazed brick	Bucket and brush hand cleaning	Wipe glazed surface with soft cloth within a few minutes of laying units. Use soft sponge or brush plus ample water supply for final washing. Use detergents where necessary and acid solutions only for *very difficult* mortar stain. For dilution rate, refer to *Bucket and Brush Hand Washing* for very difficult stains on page 2. Do not use acid on salt glazed or metallic glazed brick. Do not use abrasive powders.
Colored mortars	Method is generally controlled by the brick unit	Many manufacturers of colored mortars do not recommend chemical cleaning solutions. Most acids tend to bleach colored mortars. Mild detergent solutions are generally recommended.

Figure 14-1
Cleaning guide for new masonry

containers and water-storage tanks fixed on truck beds. To clean with high-pressure water:

- ❖ Test any solution on a sample area.
- ❖ Scrape clean, using wooden paddles or non-metallic scrape hoes or chisels. Metal objects can leave rust stains and dig into the surface. Protect metal, glass, wood, limestone and cast-stone surfaces.
- ❖ Presoak by flushing with water from the top down.
- ❖ Apply the cleaning solutions with a low-pressure orchard sprayer, 30 to 50 psi, or directly through the high-pressure cleaning unit.
- ❖ Let the cleaning solution stay on the wall about five minutes.
- ❖ Starting at the top, flush the wall with water to rinse thoroughly.

Caution: Solutions can be driven into masonry when you use high pressure to apply them. However, if you saturate the walls sufficiently before you apply the solution, the risk of penetration is less.

Type gradation	U.S. sieve size	Percent passing
Type "A" Fine Texturing[b]	30 mesh	98-100
	40 mesh	75-85
	50 mesh	44-55
	100 mesh	0-15
	200 mesh	0
Type "B" Medium Texturing[c]	16 mesh	87-100
	18 mesh	75-95
	30 mesh	20-50
	40 mesh	0-15
	50 mesh	0-15

[a] The screen analysis, as listed above, is suggested primarily for mined silica sands and crushed quartz. Reference source: "Good Practice for Cleaning New Brickwork," produced by the Brick Association of North Carolina, P.O. Box 6305, Greensboro, NC 27405.

[b] Type "A" gradation is suggested for very lightly soiled brick masonry or where very light, fine texturing of the masonry surface is permitted.

[c] Type "B" gradation is suggested for heavy mortar stains, or where a medium texture on the masonry surface is permitted.

Figure 14-2
Typical screen analysis for sandblasting sand abrasives[a]

Sandblasting

Dry sandblasting has been used for many years for restoration work and it's one method that eliminates the dangers of mortar smear, acid burn and efflorescence involved in acid cleaning. But if you're not careful you can scar the face of brick and mortar joints. Some workers like this method better than conventional wet-cleaning because it eliminates the problem of chemical reaction with vanadium salts and other foreign matter.

Sandblasting involves using a portable air compressor, blasting tank, blasting hose, nozzle, and protective clothing and hood for the operator. The air compressor should produce 60 to 100 psi at a minimum air-flow capacity of 125 cubic feet per minute. The inside orifice or bore of the nozzle may vary from $3/16$ to $5/16$ inch in diameter. The sandblasting machine (tank) should have controls to regulate the flow of abrasive materials to the nozzle at a minimum rate of 300 pounds per hour. Here's a procedure for sandblasting:

- Select sandblast materials that are clean, dust-free and abrasive. Use mined silica sand, crushed quartz, granite, white urn sand (round particles), crushed nut shells or other abrasives. Mined silica sand and crushed quartz should be of type A or B graduation. See Figure 14-2.
- Make sure the masonry is dry and well cured.
- Remove all large mortar particles.
- Protect nonmasonry surfaces. Use plastic sheeting, duct tape or other covering materials.
- Test-clean several areas at varying distances from the wall and at several angles of application. Use working distances and angles that clean best without damaging brick or mortar joints. Tell your workmen to sandblast at the brick and not the mortar joints.

Caution: Don't sandblast sanded, coated or slurry-finished brick.

Cleaning Existing Masonry

As mortar ages, it gets harder. And sometimes you don't know what's on the surface, so you may have to do some experimenting. To clean old stains, start with high-pressure steam.

High-Pressure Steam

Always use steam to clean buildings with smooth hard brick or brick with glazed surfaces. The more impervious a brick unit, the easier it should be to clean. In most cases, you can clean buildings satisfactorily with plain high-pressure steam. For stains, you may need a chemical or detergent solution.

Sandblasting

If you can't clean brick successfully with high-pressure steam, try dry sandblasting. There are abrasives softer than sand that you can use, such as crushed pecan shells. Always sandblast a sample area before proceeding with the entire job and use this method only when it won't damage the brick.

Hand Washing

Hand washing has two disadvantages: it's slow and doesn't clean as well as high-pressure steam does. Usually you do this using soap or detergent with cold water. This method generally is more costly and you probably won't use it on a job of any size.

High-Pressure Cold Water

Usually you can use this method satisfactorily if you have plenty of water. But getting rid of a lot of water can sometimes be a problem.

Chemical and Steam

Use chemicals and high-pressure steam to remove applied coatings, such as paint, to masonry. This is a highly-specialized field. It usually involves analyzing the factors you're facing before you can determine the proper cleaning agent for a particular project. I suggest that you remove some of the coating mechanically and consult a paint supplier for the chemicals to use.

Wet Sand Cleaning

This method uses a water-cushioned abrasive action. The water in the cleaning action eliminates dust. Use this method to remove paint or other surface coating from hard brick or other surfaces that won't be damaged by abrasive action.

Wet Aggregate Cleaning

This is a special process developed by Western Waterproofing Company you can use on soft brick and soft stone materials. It's a gentle but thorough process using a mixture of water and an aggregate delivered at low pressure through a special nozzle that has a scouring action. It cleans effectively without damaging the surface. Use it on surfaces with flutings, carvings and other ornamentation.

Efflorescence

Efflorescence on masonry walls usually appears as a white powder from water-soluble salts deposited on the surface when water evaporates from the brick. The salts may also be present in the mortar. Sometimes efflorescence appears as green stains on buff or gray facing brick.

When efflorescence appears repeatedly over a long time, it's usually caused by changes in climate. You'll notice efflorescence "blooms" after rains in cool weather, generally in late fall, winter, and early spring. It takes three conditions to produce efflorescence:

1) Soluble salt in the brick or mortar, or rain water (acid rain)
2) Enough moisture to cause the salt solution to seep
3) A crack or void that lets the seeping solution reach to the surface where evaporation leaves the powdery material exposed

Removing Efflorescence

Compared to removing other stains, removing efflorescent salts is easy. They're water soluble and will usually disappear of their own accord with normal weathering. This is particularly true for new deposits. Or you can remove white efflorescent salts with a dry brush or clear water and a stiff brush. Remove heavy salts by scrubbing with a solution of muriatic acid (1 part acid to 12 parts water).

Saturate the wall with water both before and after you apply the solution. These wetting steps are extremely important. If you put the acid solution on a dry wall, it may etch the joints and leave them sandy-looking. It can also carry the efflorescence dissolved by the solution back into the masonry. If you don't rinse it away completely afterward, the solution will leave runny stains and can etch the surface.

There are some proprietary compounds designed specifically to remove efflorescence. I've used Sure Klean, but you can check with masonry suppliers to find out about new cleaners.

Removing Some Common Brick Stains

Most of the methods that remove stains from brick involve making a poultice — a paste containing a solvent or agent and an inert material. The inert material may be talc, whiting, fuller's earth, bentonite or other clay. The solution or solvent you use depends on the stain you're trying to remove.

If the solvent used in preparing a poultice is an acid, don't use whiting as the inert material. Whiting is a carbonate that reacts with acids to give off carbon dioxide. While this isn't dangerous, it makes a foamy mess and negates the power of the acid.

To use a poultice, add enough of the solution or solvent to a small quantity of the inert material to make a smooth paste. Smear the paste onto the stain area with a trowel or spatula and allow it to dry. When it's dry, scrape off the remaining powder.

Poultices have a couple of advantages: They tend to keep the stain from spreading during treatment, and they pull the stain out of the pores of the brick. They work by dissolving the stain and leaching or pulling the solution into the poultice. It leaves behind a powdery substance you can simply brush off. It may take repeated applications to completely remove the stain. Poultices are normally used only for small stain spots.

Paint Stains

To remove fresh paint, apply a commercial paint remover or a solution of trisodium phosphate in water at the rate of 2 pounds of trisodium phosphate in 1 gallon of water. Let this sit to soften the paint. Remove with a scraper and wire brush. Wash with clean water. For very old dried paint, remove the paint by sandblasting or scrubbing with steel wool. There are also gel solvent paint removers you can try, but test them first on a small area.

Smoke and Dirt

Smoke is a difficult stain to remove. Try using a stiff brush with a scouring powder containing bleach. Or you can brush or spray an alkali detergent or commercial emulsifying agent. You can also use either of these with a steam cleaner. Test it on a small area first. For small, stubborn stains, make a poultice of trichloroethylene to pull the stain from the pores. Be sure to have good ventilation because trichloroethylene gives off harmful fumes.

Dirt is also sometimes difficult to remove, particularly from textured brick. If the texture isn't too rough you can use scouring powder and a stiff-bristle brush. For very rough textures, using a high-pressure steam cleaner is about the best you can do.

Plant Growth

Occasionally, an exterior masonry surface that stays damp most of the time because it doesn't get sunlight will show signs of plant growth such as moss. Apply ammonium sulfate or weed killer according to the manufacturer's directions.

Externally Caused Stains

External materials get spilled, spattered and absorbed on brick. Each is an individual case and you have to treat it accordingly. You can remove many external stains by scrubbing with a kitchen cleanser. Or you can try household bleach. Figure 14-3 lists some materials you can try — and where you can get them.

When metal is welded too close to a wall or pile of brick, some of the molten metal may splash onto the brick and meld into the surface. Use an oxalic acid-ammonium bifluoride mixture recommended for iron stains to remove welding stains.

Stains of Unknown Origin

Stains of unknown origin can be a real challenge. Appearance may be your first clue. Rust-colored stains may actually be rust. Such stains are quite common and can come from mortar ingredients, welding spatter on the back of the brick or from something placed on the pile of brick before it was laid in the wall.

Agent	Supply source
Aluminum chloride	Pharmacist.
Ammonia water	Supermarket. Household ammonia water.
Ammonium chloride	Pharmacist. Salt-like substance.
Ammonium sulfamate	Nursery and garden stores. Past use was as a base for weed killers. Not now readily avaliable. Substitute any brand weed killer solution.
Acetic acid (80%)	Commercial and scientific chemical supply firms.
Hydrochloric acid	Hardware stores. Muriatic acid is generally avaliable in 18 degree and 20 degree Baume solutions.
Hydrogen peroxide (30-35%)	Some commercial and scientific chemical supply firms.
Kieselguhr	Commercial, scientific chemical and swimming pool supply firms. Diatomaceous earth.
Lime-free glycerine	Drug stores, used as a hand lotion base.
Linseed oil	Hardware and paint stores.
Paraffin oil	Hardware stores.
Powdered pumice	Hardware stores. A sanding or polishing material.
Sodium citrate	Pharmacist. Appears like enlarged salt granules.
Sodium hydroxide	Supermarket. Available in brand name substances such as Drano.
Sodium hydrosulphite	Pharmacist or photographic stores. A white salt or "hypo" of photographic fixing agent.
Talc	Drug stores. Inert powder available as "purified talc." Bathroom talcum powder may be substituted.
Trichloroethylene	Commercial scientific chemical supply firms and possibly some service stations or supermarkets. A highly-refined solvent for dry cleaning purposes.
Trisodium phosphate	Paint stores, some hardware stores, supermarkets. Strong base type powdered cleaning material sold under brand names. Also available in brand name substance such as Calgon.
Varsol	Service stations. A refined solvent by the brand name Varsol.
Whiting	Paint manufacturers, possibly some large paint stores. A powdered chalk. Substitute kitchen flour if purchase is difficult.

Figure 14-3
Sources of cleaning and masking agents

Green stains may be grass, moss or vanadium efflorescence. Brown stains can be almost anything. Some possibilities include vanadium efflorescence, or possibly manganese staining.

One test you can use to find out if a stain is organic is to put some concentrated sulfuric acid on it. If the stain turns black, it's organic and you can use household bleach or oxalic acid to remove it.

Painting Brick Masonry

Clay masonry really doesn't need painting or surface treatment. But you can paint clay masonry walls to increase light reflection or to make them more decorative. Once painted, repaint exterior masonry every three to five years.

Selecting Brick and Mortar

Just because you're going to paint the brick you're laying doesn't mean you can use a lower grade. The brick you use should be ASTM C62 for building brick or C216 for facing brick. I recommend Grade SW. You can use different colored brick. However, make sure the bricks have similar absorption and suction characteristics, so when you paint them they'll have a uniform appearance. Follow ASTM C270, Specifications for Mortar for Unit Masonry. Select the mortar type on the basis of the structural requirements of the wall.

Don't cut any corners with workmanship, assuming you can use paint as a cover-up. For example, make sure you fill joints with mortar to keep moisture out of a wall. Any moisture trapped behind the paint will cause problems later. Water or moisture may enter masonry walls through incompletely bonded or partially-filled mortar joints, copings, sills and projections, or incomplete caulking and improperly installed or omitted flashing.

Before painting, make sure there are no efflorescent materials in the mortar, brick or the backup. Efflorescence beneath the paint film can cause problems. See the cleaning section in this chapter for how to remove efflorescence.

Preparing the Surface

Each coat of paint is the foundation for the next coat, so your success or failure with painting will depend a lot on how well you prepare each surface. The first thing to do is to thoroughly examine all surfaces. Previously-painted surfaces often require the greatest effort. Before painting, remove all loose matter. Take special care to clean any surface that you'll be covering with emulsion paint or primer. They require a cleaner surface than solvent-based paints. Be sure to follow the manufacturer's directions about applying any paint to damp surfaces.

Preparing New Masonry

Usually you don't paint new clay masonry. But if you have to, don't wash clay masonry walls with any acid cleaning solution. Acid reactions can make paint fail. Use alkali-resistant paints. Unless low-alkali portland cement was used in the mortar, neutralize the wall to reduce the possibility of alkali-caused failure with a zinc chloride or zinc sulfate solution of 2 to 3½ pounds per gallon of water.

Preparing Existing Masonry

Examine older unpainted masonry for efflorescence, mildew, mold and moss. Check all possible entry points for water. Where necessary, repair flashing and caulking, and tuck point defective mortar joints. Remove all efflorescence by scrubbing with clear water and a stiff brush.

If moss has accumulated on a damp, shaded wall, wet the wall first with clear water and then apply weed killer. Chemical weed killers may add to efflorescence or react unfavorably with paint. After you get rid of the moss, be sure to scrub the wall with a stiff brush and rinse with clear water to remove the weed killer.

Remove any mildew completely before applying any paint. Otherwise, it'll just continue to grow, damaging new paint. You can steam clean or sandblast to get rid of mildew. Or you can use this mixture:

- 3 ounces trisodium phosphate (Soilax, Spic and Span, etc.)
- 1 ounce laundry detergent (Tide, All, etc.)
- 1 quart 5 percent sodium hyperchloride (Clorox, Purex, etc.)
- 3 quarts warm water, or enough to make 1 gallon solution

Using this solution, scrub with a medium-soft brush until the surface is clean and then rinse thoroughly with fresh water. For small areas, use an ordinary household cleanser. Scrub with a medium-soft brush and rinse thoroughly. Use a mildew-proof paint to help keep molds from coming back.

Remove all peeled, cracked, flaked or blistered paint by scraping, wire-brushing or sandblasting. Paint blistering is caused by water within masonry. Find the source of the water and take the necessary steps to keep the water out of the wall.

If there's any alligatoring, remove the entire finish. There's no other means of solving this problem. Figure 14-4 shows typical types of paint failures.

If a slight chalking has occurred, brush the surface thoroughly. However, if the chalking is deep, remove it by scrubbing with a stiff fiber brush and a solution of trisodium phosphate and water. Rinse the surface thoroughly afterwards. Use a penetrating primer to help the final coat stick.

If too many coats or excessively thick coats have caused too much paint to build up on a surface, remove all the paint and treat it as a new surface.

Completely remove cement-based paints before repainting with other types. An exception to this rule is when you've used cement-based paint as a primer and you'll be doing the next coat with another paint within a couple of days. If you're going to repaint the wall with another cement-based paint, just wire brush and scrub if you don't have any mildew, efflorescence, or other problems to contend with.

Defect	Description
Alligatoring	Wrinkling of the paint surface caused by paint coats of different hardness.
Bleeding	The working up of a stain into succeeding coats, imparting a discoloration to the newly applied coat.
Blistering	Bubbles resulting from moisture trapped behind an impermeable paint film.
Chalking	Powdering at or just beneath a paint surface. Slight chalking may be normal due to weatherproofing.
Checking	A defect in organic paints, manifested by slight breaks in the film surface.
Erosion	Wearing away by weathering.
Excessive paint build-up	Result of applying too much paint or coats which are too thick.
Flaking	Detachment of small pieces.
Map checking	Breaks in paint surface extending entirely through the paint film, usually caused by shrinkage.
Mildew	Fungus growth sometimes found feeding on paint or particles adhering to the surfaces in damp places, generally black or gray in color.
Peeling	A partial detachment of paint.
Scaling	An advanced form of flaking.

Figure 14-4
Types of paint failures

Masonry Paints

In selecting paint for a brick masonry wall, your primary concern should be the characteristics of the surface and the exposure conditions of the wall. Picking a good primer coat may be of particular importance, especially where unusual or severe conditions exist. It's best to use an alkaline-resistant primer.

Masonry paint should be durable, easy to apply, and stick well. But all paints have distinct properties, and the surfaces you put them on can vary a lot . So even the most experienced paint contractors have to examine a surface carefully before deciding which paint to use. Generally though, for exterior masonry, you'll want to use a porous paint so any moisture in a wall won't be trapped behind the paint.

Fill Coats

These are base coats for exterior masonry. They're similar in composition, application, and use to cement-based paints. However, fill coats contain an emulsion paint in place of some water. This enables them to stick better and make a tougher film than unmodified cement paints. Fill coats have greater water retention and give the cement a better chance to cure. Use a fill coat in dry areas where you have trouble keeping a painted surface moist as it cures.

Cement-Based Paints

Cement-based paints are mainly portland cement, lime and pigment, with some binders and sands added. This paint is more difficult to apply than other types, but it can protect a surface very well when you apply it properly. While not completely waterproof, cement-based paints can help seal and fill porous areas. They also can make a good base for other paints if applied within a relatively short time.

You'll have better luck with white and light colors, as it's difficult to get a uniform coating with darker colors. Lighter colors tend to become translucent when wet, and dark colors become darker. Color returns to normal as the wall surface dries.

Here's how to apply cement-based paint to a properly-prepared surface:

1) Cure new masonry walls for approximately one month before you apply cement-based paints.
2) Spray wall surfaces thoroughly with water.
3) Mix powdered cement-based paint just before you apply it for the best result.
4) Apply heavy coats with a stiff brush, waiting at least 24 hours between coats. Keep the wall damp during this time by spraying it with water.
5) Apply additional coats the same way.
6) Keep the final coat damp for several days to cure properly.

Water-Thinned Emulsion Paints

Emulsion paints dry quickly, have practically no odor and present no fire hazard. And you can apply them to damp surfaces. As a group, these paints are alkali-resistant so you don't need any neutralizing washes or curing periods. They work well on brick substrates as long as you prepare the surface properly.

Emulsion paints don't stick well to moderately chalky surfaces. If possible, repaint before the previous coat chalks excessively. If it's already chalky, choose a specially-formulated paint that contains emulsified oils or alkyds that help wet the chalky surfaces. This lets the paint bond with the chalky surface and the substrate.

The most common emulsion paint is latex paint.

Latex Paint

Latex paint is relatively easy to apply with a brush, roller or spray. However, brush application is usually best, especially on coarse-textured masonry.

Butadiene-Styrene Paints

These relatively low-cost, rubber-based latex paints develop water resistance more slowly than vinyl or acrylic emulsions. Light tints are best because deep colors may chalk.

Vinyl Paints

Polyvinyl acetate emulsion paints dry faster, have improved color retention and a more uniform, lower sheen than rubber-based latex paints.

Acrylic Emulsion Paints

Acrylic emulsions set up quickly, permitting recoating in 30 minutes or less. Acrylics also resist water-spotting and alkali.

Alkyd Emulsion Paints

Alkyd emulsions are related to solvent-thinned alkyd types, but have all the general characteristics of latex paints. They penetrate more than most water-thinned emulsions and stick better to chalky surfaces. Compared to other emulsion paints, they dry slower, smell more, are less resistant to alkalis, and fade sooner. Use this paint under normal exposure conditions as a finished coat over an alkyd primer.

Multicolored Lacquer

A multicolored lacquer is a specialized paint that you apply by spray gun over another type base coat such as polyvinyl acetate or acrylic emulsion paint. The finished paint appears as a base color, with separate dots or particles of contrasting colors. Use these paints to cover surface defects and irregularities.

Solvent-Thinned Paints

Apply solvent-thinned paint only to completely dry, clean interior masonry walls that aren't susceptible to moisture. The exception to this rule is special-purpose paint, such as synthetic rubber, chlorinated rubber and epoxy paints. They're suitable for commercial building, kitchens and laboratories.

Oil-Based Paints

Oil-based paints are relatively nonporous and recommended for interior use only. They're fairly easy to apply, although you may need several coats to cover well. Allow several days of drying time between coats. As with most solvent-based paints, oil-based paints penetrate well into porous masonry and relatively chalky surfaces, but are highly susceptible to alkalis. New masonry must be thoroughly neutralized to avoid turning the paint into a soft, soapy mess.

Alkyd Paints

Alkyd paints are similar to oil-based paints. They penetrate slightly less, but are somewhat better in color uniformity. Alkyd paints are more difficult to brush, dry faster and give a harder film than oil-based paints. They're nonpermeable and recommended for interior use only. You do have to neutralize new masonry before using.

Synthetic Rubber and Chlorinated Rubber Paints

These paints are often specified for industrial applications because they:

- ❖ Penetrate well
- ❖ Resist efflorescence, alkali, corrosive fumes and chemicals
- ❖ Stick well to both previously painted, moderately chalky and new surfaces
- ❖ May be applied directly to alkaline masonry surfaces

However they're more difficult to brush on than oil paints. And darker-colored synthetic-rubber paints lack color uniformity. With either type, you have to use very strong volatile solvents which may be a fire hazard.

Epoxy Paints

Epoxy paints are synthetic resins generally made of a resin base and a liquid activator. They stick well and have good corrosion and low odor. You can use epoxy paint on alkaline surfaces. Be sure to use any epoxy paint shortly after you mix it.

Some types chalk excessively if you use them outdoors. And they're relatively expensive and somewhat difficult to apply.

High-Build Paint Coatings

Use these on interiors to give the effect of glazed brick. Some of them are based on two-component urethane polyesters and epoxies. Others are an emulsion-based coat with acrylic lacquer. These coatings usually include fillers, which smooth out surface irregularities.

Breathable Masonry Coatings

There are breathable coatings on the market that allow water vapor to pass through the wall. One of the best is from *ProSoCo* of Kansas City. Their phone number is 913-281-2700. One of their products, BMC 55, is a pigmented water-base coating you can use on interiors or exteriors. It has an acrylic binder that makes it very durable. It's designed to provide good moisture vapor permeability, which reduces the possibility of peeling or blistering. And it protects the surface from stains caused by atmospheric conditions and mildew. BMC 90 is excellent for use in extreme weather conditions, such as near the ocean.

Painting Near Unpainted Masonry

Often you'll find windows and trim of masonry buildings that have been painted with self-cleaning paints to keep the surface fresh and clean. Unfortunately, self-cleaning is generally achieved through chalking. The theory is that rain will wash away chalked paints, constantly exposing a fresh paint surface. The theory works well, but if chalk-contaminated rainwater gets on a masonry surface, the result is usually more unsightly than dirty paint on trim or windows. Avoid this staining by choosing nonchalking paints for windows and trim and making sure water drains away from wall surfaces.

15

Flashing and Moisture Control

In this chapter, we'll discuss the flashing you use when you build masonry walls and buildings. You'll find information on flashing for chimneys and fireplaces in Chapter 16.

Flashing is a membrane you install in a masonry wall system to keep water from penetrating into the wall. I recommend flashing every barrier wall, parapet, sill, projection, recess and intersection. It's essential to flash drainage walls. Drainage-type walls, such as brick veneer and cavity walls, will deteriorate quickly if you don't install flashing.

However, if you're building in an area where the amount of rainfall is slight, you don't need internal flashing at spandrels, lintels, sills, etc. Also, you can reduce external flashing to a minimum. I've made recommendations in this chapter for areas which have severe or moderate amounts of rain.

Flashing Materials

You can use bituminous membranes, plastics, sheet metals or combinations of these as flashing material. It's important to use only superior materials because it'll be expensive or even impossible to replace flashing later if it fails. I don't recommend using asphalt-impregnated felt paper as a flashing material. It tears too easily.

Copper

One of the best types of flashing material is copper. Copper flashing is sheet copper with a layer of textured fabric attached to both sides. It comes in 2, 3, 5 and 7 ounce weights, packaged in rolls usually 60 inches wide by 25 feet long. But you can also get other combinations of copper and lead. Copper has several advantages that make it a good choice for flashing:

- ❖ It resists harmful acid and alkali in fresh mortar.
- ❖ It's waterproof.
- ❖ It has a high enough tensile strength to resist stretching, and is tough enough to bear the compressive force of masonry walls.
- ❖ It's easy to trim to any shape on the job site.
- ❖ It's as durable as a masonry wall.

Figure 15-1 shows York copper flashing on a lintel. Here you apply sheathing over the top of the flashing to keep moisture from getting behind the flashing.

Of course, copper costs more than most other flashing materials. When using copper flashing, don't use chloride-based additives in mortar. Also, exposed copper may tend to stain adjacent masonry. Don't extend the flashing out of the edge of the mortar joints.

Zinc

Galvanized coatings may corrode in fresh mortar. Although the coating products seem to form a very compact film around zinc, and a good bond with the mortar, you can't predict how bad the corrosion will get over time. Galvanized steel will also corrode if it's bent or the coating is cracked.

Figure 15-1
Using York copper flashing

Aluminum and Lead

Aluminum flashing is popular because it's cheap. But it's also susceptible to the alkali in fresh mortar. Once mortar is dry, it doesn't affect aluminum, but if it gets wet later, it can corrode the aluminum. I don't recommend aluminum as a flashing material in masonry construction.

Lead, like aluminum, will corrode in fresh mortar. Furthermore, if you only partially embed lead in mortar, moisture will cause an electrolytic action that will gradually disintegrate the embedded lead.

Plastics

As a group, plastics are becoming one of the most widely-used flashing materials. High-quality plastic flashings are tough, resilient, and highly resistant to corrosion. However, some plastics won't withstand the corrosive effects of masonry mortars. Study the product literature or test data to make sure any plastic you choose will perform well.

Wascoseal is a one good plastic product that's easy to use. This polyvinyl chloride sheeting comes in four thicknesses, in rolls 50 or 144 feet long. It's tough, so it resists puncturing, ripping, tearing, cracking and flaking when you install it. It won't crack or flake when you bend it 180 degrees over a 1/32-inch mandrel, then 360 degrees in the opposite direction. Use a rubber base adhesive to join this material.

Bituminous Flashings

You can use fabrics saturated with bitumen as damp checks (to prevent moisture entering the masonry by capillary action). It's also a low-cost substitute for metal flashing at the heads of openings at spandrels and window sills. They're effective if they're permanently insoluble in water and you can install them without breaking the fabric skin. But they're not as permanent as good metal flashing. I don't recommend asphalt-impregnated felt for flashing masonry because it's sometimes brittle and easily damaged during installation.

Stainless Steel

Stainless steel is an excellent flashing material because it resists chemicals. It costs about the same as aluminum but it's more durable. Some masons don't use it because it will fail if it comes in contact with other ferrous metals. It also tends to tear when you cut it.

Combination Flashings

Often, you'll find a combination of materials that utilize the best properties of each material. For example, you can use sheet metal coated with plastic or bituminous materials for a good material that's not too expensive.

How to Install Flashing

Where flashing extends to the interior, place its end between the furring and the interior finish and turn it up at least 1 inch to collect moisture that may get through the wall.

Unless coping has watertight joints, place through flashing in the mortar bed beneath it. That's flashing that extends from one side of the wall across to the other side. Where coping provides an adequate drip, you can stop the flashing at the wall surface. However, where copings are flush with the surface of a wall, extend the flashing at least ¼ inch on both sides and turn it down to provide a drip. Follow the same procedure when topping out piers, pilasters, and so on.

Flashing for horizontal masonry surfaces should be laid in and topped with a soft bed of mortar. Start horizontal flashing about ½ inch from the face of a wall and carry it through the wall. Turn up a lip at the back of the wall to encourage moisture to flow through weep holes to the outside.

On vertical masonry surfaces, coat flashing with an asphalt mastic to hold it in place until the masonry is laid up to hold it permanently.

At heads, sills and cavity walls, turn the flashing up at the ends to form a pan. Fold the corners, don't cut them. Weep holes (described in the next section) will then carry moisture away to the outside.

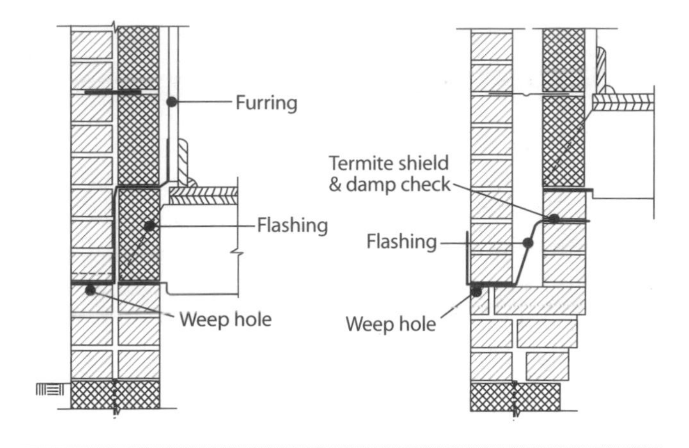

Figure 15-2
Flashing at the base of masonry walls

Where to Install Flashing

Now that we've discussed how to install flashing, let's look at places that need it. Moisture may enter masonry at vulnerable spots. You need flashing to divert this moisture to the outside. In areas of severe or moderate moisture, provide flashing in these areas:

- ❖ Under horizontal masonry surfaces, such as roof and parapet or roof chimney
- ❖ At overheads of openings, such as doors and windows
- ❖ At floor lines in exterior construction, install the flashing on top of a metal supporting angle attached to the superstructure or on a projecting concrete shelf or ledge

Wall Base

Any moisture which does enter a wall generally travels downward. Place flashing above grade at the wall base to divert this water to the exterior. To stop ground moisture from soaking upward, especially if there's no basement, place damp checks about 6 inches above grade. In areas where first-floor wood joists require protection from termites, use a metal damp check that projects at least 2 inches past the inside face of the wall and bend it down at an angle, as shown in Figure 15-2.

Base flashing, installed at a little above ground level, also helps keep water from entering a wall. Through-wall flashing is laid in and topped with a soft bed of mortar which is placed directly on the precast or steel lintel of the wall opening.

Window Sills

Place through-wall flashing under and behind all sills. Extend the ends of the sill flashing beyond the jamb line on both sides and turn them up at least 1 inch into the wall. Slope all sills to drain water away from the building. Where the undersides of sills don't slope away from the building, provide a drip notch, or else extend the flashing and bend it down to form a drip, as shown in Figure 15-3. If water runs down windows, over sills and continues down a building face, it's likely to stain the building.

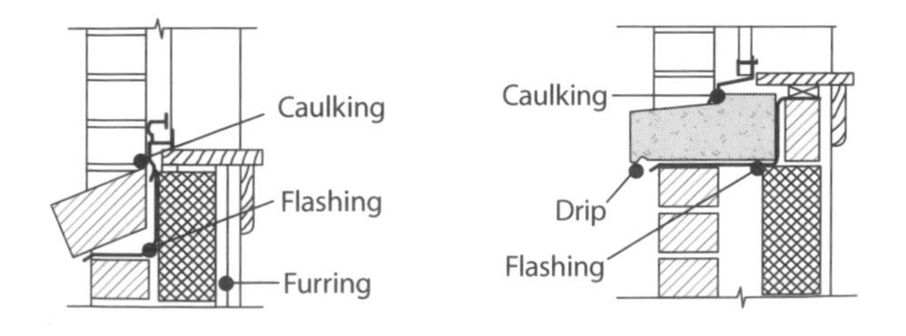

Figure 15-3
Flashing window sills

Opening Heads

Install through flashing over all openings except those completely protected by overhanging projections. At steel lintels, place flashing under and behind the facing material and over the top of the lintel. Bend its outer edge down to form a drip. See Figure 15-4.

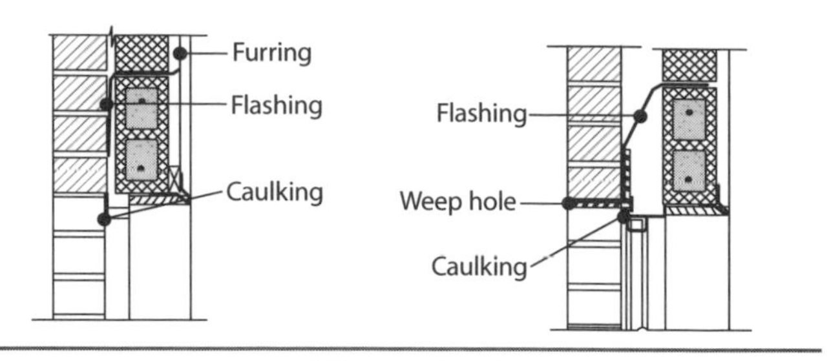

Figure 15-4
Flashing opening heads

Spandrels

In skeleton-frame structures, flash spandrels continuously at beams or with a reglet. When you're flashing the entire spandrel, use two-piece flashings lapped at least 4 inches. When cavity walls are supported by concrete spandrel beams, place flashing on the shelf angle and extend it at least 8 inches up the beam and anchor it into a reglet. If you use galvanized or stainless steel-covered angles, you may not need flashing except to cover between lengths. See Figure 15-5.

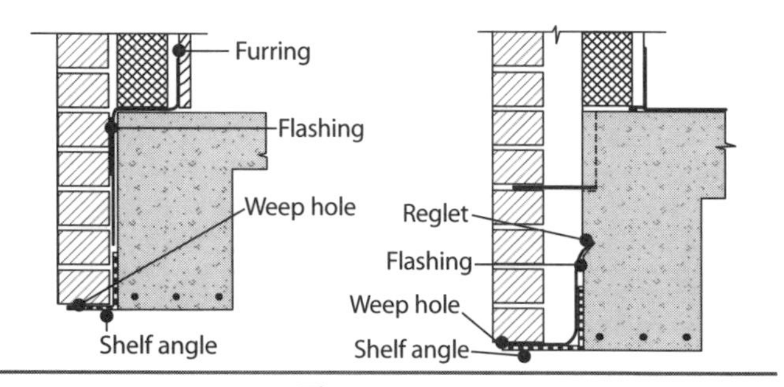

Figure 15-5
Spandrel flashing

Projections and Recesses

Projections and recesses tend to hold rainwater and snow. Slope flashings at such locations away from walls. Make them with a top slope for drainage and, if possible, a drip to keep water away from the wall surface below. Place flashing over the tops of projections and recesses, with the outer edge bent down to form a drip and the back edge turned up at the inner face of the wall. See Figure 15-6. Put a cap or coping at the top of any wall. For a cavity wall, install through-wall flashing beneath the cap or coping.

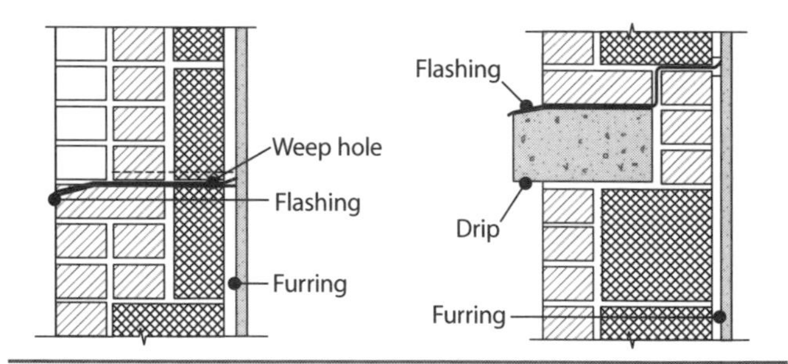

Figure 15-6
Flashing projections and recesses

Roof Flashing

Roof flashing occurs at very vulnerable points, so you need to design and install it with great care. Choose a flashing material based on the type of roofing material. You may have to work with the roofing contractor to make the flashing work with the roof covering. If you use metal base flashing, use metal counterflashing, extended into the wall and overlapping the metal base flashing. Dissimilar metals can fail due to electrolytic action that gradually disintegrates the metal. Check the blueprints to see if the flashing material is specified. If not, ask your supplier for compatible flashing materials you can choose from. At chimney walls, embed counterflashing securely in mortar joints. See Figures 15-7 and 15-8. For more information on chimney flashing, see Chapter 16.

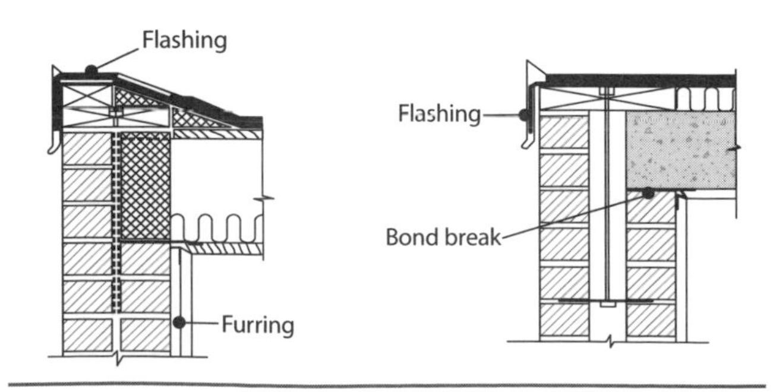

Figure 15-7
Roof flashing

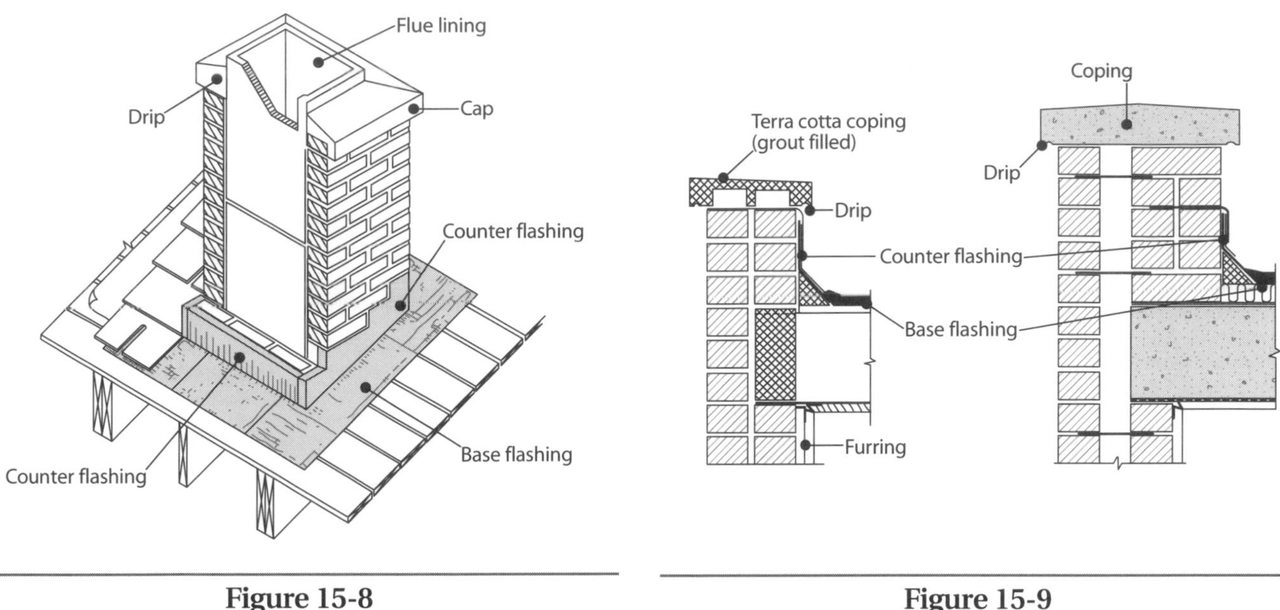

Figure 15-8
Chimney flashing

Figure 15-9
Flashing copings of parapet walls

Parapet Walls

Face both sides of parapet walls using the same durable masonry materials. Using an inferior material on the side not exposed to view is a frequent cause of future trouble. The amount of money you save by doing this is nowhere near worth the amount of trouble it can bring you. Don't paint or coat the backs of parapet walls. Leave them free to dry rapidly. Figure 15-9 shows how to flash copings of parapet walls.

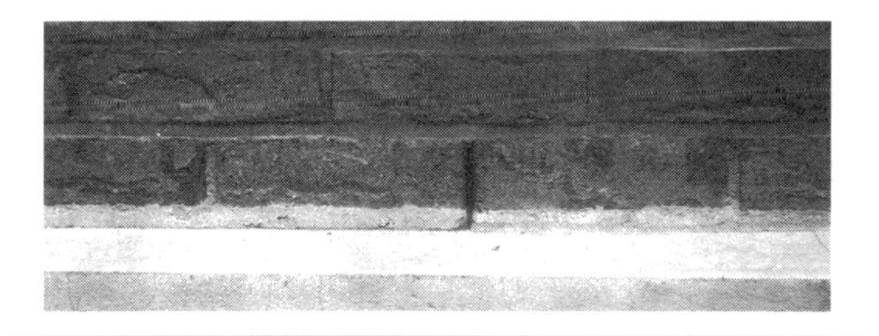

Figure 15-10
Weep hole

Weep Holes

A weep hole is an opening at the level of a flashing that lets moisture escape from inside a wall, as shown in Figure 15-10. You make a weep hole by inserting a forming material into the mortar joint or by omitting all or part of the head joint. If you use a well-oiled rod as a forming material it will leave an unobstructed opening when you take it out. With other forming materials, such as plastic tubes or rope wicks, you leave them in place. But if you put any metal screening, fibrous glass or other material in an open weep hole, be careful. Materials such as metal screening can corrode and stain the masonry.

Generally you need a weep hole every 16 to 24 inches to let the water accumulated on flashing drain off. Put weep holes on 24-inch centers at sills, lintels and base courses. Leave a void at every third head joint on the flashing course, or use some type of rope wick. Install the weep holes or wick immediately above the flashing at all flashing locations. Figure 15-11 shows rope wick material at the top and bottom of a window. Figure 15-12 shows wick material at a base course flashing.

If you have to hide some flashing to improve the look of an area, don't space the weep holes more than 16 inches on center. Concealed flashing with tooled mortar joints and no weep holes can retain water in a wall for longer periods of time. Moisture may concentrate in a low spot spot.

Figure 15-11
Rope wick above and below a window

Figure 15-12
Rope wick at a base course flashing

Controlling Moisture in Brick Veneer and Cavity Walls

Good flashing details are essential for brick veneer construction. Look back to Figures 12-3 and 12-16 in Chapter 12. To divert the moisture out of the air space through the weep holes, install continuous flashing at the bottom of the air space. The flashing must be at or above grade. Where the veneer continues below grade, completely fill the space between the veneer and the existing structure with mortar, grout or concrete to avoid a place for moisture to accumulate. Apply waterproofing on the exterior of the wall down from the grade line to prevent capillary action. Extend the flashing through the face of the brick veneer to form a drip. Where the flashing isn't continuous, such as at heads and sills, turn the ends up about 1 inch.

Locate weep holes in the head joints immediately above all flashings. Don't space the weep holes more than 24 inches on center.

Form weep holes by omitting mortar from all or part of the head joint when you construct the veneer. Or you can get a manufactured weep hole that's just a plastic tube slightly longer than the width of a brick. Insert the tube in the bottom of a weep hole joint to make a more reliable weep hole. Usually you make a weep hole at every third brick.

If you use wick materials in the weep holes or the flashing doesn't extend through the face of the brick veneer, don't space the weep holes more than 16 inches on center.

All flashing must be drained to the outside. Tests at the National Bureau of Standards show that concealed flashings in tooled mortar joints aren't self-draining without weep holes. Rather, they collect moisture. Put weep holes every 24 inches in head joints immediately above all flashings. When building, keep weep holes free of mortar droppings.

Weep holes formed with wick material shouldn't be spaced more than 16 inches on center. If the veneer continues below the flashing at the base of the wall, grout the space between the veneer and the backup to the height of the flashing. Securely fasten the flashing to the backup system and extend it through the face of the brick veneer. Install the flashing carefully so it isn't punctured or torn. Where you need several pieces of flashing to flash a section of the veneer, lap the ends of the flashing and properly seal the joints.

For a cavity wall you need to keep moisture from infiltrating into the cavity. And you want to make the cavity wall drain as well as possible. To accomplish this it's best to place the bottom of a cavity wall above the finished grade. With basement construction, use through-wall flashing at the bottom of the cavity to pre-

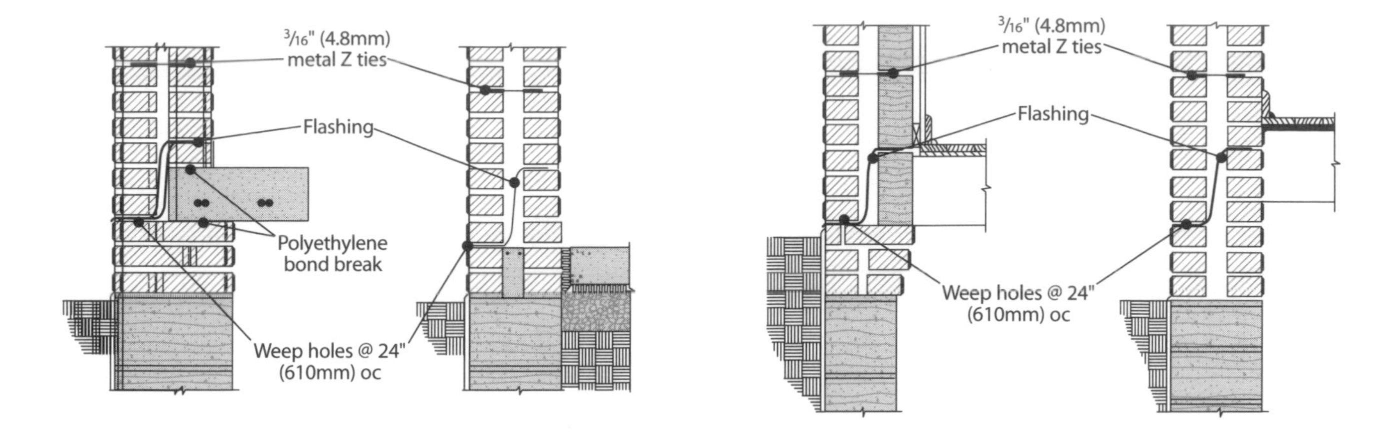

Figure 15-13
Flashing and weep holes in basement walls

vent moisture from getting to the inside surface of the basement wall. See Figure 15-13. Also be careful not to put earth over the weep holes during landscaping.

Put flashing and weep holes above and as near to grade as possible at the bottom of the wall and at all openings so that the wall will drain properly. Locate weep holes in the head joints immediately above all flashings, spaced no more than 24 inches on center. Be careful not to let mortar fall into the cavity wall and block the weep holes from the inside.

Sealants and Caulking

Almost all openings in masonry walls require a sealant or some kind of caulk to keep moisture out. At windows and doorways, use caulking containing a solvent-based acrylic sealant or a butyl caulk composition. In large areas that have expansion joints, use an elastomeric sealant at the joints. Urethanes, silicones and polysulfides are elastomeric sealants. Use Styrofoam rope as a backer rod to complete the application.

16

Chimneys and Fireplaces

Let's begin by taking a look at some details about the important parts of a chimney and how to construct them. Be sure to check all the building code regulations before you actually begin construction, of course.

Flues

A flue is the passage in a chimney that air, gases and smoke travel through. Figure 16-1 shows a typical chimney design. The size (area), shape, height, tightness and smoothness of the inside of a flue determine how effective the chimney is, because they all affect its draft. Overheated or defective flue linings are a major cause of house fires, especially now when so many people use firewood for heat. Most stove manufacturers recommend what size flue to use and give suggestions for safe hookups.

Flue Linings

Years ago chimneys were built without clay flue linings. But now we know that a chimney with a liner is safer and more efficient. Lined flues are definitely recommended for use with brick chimneys because bricks exposed to flue gases will disintegrate. This disintegration can open cracks in the masonry, reducing draft and increasing fire hazard.

Use linings made of vitrified fire clay at least $^5/_8$ inches thick. The ASTM specification Clay Flue Linings C315-78c defines acceptable flue linings as rectangular nonmodular, rectangular modular, round or oval. A round flue lining is the most efficient, but rectangular flues fit better in brick construction.

To line a flue, place each length of lining in position. Use fire clay cement to join the liner sections because regular mortar may disintegrate under the high temperatures generated in the flue. Then strike the inside of the liner joint smooth. In chimney block construction, lay the blocks one or two ahead of the tile. In brick construction, lay the lining first and then lay the brick around it. Placing the lining after you lay the brick can cause an air leak — a potential fire hazard. In masonry chimney construction with walls less than 8 inches thick, leave a space between the lining and the chimney walls. Don't fill the space solid with mortar. Use only enough mortar to make good joints and hold the lining in position.

Some building codes now specify solid chimney blocks instead of hollow core blocks. Both are shown in Figure 16-2. In wood-frame construction, the codes also require you to tie a chimney to a structure with metal wall ties, which are ribbed to hold well in the mortar. Embed one end of the tie in the mortar at

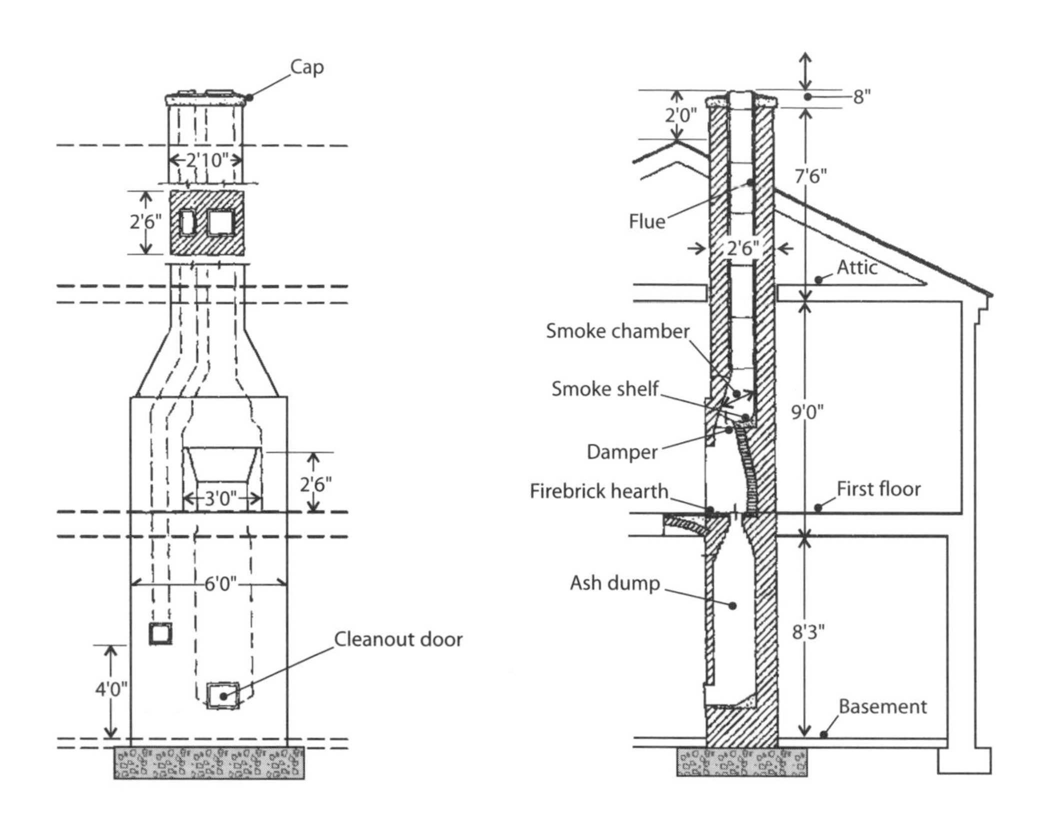

Figure 16-1
Typical chimney design

A Solid chimney blocks

B Hollow core chimney blocks

Figure 16-2
Typical chimney blocks

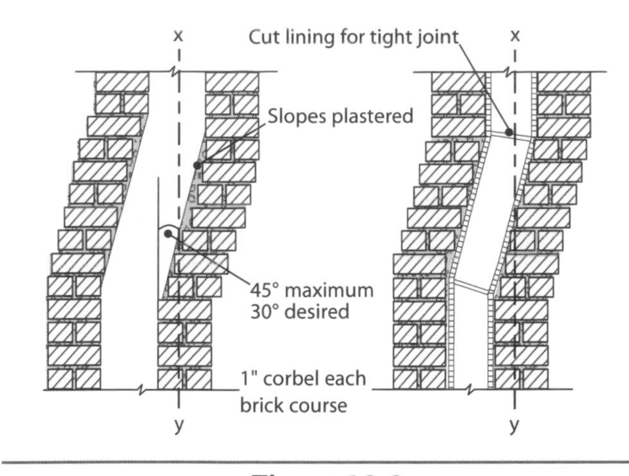

Figure 16-3
Offsetting a chimney

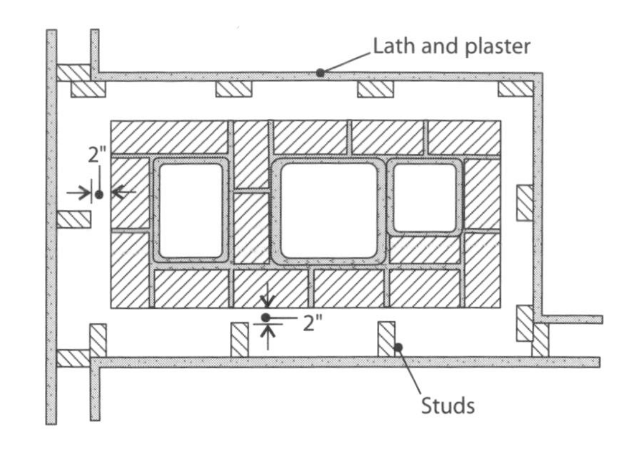

Figure 16-4
Multiple flue arrangement

each joint, and nail the other end to the building exterior. Use nails made of the same material as the ties to prevent chemical reactions and failure. Check with your local code officials for the minimum distance between the chimney and the wood frame.

Unless it rests on solid masonry at the bottom of the flue, support the lower section of lining on at least three sides by a brick corbeled to the inside of the chimney.

Always place flues as plumb as possible. If a projection makes this impossible, try not to make the angle of curve around the projection more than 30 degrees. See Figure 16-3. Sharp turns in a flue lining set up eddies which can affect the motion of gas and smoke.

Where a flue changes directions, make the lining joints as tight as possible by mitering the ends of the tiles. Cut the lining before it's built into the chimney. If you cut it afterwards, it might break and fall into the chimney. Getting it out can be quite a job. A diamond blade is the best way to cut a clay flue lining, but a small cutoff blade and an electric handsaw will work. Another way to cut a lining is to fill the tile full of damp sand. Then tap around the tile with a sharp chisel until it breaks. The compacted sand in the tile will absorb some of the shock, and help protect the tile. Because flue tiles are very brittle, it takes a lot of skill to cut a tile with a hammer and chisel.

As you lay tile in either a brick or a block chimney, clean the inside of the chimney as the work progresses. You want to remove any extra material and keep the inside as smooth as possible.

A chimney can contain more than one flue. If you're building one of these, check the building code carefully. Building codes generally require a separate flue for each fireplace, furnace or boiler. If a chimney contains three or more lined flues, separate each group of two flues from the other single flue or group of two flues by a brick division. Figure 16-4 shows a multiple flue arrangement. If two flues are grouped together with no dividing wall, stagger the lining joints at least 7 inches and completely fill the joints with mortar. If the chimney has two or more unlined flues, separate them with a well-bonded wythe at least 8 inches thick.

Don't connect any range, stove, fireplace or other equipment to the flue for the central heating unit. Connect each unit to a separate flue. Sparks passing into one flue opening and out through another connection to the same flue may cause a fire.

Chimney Construction

A chimney should extend about 36 inches above a flat roof and about 24 inches above the peak of a pitched roof. If the chimney isn't near the peak, it

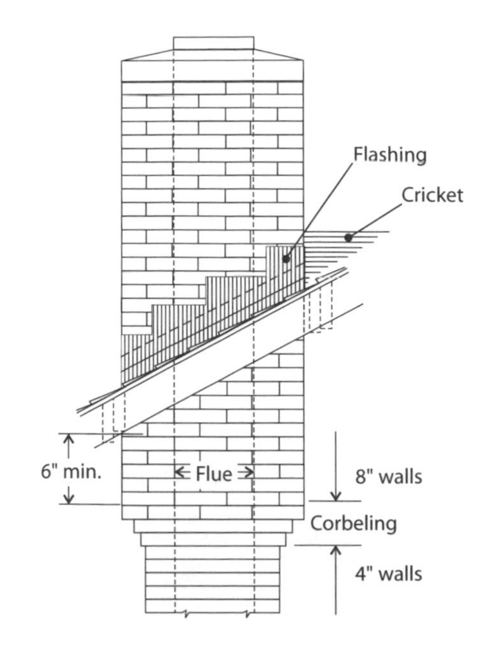

Figure 16-5
Corbeling a chimney to make the outside walls 8 inches thick

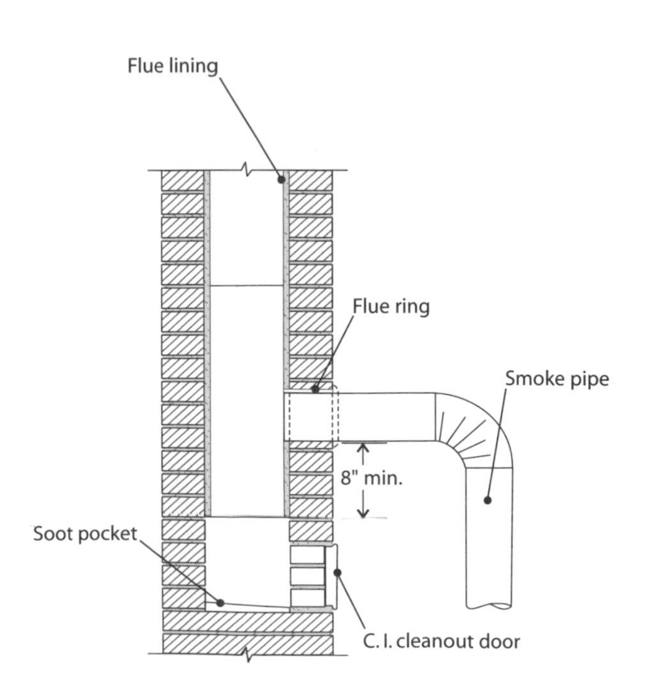

Figure 16-6
Making a soot pocket and cleanout

should be at least 24 inches above any part of the roof within 10 feet of the chimney. Install a cap or hood if you can't build a chimney high enough above the ridge of a roof to keep wind from getting into the chimney. The open end of the hood or cap should be parallel with the ridge of the roof.

Support

A chimney is usually the heaviest part of a building. Always build it on a solid foundation. Use concrete footings wide enough to distribute the load of the chimney so the weight of the chimney doesn't exceed the safe load-bearing capacity of the soil. Make footings at least 12 inches thick and about 8 inches past the outer edges of a chimney on all sides.

Walls

When you're building a brick flue-lined chimney less than 30 feet high, the walls should be at least 4 inches thick. For a stone chimney less than 30 feet, make the walls at least 12 inches thick. The outside wall of any chimney exposed to the elements should be at least 8 inches thick. Brick chimneys that extend up through a roof may sway enough in heavy winds to open up mortar joints at the roof line. Openings to the flue at this point are dangerous because sparks from the flue may start fires in the woodwork or roofing. It's good practice to make the chimney wider through corbeling. Make the upper walls 8 inches thick by starting to offset the bricks at least 6 inches below the underside of roof joists or rafters, as shown in Figure 16-5.

Soot Pocket and Cleanout

Make a soot pocket and a cleanout for each flue. Deep soot pockets can accumulate soot, which may catch fire, so make the pocket just deep enough to install a cleanout door below the smoke pipe connection. Fill the lower part of the chimney — from the bottom of the soot pocket to the base of the chimney — with solid masonry. Use a cleanout door made of cast iron. Make sure it fits snugly and closes tightly to keep air out. Figure 16-6 shows a typical soot pocket and cleanout.

Smoke Pipe Connection

Make sure the smoke pipe enters the chimney horizontally, as shown in Figure 16-6 and that it doesn't extend into the flue. Line the hole in the chimney wall with fire clay, or build thimbles tightly into the masonry. A thimble is the round clay tile or metal pipe that extends from the stove pipe to the interior of the chimney at the flue. I recommend using clay thimbles for the entrance to the flue liner. Some metal stove pipe can react with wet mortar, causing it to disintegrate over time. Usually the stove pipe will fit inside the thimble, which should protrude slightly from the outer edge of the chimney.

Set the thimble (made of the same material as the flue) level in the flue. Enclose the area surrounding the thimble with brick and fire clay to cement the thimble to the flue lining, as shown in Figure 16-7. To be extra safe, lay fire brick in fire clay around the thimble.

Figure 16-8 shows the inside of the flue lining with the thimble cemented to the lining with fire clay. Note how the thimble doesn't extend into the chimney. Smooth the fire clay around the thimble for more efficient operation and a better seal.

Many building-supply dealers sell flue tile that already has the hole cut for the thimble as shown in Figure 16-9. This is a great time and money saver, but it does require more planning. You have to make sure the flue tile will be at the right height for the thimble. Drill a hole through the wall at the center of the thimble and measure down from this point to get the correct alignment.

When building a block chimney in an existing structure, you'll have to cut through the soffit and fascia on the edge of a roof. It's quick and easy to use a chain saw for this, but it's also dangerous. I recommend using a reciprocating saw. See Figure 16-10. Never cut with the saw higher than your head. Draw plumb lines on the siding and cut the hole just before you lay the blocks.

Footings

When you build a new foundation, try to pour the footing for the fireplace or chimney at the same time as the foundation (Figure 16-11). Make the footing

Figure 16-7
Set the thimble level into the flue using fire brick and fire clay

Figure 16-8
Inside of flue lining showing thimble entrance

Figure 16-9
Flue tile with precut thimble hole

Figure 16-10
The reciprocating saw is an excellent tool for chimney construction

Figure 16-11
Pour the chimney or fireplace footing at the same time as the foundation

about 12 inches thick and reinforce it. If a building has no basement, pour the footing for the chimney below the frost line. Get information on the frost line from a Cooperative Extension Office of the Department of Agriculture.

If a house is built of solid masonry (walls at least 12 inches thick), you can build a chimney integrally with the wall. Instead of carrying the chimney to the ground, use corbeling to offset it enough from the wall to make space for the flue. Don't extend the offset more than 6 inches from the face of the wall. Make sure each course projects not more than 1 inch, and don't build the corbel more than 12 inches high. You can only do this on a single-story house.

In a frame building, build a chimney from the ground up, or rest it on the building foundation. If the basement walls are at least 12 inches thick and have adequate footings, you can use them as a base.

Connecting with the Roof

Where a chimney passes through a roof, make a 2-inch clearance between the wood framing and the chimney for fire protection. This will also leave room for expansion due to temperature changes, settlement and slight movement during heavy winds.

Flash and counterflash chimneys to make the junction with the roof watertight, as shown in Figures 16-12 and 16-13. Begin with the base flashing at the downslope side of the chimney. Make the flashing 12

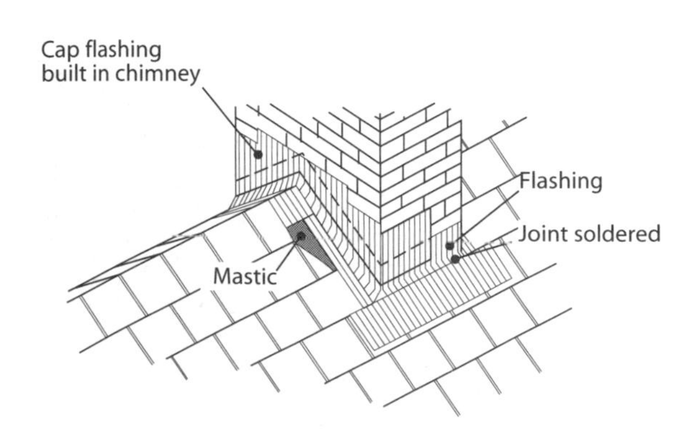

Figure 16-12
Flashing a chimney on a roof ridge

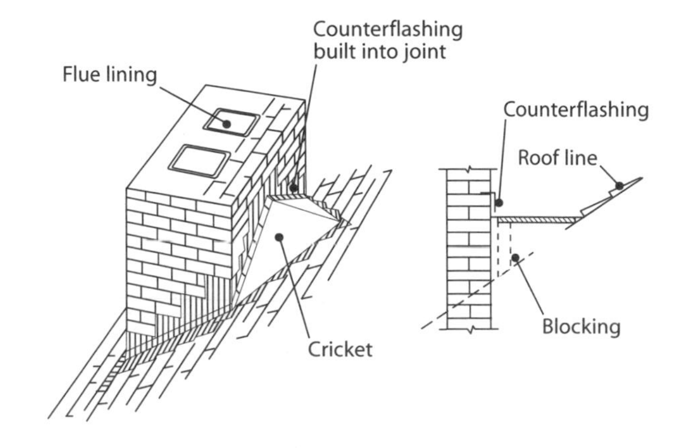

Figure 16-13
Counterflashing and cricket behind a chimney

inches longer than the width of the chimney and at least 24 inches wide. Cut and install the base flashing as shown in Figure 16-14.

Next, cut pieces of step flashing 10 inches square. Bend them more than 90 degrees so they'll press against the chimney. Cut and fold the first piece of flashing so it wraps 1 to 2 inches around the front of the chimney. Work up the chimney, lapping each piece of flashing at least 2 inches over the one below. Fasten the flashing with $1^1/_2$-inch galvanized nails. See Figure 16-15.

When you reach the high side of the chimney, install another base flashing. Again, cut it 24 inches longer than the width of the chimney, and 12 to 24 inches wide. Cut the flashing to fit around the chimney, as shown in Figure 16-16.

Embed 10-inch-wide counterflashing about 1 inch into a mortar joint in the chimney. If you use aluminum flashing which reacts with chemicals in wet mortar, cut in the flashing after the chimney is built and seal it with caulking, as shown in Figure 16-17.

When a chimney is located on the slope of a roof, build a cricket (or saddle) high enough to shed water around the chimney. Also install a water diverter on chimneys on the edge of the roof line. In snow- and ice-prone areas of the country, the cricket will help prevent buildup behind the chimney during freeze-thaw cycles. See Figure 16-18.

To build the cricket, begin by cutting a 2 x 4 to match the roof slope and tack it at the center of the chimney. See Figure 16-19. Cut plywood triangles to

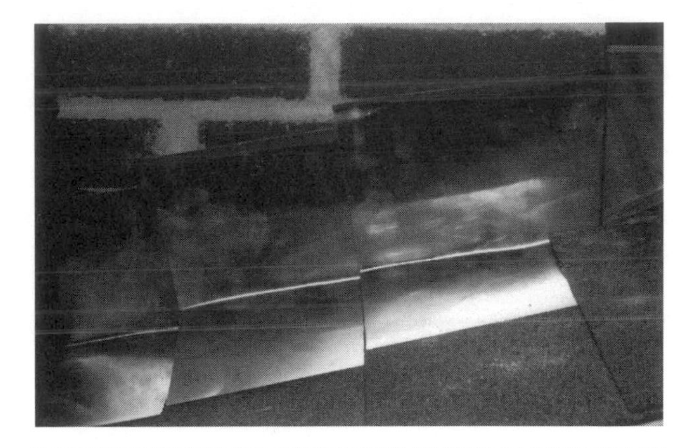

Figure 16-15
Install step flashing up the slope

Figure 16-16
Installing base flashing on high side of chimney

Figure 16-14
Installing the base flashing

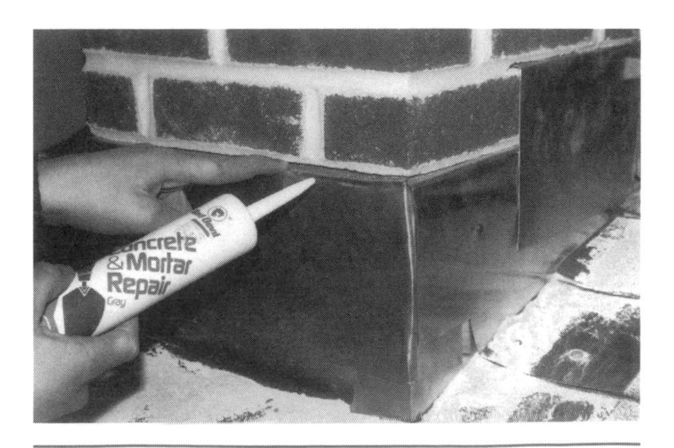

Figure 16-17
Install and caulk the cap flashing

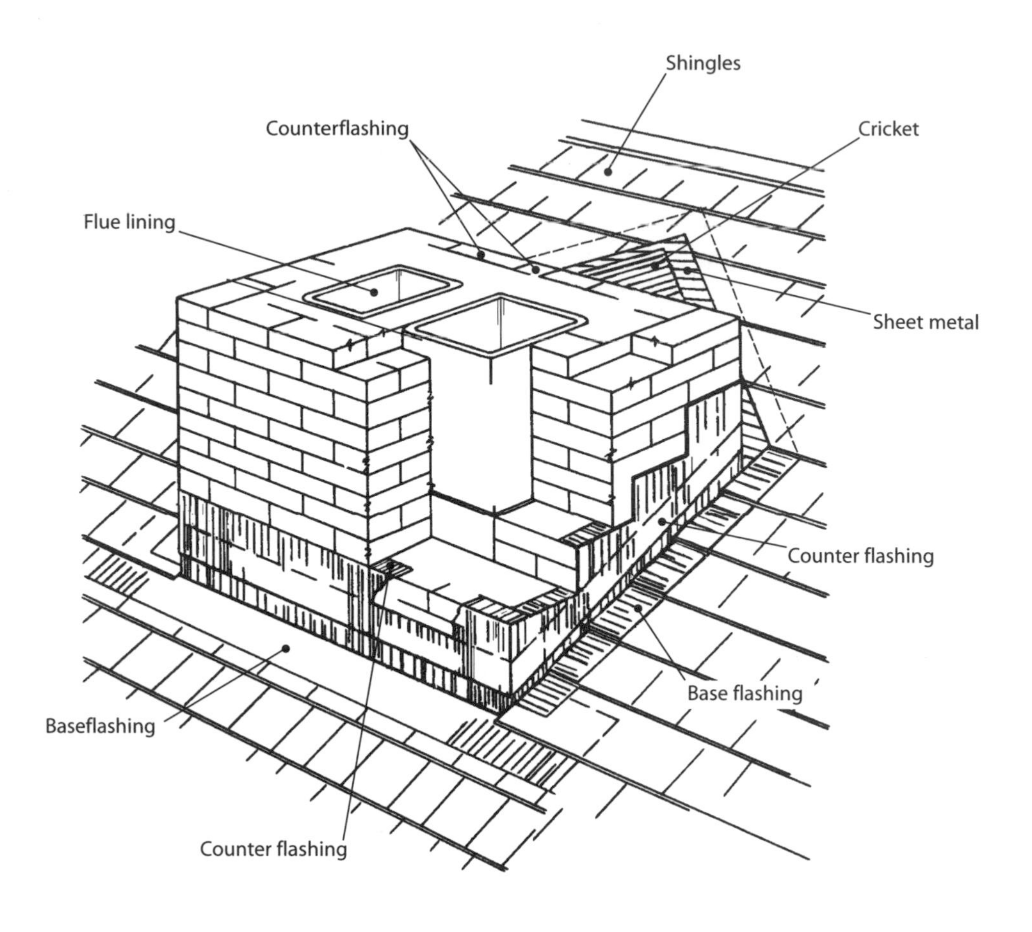

Figure 16-18
Flashing a chimney on a sloped roof

make the top of the cricket (Figure 16-20). Cover the cricket with flashing cut from the pattern, like the one in Figure 16-21. Use corrosion-resistant metal, such as copper, zinc or lead for flashing. Galvanized or tinned sheet steel isn't as good because it has to be painted occasionally. Make sure the sheet metal extends at least 4 inches under the shingles, and is counter-flashed at the joint. Lap the cap flashing over the base flashing to provide waterproof construction. Provide a full bed of mortar where cap flashing is inserted in the joints. Figure 16-22 shows a finished cricket.

Finishing the Top of a Chimney

Figure 16-23 shows three good ways to finish the top of a chimney. In the first illustration, a concrete wash is sloped to shed water. Extend the flue lining at least 4 inches above the cap or top course of brick and surround it with at least 2 inches of cement mortar. Finish the mortar with a straight or concave slope to direct air currents upward at the top of the flue and to drain water from the top of the chimney.

Another option is to use a hood to keep rain out of the chimney and to prevent downdrafts caused by nearby buildings, trees or other tall objects. Common types are the arched brick hood and the flat stone or cast-concrete cap. If a hood covers more than one flue, divide it by wythes so that each flue has a separate section. The area of the hood opening for each flue must be larger than the area of the flue.

A spark arrester won't entirely eliminate the discharge of sparks, but if you build and install it properly, it'll greatly reduce the hazard. Your building code may require a spark arrester, so be sure to check on that. A spark arrester should be made of rust-resistant

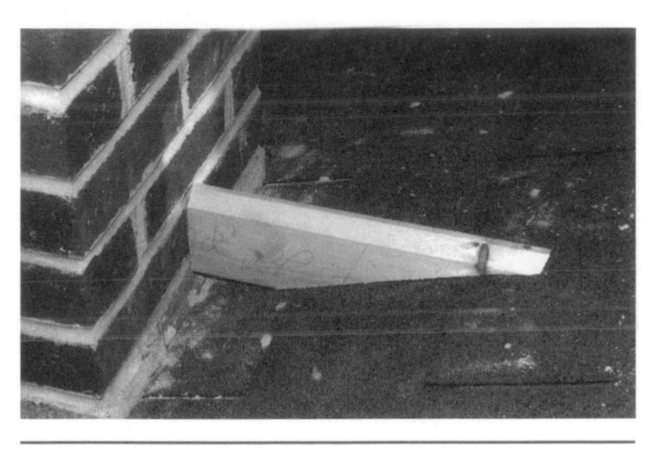

Figure 16-19
First, place a 2 x 4 brace for the cricket

Figure 16-20
Put plywood triangles in place to form cricket

material with openings not larger than 5/8 inch, nor smaller than 5/16 inch. Enclose the flue discharge area completely and fasten the spark arrester securely to the top of the chimney.

Fireplaces

As a mason you probably won't have to do much in the way of fireplace design. But if one that you build doesn't work right, or looks bad, it's been known for the designer to let the builder take the blame. Right or wrong, you're going to get bloodied. So here are a few basics on the subject:

- ❖ A fireplace should harmonize in detail and proportion with the room it's in.
- ❖ Don't put a fireplace near a door.
- ❖ Make a second-floor fireplace smaller than you would a first-floor one because of the reduced flue height.

You can make a fireplace opening from about 2 to 6 feet wide. You might want to consider the size of the wood that will be burned in the fireplace. For example, for cordwood (4 feet long) cut in half, an opening 30 inches wide is pretty good. Make the fireplace opening 18 inches high for an opening 2 feet wide, and up to 28 inches high for a fireplace up to 6 feet wide. The higher the opening, the more chance of a smoky fireplace. In general, the wider the opening, the greater the depth. A shallow opening throws out

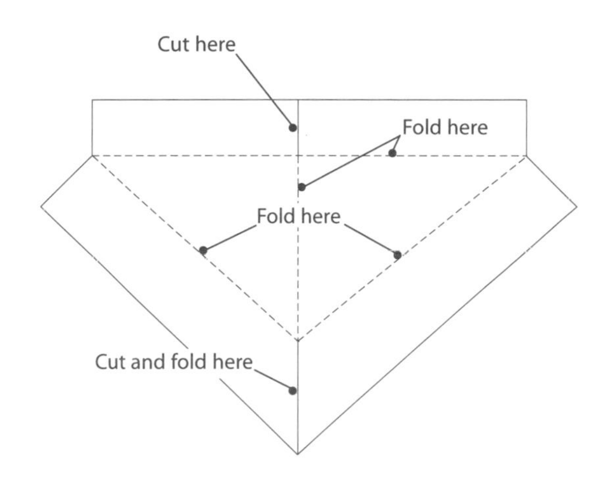

Figure 16-21
Typical pattern for cutting flashing

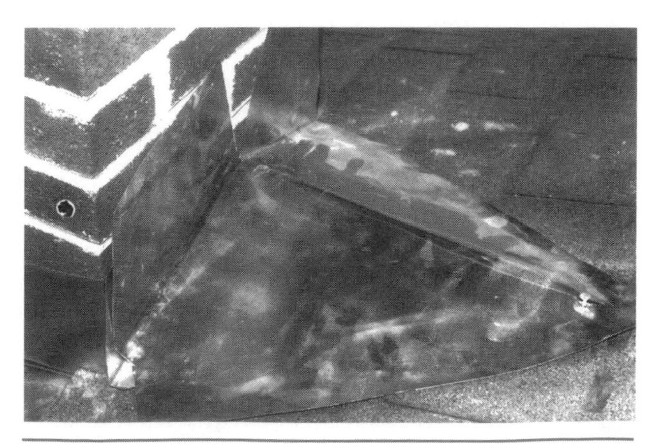

Figure 16-22
Cover plywood with metal flashing

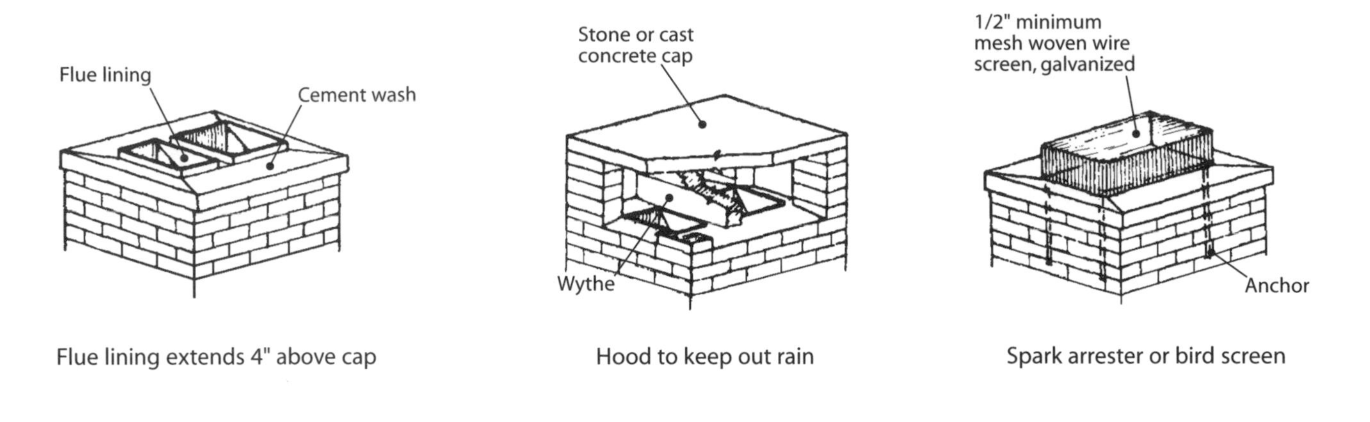

Figure 16-23
Building the top of a chimney

relatively more heat than a deep one, but it holds smaller pieces of wood. In a small fireplace, a depth of 12 inches may make good draft, but I recommend a minimum depth of 16 inches to lessen the danger of sparks falling onto the floor.

Building a Fireplace

Basically there's just one way to build a fireplace, regardless of its design. Figure 16-24 shows construction details of a typical fireplace. Figure 16-25 gives the recommended dimensions for essential parts or areas of various size fireplaces.

Footings

Foundation and footing construction for chimneys with fireplaces is similar to that for chimneys without fireplaces. Always make certain footings rest on firm ground, since chimneys are very heavy.

Hearth

Make the fireplace hearth of brick, stone, terra cotta or reinforced concrete at least 4 inches thick. It should project at least 20 inches from the top of the opening of the fireplace and be 24 inches wider than the fireplace opening (12 inches on each side). If there's a basement, it's a good idea to build an ash dump under the back of the hearth, as shown in Figure 16-26.

In buildings with wooden floors, support the hearth in front of a fireplace with a masonry trimmer arch or other fire-resistant construction. Look at the section showing the alternative hearth in Figure 16-24. Make sure you remove wood centering under the arches used during construction when construction is completed. Figure 16-27 shows the recommended method of installing the floor around the hearth.

Walls

Building codes generally require that you make the back and sides of fireplaces of solid masonry or reinforced concrete at least 8 inches thick, and lined with firebrick, approved noncombustible material at least 2 inches thick, or steel lining at least $^{1}/_{4}$ inch thick. You don't need this lining if you make the walls of solid masonry or reinforced concrete at least 12 inches thick.

Jambs

For a fireplace opening 3 feet wide or less, make the jambs 12 inches wide for a wood mantle or 16 inches wide if the jambs are exposed masonry. For wider

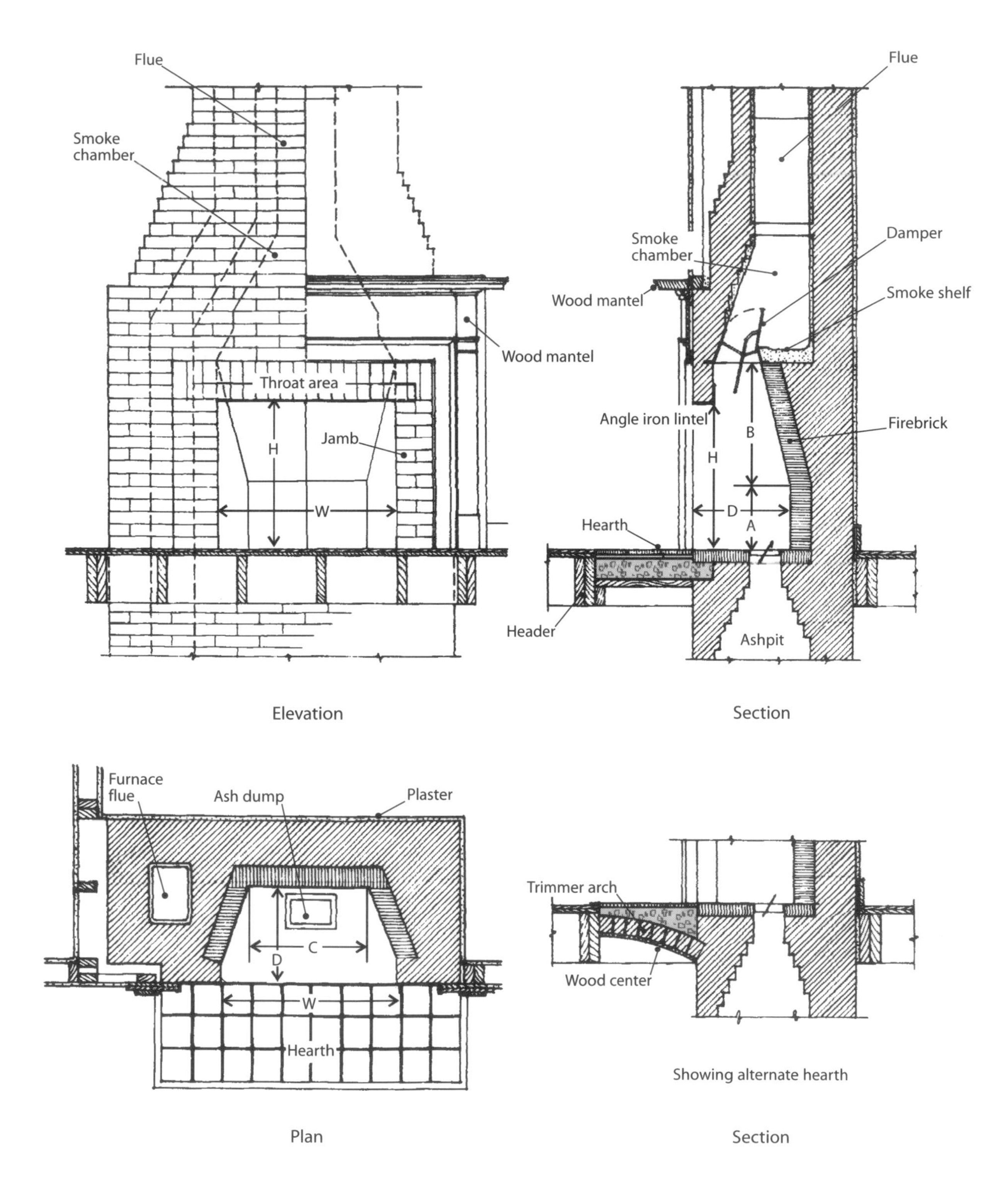

Figure 16-24
Construction details of a typical fireplace

Size of fireplace opening						Required flue lining size	
Width W (in.)	Height H (in.)	Depth D (in.)	Minimum width of back wall C (in.)	Height of vertical back wall A (in.)	Height of inclined back wall B (in.)	Standard rect. outside dimensions (in.)	Standard round inside diameter (in.)
24	24	16-18	14	14	16	$8^1/_2$ x $8^1/_2$	10
28	24	16-18	14	14	16	$8^1/_2$ x $8^1/_2$	10
30	28-30	16-18	16	14	18	$8^1/_2$ x 13	10
36	28-30	16-18	22	14	18	$8^1/_2$ x 13	12
42	28-32	16-18	28	14	18	13 x 13	12
48	32	18-20	32	14	24	13 x 13	15
54	36	18-20	36	14	28	13 x 18	15
60	36	18-20	44	14	28	13 x 18	15
54	40	20-22	36	17	29	13 x 18	15
60	40	20-22	42	17	30	18 x 18	18
66	40	20-22	44	17	30	18 x 18	18
72	40	22-28	51	17	30	18 x 18	18

Figure 16-25
Recommended fireplace dimensions and required flue lining sizes

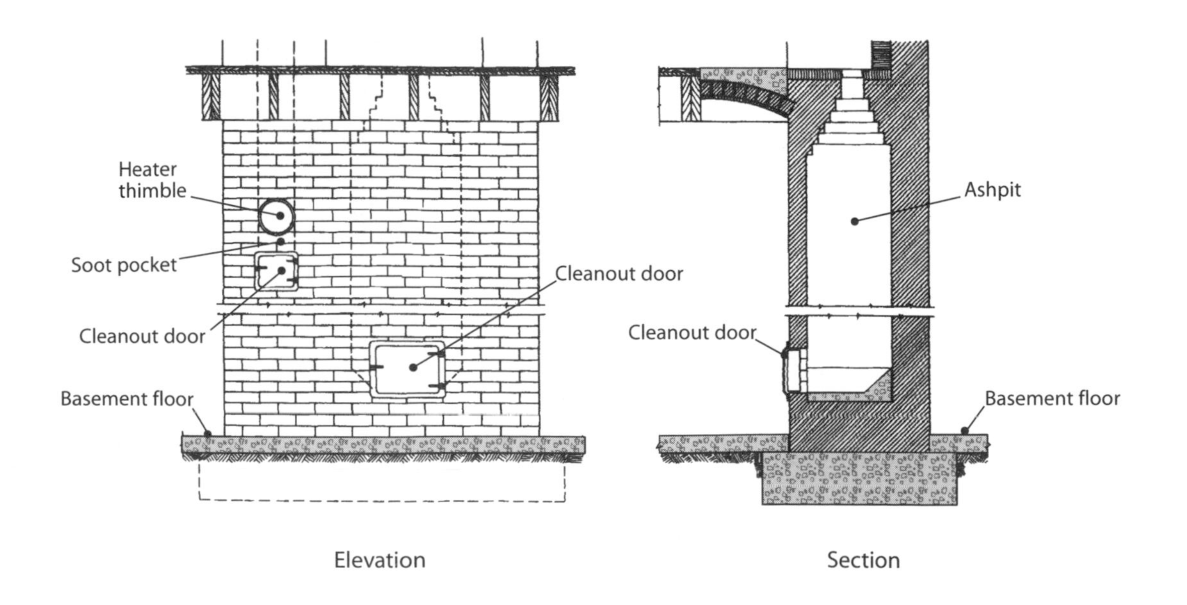

Figure 16-26
Construction details of a typical ashpit

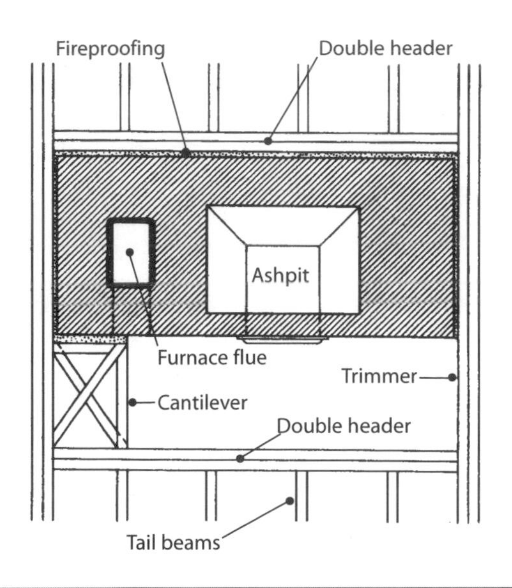

Figure 16-27
Building a floor around a chimney and hearth

fireplace openings or a fireplace in a large room, make the jambs proportionately wider. Often you'll want to face fireplace jambs with ornamental brick or tile.

Don't put any woodwork within 6 inches of the fireplace opening. Place any woodwork above or projecting more than 1½ inches from a fireplace at least 12 inches from the top of the fireplace opening.

Lintel

Install a lintel across the top of the fireplace opening to support the masonry. For fireplace openings 4 feet wide or less, use ½ by 3 inch flat steel bars, 3½ by 3½ by ¼ inch angle irons, or specially-designed damper frames. Wider openings require heavier lintels. If you use a masonry arch over the opening, make sure the fireplace jambs are heavy enough to resist the thrust of the arch.

Throat

To build a good fireplace you must pay attention to its throat area. The sides of the fireplace must be vertical up to the throat, which should be 6 to 8 inches or more above the bottom of the lintel. The area of the throat must be at least as big as the flue. The length must be equal to the width of the fireplace opening, and that width depends on the width of the damper frame (if you've installed a damper). Five inches above the throat, start sloping the side walls inward to meet the flue. See Figure 16-24.

Damper

A damper is a cast iron frame with a hinged lid that opens or closes the throat opening. Ask the building code officer if a damper is required. Even if it's not, I definitely recommend you install one, especially in cold climates. Dampers control the draft and shut off the opening when the fireplace isn't in use.

Some dampers are designed to support the masonry over fireplace openings, replacing ordinary lintels. Consult a good manufacturer of fireplace equipment for help when you pick a damper for a fireplace. Make sure the full-damper opening is the same as the area of the flue.

Smoke Shelf and Chamber

A smoke shelf prevents downdraft. To make a smoke shelf, set the brickwork at the top of the throat back to the line of the flue wall for the full length of the throat. Make the shelf 6 inches or deeper, depending on the depth of the fireplace. The smoke chamber is the area from the top of the throat to the bottom of the flue. See Figure 16-24. Plaster the smoke shelf and the smoke chamber with cement mortar at least ½ inch thick.

Fireplace Flue

The relationship between the area of a fireplace opening and the area and height of its flue must be perfect if the fireplace is to function properly. For a lined flue 22 feet high, make sure the area of the flue is at least 1/12 the area of the fireplace opening. For an unlined flue or a flue less than 22 feet high, make the area of the flue 1/10 the area of the fireplace opening. Figure 16-25 lists the dimensions of fireplace openings shown in the construction details in Figure 16-24. The last two columns tell you what size flue you need for each dimension. Use this table to figure what size lining you need for a fireplace opening or what size opening to use with an existing flue. Figures 16-28, 16-29, 16-30 and 16-31 show different sizes of flue liners and their dimensions.

Outside dimension in. (mm)	Nominal wall thickness in. (mm)	Outside corner radius maximum, in. (mm)
4 1/2 x 8 1/2 (115 x 215)	5/8 (16)	1 (25)
4 1/2 x 13 (115 x 330)	3/4 (19)	1 (25)
8 1/2 x 8 1/2 (215 x 215)	3/4 (19)	2 (50)
8 1/2 x 13 (215 x 330)	7/8 (23)	2 (50)
8 1/2 x 17 1/4 (215 x 450)	1(25)	2 (50)
13 x 13 (330 x 330)	7/8 (23)	3 (75)
13 x 17 3/4 (330 x 450)	1 (25)	4 (100)
17 3/4 x 17 3/4 (450 x 450)	1 1/4 (32)	4 (100)
20 x 20 (500 x 500)	1 3/8 (35)	5 (125)
20 x 24 (500 x 600)	1 1/2 (38)	5 (125)
20 x 24 (600 x 600)	1 5/8 (41)	6 (150)

Figure 16-28
Standard dimensions of rectangular nonmodular clay flue linings

Outside dimensions in. (mm)	Nominal dimensions in. (mm)	Nominal wall thickness in. (mm)	Outside corner radius maximum, in. (mm)
3 1/2 x 7 1/2 (90 x 190)	4 x 8 (100 x 200)	5/8 (16)	1 (25)
3 1/2 x 11 1/2 (90 x 290)	4 x 12 (100 x 300)	5/8 (16)	1 (25)
7 1/2 x 7 1/2 (190 x 190)	8 x 8 (200 x 200)	5/8 (16)	2 (50)
7 1/2 x 11 1/2 (190 x 290)	8 x 12 (200 x 300)	3/4 (19)	2 (50)
11 1/2 x 11 1/2 (290 x 290)	12 x 12 (300 x 300)	3/4 (19)	2 (50)
11 1/2 x 15 1/2 (290 x 390)	12 x 16 (300 x 400)	1 (25)	3 (75)
15 1/2 x 15 1/2 (390 x 390)	16 x 16 (400 x 400)	1 1/8 (29)	4 (100)
15 1/2 x 19 1/2 (390 x 490)	16 x 20 (400 x 500)	1 1/4 (32)	4 (100)
19 1/2 x 19 1/2 (490 x 490)	20 x 20 (500 x 500)	1 3/8 (35)	5 (125)
19 1/2 x 23 1/2 (490 x 590)	20 x 24 (500 x 600)	1 1/2 (38)	5 (125)
23 1/2 x 23 1/2 (590 x 590)	24 x 24 (600 x 600)	1 5/8 (41)	6 (150)

Figure 16-29
Standard dimensions of rectangular modular clay flue linings

Nominal inside diameter in. (mm)	Permissible variation in inside diameter ± in. (±mm)	Nominal wall thickness in. (mm)
6 (150)	$\frac{1}{4}$ (6)	$\frac{5}{8}$ (16)
8 (200)	$\frac{1}{4}$ (6)	$\frac{3}{4}$ (19)
10 (250)	$\frac{5}{16}$ (8)	$\frac{7}{8}$ (23)
12 (300)	$\frac{3}{8}$ (10)	1 (25)
15 (375)	$\frac{3}{8}$ (10)	$1\frac{1}{8}$ (29)
18 (450)	$\frac{7}{16}$ (11)	$1\frac{1}{4}$ (32)
21 (525)	$\frac{7}{16}$ (11)	$1\frac{5}{8}$ (41)
24 (600)	$\frac{1}{2}$ (13)	$1\frac{5}{8}$ (41)
27 (675)	$\frac{9}{16}$ (14)	2 (50)
30 (750)	$\frac{5}{8}$ (16)	$2\frac{1}{8}$ (54)
33 (825)	$\frac{11}{16}$ (17)	$2\frac{1}{4}$ (57)
36 (900)	$1\frac{1}{4}$ (32)	$2\frac{1}{2}$ (64)

Figure 16-30

Standard dimensions of round clay flue linings

Outside dimensions in. (mm)	Nominal wall thickness in. (mm)	Nominal outside corner radius in. (mm)
$8\frac{1}{2}$ round, (215) round	$\frac{3}{4}$ (19)	$4\frac{1}{4}$ (105)
$12\frac{3}{4}$ round, (325) round	1 (25)	$6\frac{3}{8}$ (160)
$8\frac{1}{2}$ x $12\frac{3}{4}$ (215 x 325)	$\frac{3}{4}$ (19)	$4\frac{1}{4}$ (105)
$8\frac{1}{2}$ x $16\frac{3}{4}$ (215 x 425)	1 (25)	$4\frac{1}{4}$ (105)
10 x $17\frac{3}{4}$ (250 x 450)	1 (25)	5 (125)
$12\frac{3}{4}$ x $16\frac{3}{4}$ (325 x 425)	1 (25)	$6\frac{3}{8}$ (160)
$12\frac{3}{4}$ x 21 (325 x 525)	$1\frac{1}{8}$ (29)	$6\frac{3}{8}$ (160)
$16\frac{3}{4}$ x $16\frac{3}{4}$ (425 x 425)	1 (25)	$6\frac{3}{8}$ (160)
$16\frac{3}{4}$ x 21 (425 x 525)	$1\frac{3}{16}$ (30)	$6\frac{3}{8}$ (160)
21 x 21 (525 x 525)	$1\frac{1}{4}$ (32)	$6\frac{3}{8}$ (160)

Figure 16-31

Standard dimensions of oval clay flue linings

Air Supply

You have several options for laying out and building the air passageway and ashpit. Since new buildings are designed to minimize air infiltration, the fireplace may have trouble getting enough air for efficient combustion. A large fireplace may deplete the oxygen in the building. Some codes require a method of directing outside air to the fireplace. Check the code in your area before starting construction.

An exterior air system makes a fireplace work better by cutting down the amount of air drawn from the building for combustion and draft. The air source can be either outside the building, or an unheated area of the building such as a crawl space.

There are several products on the market that you can install during construction to provide the outside air needed for combustion. Check with your masonry supplier. One product takes the place of one modular 4 x 8 brick. You just lay it in the wall and build a channel to the fireplace with galvanized heat pipe to form an air passage. Your supplier should be able to advise you on the best size to use for the size of the fireplace.

When the fireplace design doesn't allow for an air passageway in the base, you can put the intake on any exterior wall. No matter where you put it, provide the intake with a screen-backed, closeable louver. If possible, make the louver operable from inside the building. Check your local building codes for restrictions on air intake locations. For instance, many codes prohibit air from a garage to be vented into habitable space.

It's best to incorporate an air passageway into the base assembly. Sometimes that's not practical due to the fireplace configuration or the outside ground level at the fireplace. In that case, you might be able to change the fireplace design by raising the hearth. Or you can make the passageway of ductwork attached to, or incorporated into, the floor. Insulate the air passageway, especially if it adjoins or passes through heated areas.

The best place for the air inlet is in the base near the front of the combustion chamber so the fire burns evenly and toward the rear of the chamber. You can also install it in the side walls or even the rear wall, but there's a danger that a surge of air will force smoke and gases into the room. Equip the inlet with both directional and volume controls.

It's important to take time to build any fireplace carefully, both for efficiency and aesthetics. The fireplace must be designed for the room size, with the correct size opening and flue. And since it will probably be the focal point of a living room, make it pleasing to the eye.

17

Brick Floors and Pavements

You can use brick to make attractive, long-lasting floors and pavements. But you do have to be careful to work out a design that's practical and effective when you select materials.

Selecting the Right Brick for the Job

When you select bricks, there are four factors to consider:

1) Traffic on the floor or paving
2) Exposure to moisture and freezing cycles
3) Resistance to chemicals and acids (often a major problem for industrial paving)
4) How you want the finished product to look

Vitrified Clay Brick

There are two types of paving brick. One is vitrified clay brick that's kiln dried. Vitrified clay bricks for paving are covered by ASTM C410 and C902. They're naturally durable and abrasion resistant, as well as dense, low absorbing, and easy to clean and maintain. I recommend using ASTM 62 SW (severe weathering) brick for paving. You can use MW (moderate weathering) indoors where there's no moisture present. Make sure you know whether you're getting extruded or molded brick. It's important to know the compressive strength and the water absorption rate for pavers you'll use outdoors. Brick manufacturers and their distributors can help you select the right brick for the job. Look for:

- Extruded brick — Minimum average compressive strength of 8000 psi, maximum average cold-water absorption of 8 percent, and maximum saturation coefficient of 0.78
- Molded brick — Minimum average compressive strength of 4500 psi, maximum average cold-water absorption of 8 percent, and a maximum saturation coefficient of 0.78

These minimums are purposely conservative. I'm sure there are brick that don't meet these requirements that you could use with no problems. But, until you have experience to go by, you know brick that meets these standards will do the job.

Don't use salvaged brick for brick paving. You need uniform durable material for this kind of job and, generally speaking, used brick aren't uniformly durable when exposed to the weather. For example, they may spall, flake, pit and crack if they freeze.

Concrete Pavers

The second kind of paver is concrete brick made from a "zero slump" concrete mix with ¼-inch washed aggregate, sand and portland cement. They're manufactured under great pressure and high frequency vibrations. They have a compressive strength of 8,000 psi and a very low water absorption rate of 5 percent.

Concrete pavers are manufactured in a wide variety of colors, and the colors tend to be more standard than with brick pavers. Additives like air-entraining agents and water repellents can make them look and perform better. And since they're cast rather than kiln fired, their sizes tend to be more uniform.

Most concrete pavers have spacers on their edges to help you lay them uniformly. If they don't have spacers, lay them with an ⅛-inch space between. Place all pavers carefully so the setting material does not get into the space between pavers.

Most concrete pavers are laid in a gravel and sand base. The depth of excavation for concrete pavers on gravel and sand is 7½ inches for sidewalks and 10½ to 14 inches for driveways. Use a minimum paver thickness of 2⅜ inches for walks and patios and 3⅛ inch for drives. The minimum thickness of the subbase for walks and patios is 4 inches, and for driveways, 10 inches.

Use these figures for estimating the materials for setting concrete pavers:

- Crusher run gravel: 1 ton = ⅔ cubic yard. One cubic yard 4 inches deep covers 81 square feet.
- Bedding sand: Use coarse concrete type. One ton of sand = 1⅓ cubic yards. One cubic yard 1 inch deep covers 300 square feet. To cover 100 square feet, you'll need ⅔ cubic yard of sand that's 2 inches deep, and 1¼ cubic yards for 4 inches deep.
- Finishing sand: Use fine mason's type that's sold in 100 square foot bags. Divide the square feet of area by 100 to find the number of bags you'll need.

Estimating Pavers

Begin by estimating the area to be paved in square feet:

- Squares and rectangles: multiply length × width in feet
- Circles: multiply the square of the diameter by 0.7854
- Triangles: multiply the base by half the perpendicular height

Then check the manufacturer's recommendations for the number of units per square foot. For standard paving brick, you can use Figure 17-1. To figure the edging, divide the length of the perimeter in linear feet by the length of the edging material.

Planning the Job

Unless you're working from architectural blueprints, you'll have to do some preliminary design work. Begin by sketching your project on paper, including all the measurements. Include all the diameters of circles and the base and height of any triangles. Then try to adjust the size of the paved area to eliminate as much cutting as possible. It's best not to use any pieces smaller than a half unit.

When you have a final design, lay out the area with a line and stakes. You can lay out circular areas with a hose. Always check the diagonals to make sure the layout is square.

Width × length	Number per SF
4 × 8	4.5
8 × 8	2.25
8 × 12	1.5
6 × 6	4.13
6 × 12	2.07
Allow 5% for waste. Add 19% for curves and circles.	

Figure 17-1
Estimating standard paving brick

Traffic

Generally, the first design consideration is to figure out how much traffic the area will get. If heavy vehicles will use the area, you'll need a rigid-base diaphragm (reinforced concrete slab on grade) or a semi-rigid continuous base. You can use rigid brick paving over this type of base. Rigid brick paving is brick laid in a bed of mortar with mortar joints between the units. Use pavers a minimum of 3⅛-inches thick for driveways, and up to 6 inches for roadways.

For an area with light vehicular traffic, such as a residential driveway, you can use a flexible base and flexible paving. Flexible brick paving has no mortar below or between the units. For an area with only pedestrian traffic you can use either of these.

Consider the Site

Whether the site is a small residential patio or a major urban-renewal project taking in several city blocks, you have to prepare it carefully. If you're in the planning stages of a large project, make sure you consider the placement of underground utilities. In some states it's mandatory to contact utility companies to find out if there are underground utilities in the area. Cutting into a utility line can cause an explosion or electrocution.

Before excavating the site, it's a good idea to have the materials delivered and stacked in convenient spots for installation. The weight of the delivery truck could cause problems if it had to drive over an excavated area. Have a dumpster delivered before work starts. Spread plastic on the ground before placing the sand to prevent waste and make cleanup easier. Remember to cover the sand, as wet sand is hard to use for installing the pavers, and for sweeping into the cracks.

The subgrade is important. Remove all vegetation and organic material from the area you're going to pave. Remove soft spots with poor base material and refill with suitable material that's properly compacted. If the area to be paved is large and covered with good sod, you may want to see if it can be sold or stored for later use on some other project. If you store it, cover it with plastic so it doesn't turn to mud.

Drainage

Surface and subsurface drainage are of major importance. Generally, you need to slope exterior brick paving ⅛ to ¼ inch per foot. For large paved areas for malls and vehicular parking, use more slope. Slope all paving away from buildings, retaining walls and other areas that collect water. In high water table areas, use a porous base and a cushion of gravel. In an area where the soil is so hard packed that water won't drain through it, you may have to add a subsurface drainage system.

Edging

Use some type of edge restraint for all paving, installed before you start laying the brick. You can choose from aluminum and plastic restraints, treated wood or precast edge pavers. One of the best types of plastic edging is called Snap Edge. It comes in 6'8" lengths packed in bundles of 24 pieces. You use 12-inch spikes to install it. Or use an edge made of a brick soldier course set on concrete with mortar. If you're lucky, you may be able to use existing concrete curbing, a building or a retaining wall.

Selecting the Color, Texture and Bond Pattern

The way brick paving looks depends on its shape, size, color, texture and the pattern you lay it in. Most of the time you'll place solid, uncored brick flat. However, you may also want to consider using cored or uncored brick placed on edge. Before you begin designing pavement, get familiar with the various types of brick paving. The Brick Institute of America says there are about 38 sizes and shapes of brick paving available. Figure 17-2 shows the three most common sizes. A few manufacturers make special molded shapes.

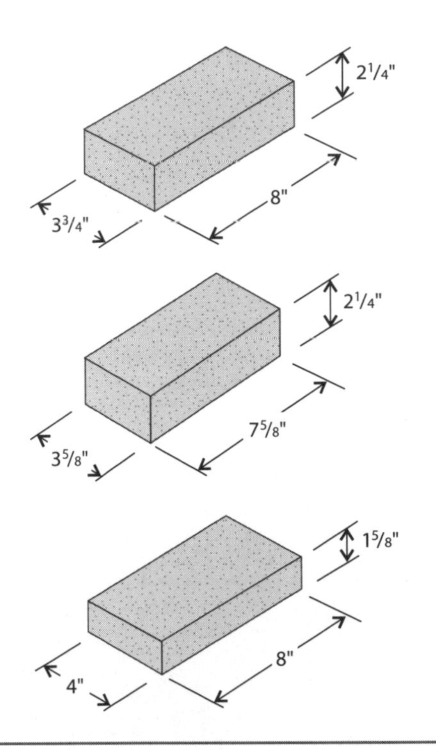

Figure 17-2
Common brick paving units

Color and Texture

Brick pavers come in many colors and textures. Red is the most popular color and it comes in several shades. You can also get buff, gray or brown brick pavers.

Before installing pavers, refer to the manufacturer's recommendations. The colors vary and are randomly mixed at the factory. Don't pick out colors — stay with the manufacturer's mix. In large projects, take random units from different packages to vary different factory runs.

The texture of brick affects how it will perform, how you install it, and how to maintain it. For example, the coarser the texture of a brick, the better its slip resistance. Rough brick are generally more suitable for exterior use where good slip resistance is important. For interiors, consider using smooth brick pavers. They take sealers, coatings and waxes better so they're easier to maintain and clean. Generally, smooth low-maintenance brick pavers are good to use in high-traffic areas such as lobbies and foyers.

Be sure to specify the proper size brick, especially for pattern bonds.

Bond Patterns

There's just about an endless variety of patterns you can make with brick paving. Some of the most common are shown in Figure 17-3. The running bond and the basketweave are the easiest to lay. If you can lay out the area to eliminate cuts, the basketweave is pretty easy. You've got to lay out the running bond in full units to eliminate excessive cuts or small cuts. Don't start your cuts until all full units are in place. Make the cuts with a mason's hammer and chisel or with a masonry saw. Place the cuts against the outside edging.

Joints

Here are the three basic ways to install brick paving with mortar joints:

1) *Conventional use of mortar and a trowel* — Butter brick pavers with mortar and lay them in a leveling base of mortar.

2) *Laying each brick in a mortar leveling bed* — Leave ⅜ to ½ inch space between the units. Pour a grout mixture into these spaces. Generally, grout proportions of portland cement and sand are the same as for mortar, except that you can leave out the hydrated lime. When you pour grout into the joints, be careful to protect the brick so they won't be so difficult to clean later.

3) *Laying each brick in a dry mixture of portland cement and sand* — Using the same proportions as for grout, lay brick pavers on a cushion of damp mixture. Put the same mixture between the pavers. Clean the extra material from the paving surface and spray with a fine mist of water until the joints are saturated. Keep the pavement damp for two or three days.

For brick paving without mortar joints, just sweep the paving with plain dry sand or a mixture of portland cement and sand. There's more on the proper proportions of portland cement and sand later in this chapter.

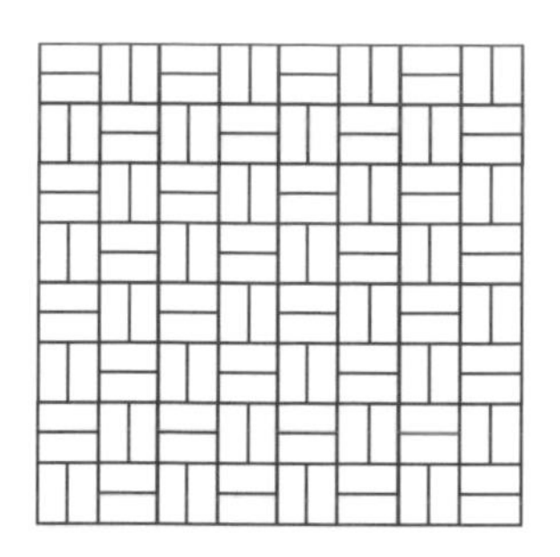

Basketweave

Herringbone

Runnng bond

Variation of basketweave

Variation of basketweave

Variation of basketweave

Variation of basketweave

¼ running bond

Circular and running bond mixed

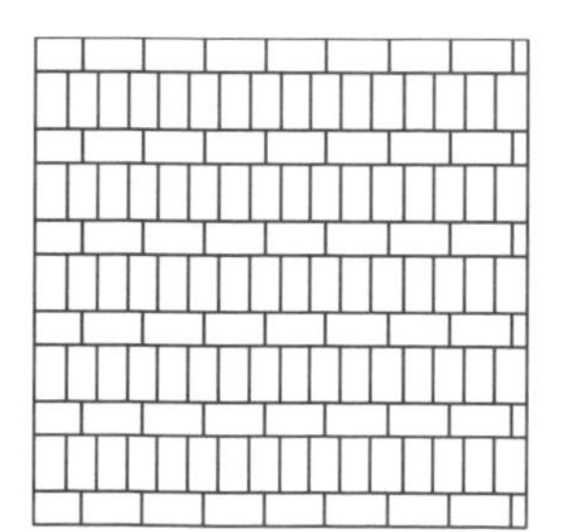

Running and stack bond mixed

Stack bond

Figure 17-3
Brick paving patterns

Expansion Joints

Temperature changes and moisture can make brick paving expand and contract. Use expansion joints to allow for this movement. Always install expansion joints in rigid paving systems to allow for thermal movement. There are no standard joint locations because of the wide variety of situations and patterns. But it's always a good idea to put joints parallel to edges and at any angle, 90-degree turn or other changes in the straight, flat surface. Expansion joints are made of butyl rubber or neoprene.

You may also need expansion joints in the supporting base because it can be affected by moisture and temperature changes.

Membrane Materials

Use a membrane under the brick paving to:

- ❖ Control or reduce moisture getting into the paving
- ❖ Discourage weed growth
- ❖ Separate layers to accommodate differential movement

Membranes are usually made of sheet materials that resist rot and decay. One of the best is called Geotextile, used mainly for driveways.

Be careful not to damage the membrane during construction. Some membranes resist abrasion better than others. Protection is especially important for roof deck construction — a deck over a garage, for instance.

Base Materials

Use a base material for:

- ❖ Support
- ❖ Drainage
- ❖ Ground-swell protection

Gravel Bases

Use clean, washed gravel to provide the best drainage for pavement. The size of the gravel will depend on the installation. Pea gravel is self-compacting and you can easily screed it to a finished grade. Where you to build up a thick base for drainage purposes, a larger stone size may be more economical.

Stone screenings (pieces that fall through a ¼-inch screen) and bank run gravel are both readily available and compact well. Stone dust (residue from limestone materials) compact well, but will harden if exposed to moisture. And stone dust that gets wet may cause water-soluble salts to deposit on the surface of the pavers. See Figure 17-4.

Concrete Bases

You can use either an existing or a new concrete base for brick paving. For a new base, install concrete following recommended concrete practices. If you use a mortar bed, give the slab surface a raked or floated finish to make a good bond. If you use a

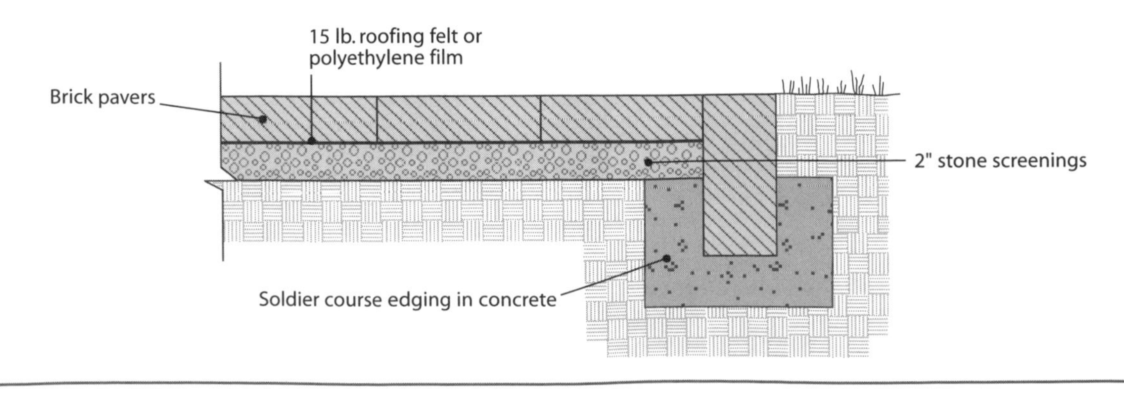

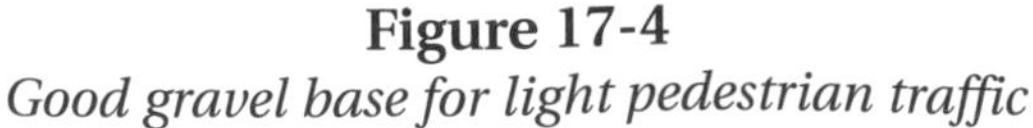
Figure 17-4
Good gravel base for light pedestrian traffic

noncementitious type of leveling bed or cushion, just screed the slab. You don't need any other finishing. Figure 17-5 shows a concrete base with mortarless pavers.

Asphalt Bases

You can use new or existing asphalt paving bases to support brick paving. You can also use a mortar bed with an asphalt base, but the mortar and asphalt won't bond well. Avoid placing a mortar bed on hot asphalt because the mortar will set too quickly. Repair major defects in an existing asphalt pavement before you install any brick paving.

Flexible Bases

A flexible base is made of compacted gravel or a damp, loose, sand-cement mixture tamped in place. You can only use flexible paving over a flexible base. Generally, a flexible base is the most economical type of base to install. It's usually made of layers of gravel, damp loose sand or a mixture of sand and cement.

You can make a porous gravel base of graded or ungraded stone. Ungraded gravel or bank run gravel is gravel that's not screened to separate it by size. Because it has many different sizes of particles, it interlocks better, for more stability. Graded gravel that's classified No. 1 (½ inch) or No. 2 (1 inch) used alone has a tendency to roll under foot. Stone screenings containing finer particles generally compact better than pea gravel. However, a pea gravel base drains better.

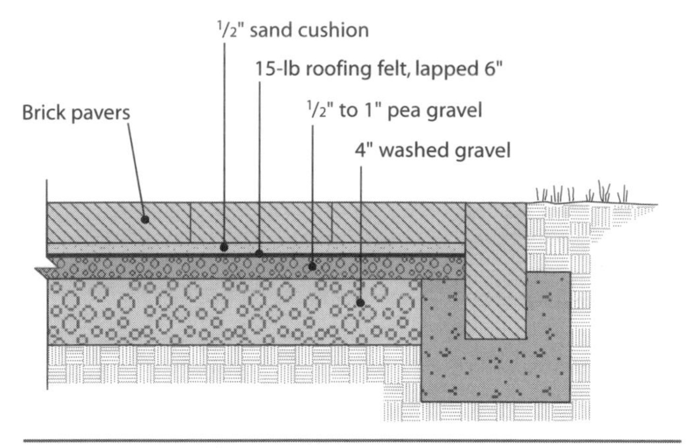

Figure 17-6
Gravel base for a residential patio

I recommend a ½-inch sand cushion, if you use pea gravel under it, as shown in Figure 17-6. This is a good base for areas with a high groundwater table, heavy rain, or problem drainage areas. You have to use a membrane or the sand will eventually settle through the pea gravel. The sand layer accomplishes two basic purposes:

1) You can lay the brick pavers more efficiently to the desired grade.
2) The sand cushion makes the brick paver more stable.

Figure 17-7 shows a typical residential driveway with a gravel base. The gravel for the base should include stones ranging from 1 inch to bank run gravel of various sizes. This is often referred to as crusher run.

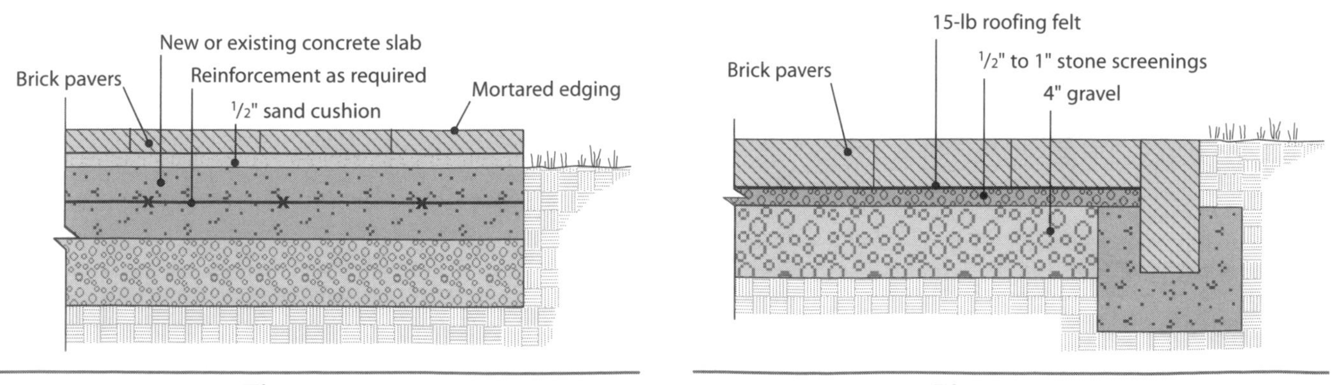

Figure 17-5
Concrete base used with mortarless pavers on a residential patio

Figure 17-7
Gravel base for a residential driveway

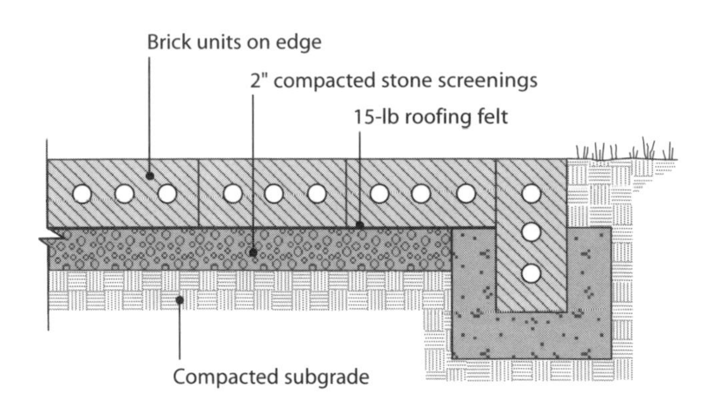

Figure 17-8
Another gravel base suitable for a residential driveway

Figure 17-8 is an alternative gravel base for the residential driveway. It includes a 2-inch stone base which is about half of the base thickness shown in Figure 17-6. You can use this thinner base if you use a thicker brick paver along with it. To do this, lay the pavers on edge. Laid this way, you can use cored brick pavers instead of solid ones. Either of the bases shown in Figures 17-7 and 17-8 will work for residential driveway traffic.

Look back to Figure 17-4. It shows a gravel base of 2-inch stone screenings for light pedestrian traffic. Spread the screenings to a uniform depth after you've done the necessary grading and edging. Then moisten and thoroughly compact the screenings by hand or mechanical means. Rescreed and compact more if necessary to get the specified grade. Add a membrane layer of asphalt-impregnated roofing felt or polyethylene plastic sheeting over the screenings to prevent efflorescence and to protect the surface.

Close to buildings where you want a slope to drain water away, add portland cement to the stone screenings to provide stabilization. Add 1 part portland cement to 6 parts stone screenings by volume and mix with only enough water to form a ball. Screed this mixture into place and let it set up. This makes a stable base that still lets moisture drain through the screenings.

For a large area, such as a pedestrian mall, you can use the installation shown in Figure 17-9 if the subsurface soil of the area drains well. Excavate a grid system of earth trenches to hold reinforced concrete grade beams. After the concrete has cured, strip the subgrade away and replace it with a sand-cement base (3 to 6 parts sand to 1 part portland cement). Spread a 1-inch sand cushion over the compacted base to make it easier to lay the pavers and separate them from the base.

Figure 17-10 shows the most economical way to lay a residential patio in an area with a low groundwater table and moderate rain. Don't use this method unless the subsurface soil drains well.

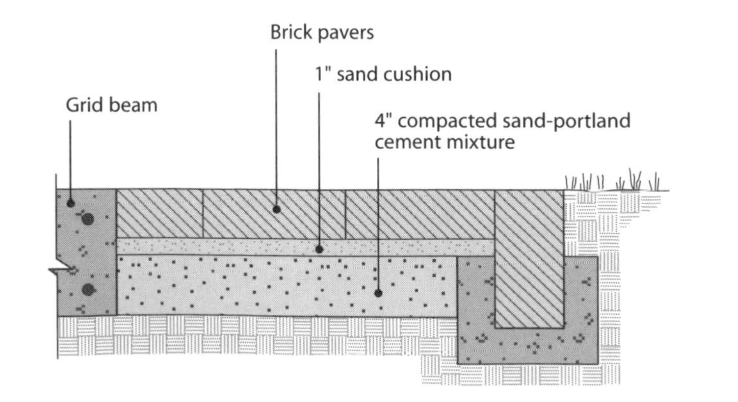

Figure 17-9
Sand and portland cement base for pedestrian mall traffic

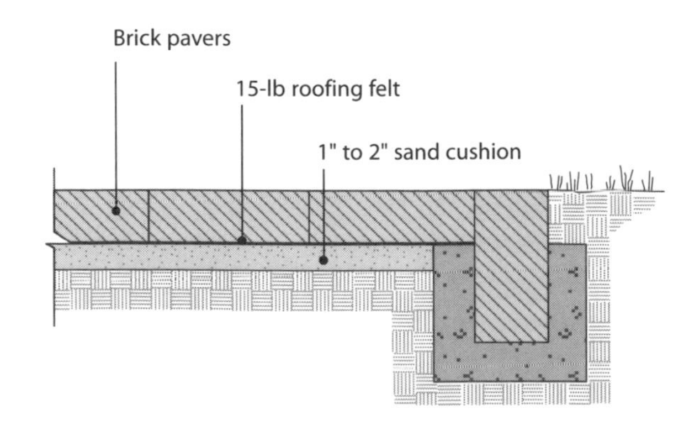

Figure 17-10
Sand base for economical residential patio

Compact the subgrade before installing subbase materials. If the subbase is over 4 inches thick, place it in layers not more than 4 inches thick. Compact each layer so the base is smooth and flat, and at the desired slope. Inadequate or uneven compacting of the subbase can cause irregular settlement of the pavers.

Suspended Diaphragm Base

Suspended diaphragm bases are structural roof or floor deck assemblies. The composition varies depending on the design. Either flexible or rigid brick paving is suitable for this type of base.

Setting Beds

The usual types of setting beds are mortar or bituminous.

Mortar Setting Beds

You can use a mortar setting bed with concrete and asphalt bases. For exterior paving, use a Type M portland cement and lime mortar to make a mortar bed. For interior applications, use Type S or N. The thickness of the bed may vary from ½ inch to 1 inch.

You can also get mortars with high-bond additives or latex-modified portland cement. Be careful to mix and use these mortars according to the manufacturer's recommendations. High-bond mortar bonds well to some brick — but not all. Test it carefully to make sure it works well with the brick you're using.

Here are the types of mortar, their uses and specifications:

- *Portland cement-lime mortars* — For exterior mortared paving on grade, use Type M (1:¼:3) portland cement-lime-sand. It has high durability and is specifically recommended for masonry in contact with the earth. For interior mortared paving, use Type S (1:½:4¼) or Type N (1:1:6) portland cement-lime-sand mortars. Use Type S mortar for both reinforced and unreinforced masonry where you need maximum bond strength. It's also suitable for outdoor use, provided the brick slab isn't in contact with the ground. Type N is a medium-strength mortar suitable for interior use. Don't use masonry cement mortars.
- *Dry-mixed or grout-type mortar* — You can mix sand and dry cement and sweep it between the brick pavers without mortar joints. Then fog the pavement down with water to set the mixture. For exterior use, follow Type M mortar proportions. For interior use, use Type S or N mortar. Some installers prefer a soupy, grout-type mixture to pour between the mortar joints. The main difference between grout and mortar is that there's no hydrated lime in the grout. To make cleanup easier, coat the brick with melted paraffin before you lay them. Then any grout that spills on the face of the coated units won't stain the surface. *But make sure the paraffin doesn't coat the joint side of the brick.* After you've done all the grouting, clean off the paraffin coating with a steam jenny. You can also remove paraffin by hosing the surface off with hot water or cold water under pressure.
- *Latex-portland cement mortars* — Latex materials for cement mortar vary among manufacturers. Follow the manufacturer's directions carefully. For material and installation specifications see ANSI A119.4-1973, Specification for Latex-Portland Cement Mortar. According to the Tile Council of America, latex mortars are somewhat more flexible than conventional mortars, resist water well, and clean up easily. This type of mortar is good to use for mortared paving supported by a flexural structural system, such as a wood-joist floor assembly.
- *High-bond mortar* — SARABOND® high-bond mortar additive is a liquid saran polymer that greatly improves the bonding, compressive and tensile-strength characteristics of mortar.

Figure 17-11 shows mortared brick paving for interior or exterior use.

Bituminous Setting Beds

You can use a bituminous setting bed of aggregate and asphaltic cement to support brick paving. Usually the mix is delivered hot from the plant and rolled to a ¾-inch depth.

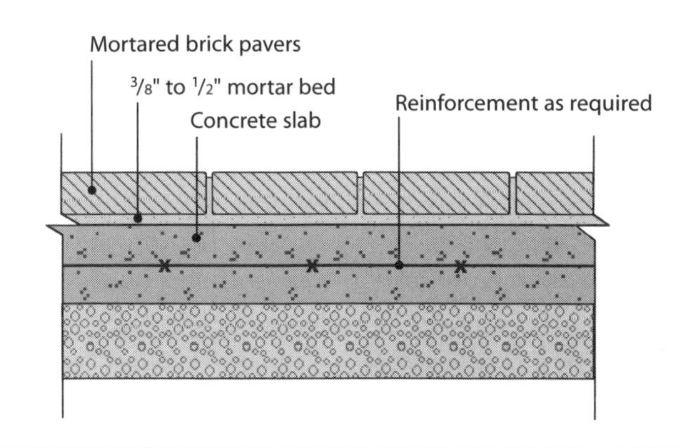

Figure 17-11
Mortared brick paving for interior and exterior use

Bituminous setting beds are usually used on rigid or semi-rigid bases as shown in Figure 17-12. Typically, you'll use a concrete slab or asphalt pavement as a base for a bituminous setting bed. The specialty contractor on the job will probably be the one to specify the proportions of asphalt and aggregate so I won't discuss them here. You'll want to make the supporting base about 4 to 6 inches thick, depending on how much traffic the area will get and the dead weight of the brick pavers and other materials. Install brick pavers hand-tight on a bituminous setting bed.

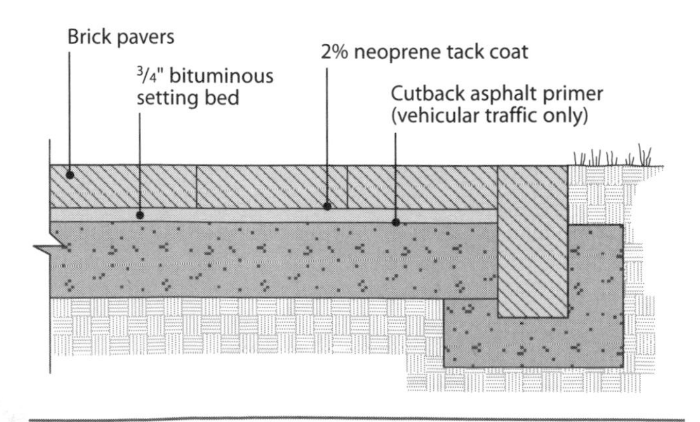

Figure 17-12
Concrete or asphalt base with flexible brick paving for pedestrian malls or slow vehicular traffic

To accommodate light vehicular traffic, apply a primer coat of rapid-curing cut-back asphalt to the concrete slab or asphalt pavement.

Cushion Materials

A layer of cushion material helps make up for irregularities in the surface of the base and the paver. You can use a 1- or 2-inch layer of sand, pea gravel, stone screenings or even several layers of roofing felt. Under extremely wet conditions, avoid fine-particle cushions such as sand or stone screenings because they may make moisture drain off too slowly.

Sand and Cement Mixtures

Any sand you use for cushion material (or bases, joints or mortar) should conform with ASTM C144, Aggregate for Masonry Mortar. Make sure the sand is free of clay to avoid scumming when you sweep it over the face of brick pavers.

You can also use a dry mixture of sand and cement as a base, or cushion. Mix 1 part portland cement with 3 to 6 parts damp, loose sand. It's easiest to use your shovel as a measure. The brick and the leveling bed won't bond well if there's too much sand in the mixture.

Roofing Felt

Generally you can use 15- to 30-pound weight roofing felt as a cushion between brick pavers and concrete or asphalt paving. This material goes down fast and it covers up minor irregularities between the base and the brick pavers. Roofing felt may also make the paving more resilient. Use two layers of 15-pound roofing felt over a rigid or semi-rigid base as a cushion for regular flat brick pavers. Figure 17-13 shows two layers of roofing felt on a concrete or asphalt base.

Consider using a depressed concrete slab to avoid an abrupt change in floor finishes, as shown in Figure 17-14. If you overlap the felt, it can make the units over the overlap higher than the other brick. Avoid this by abutting the edges before you lay the pavers.

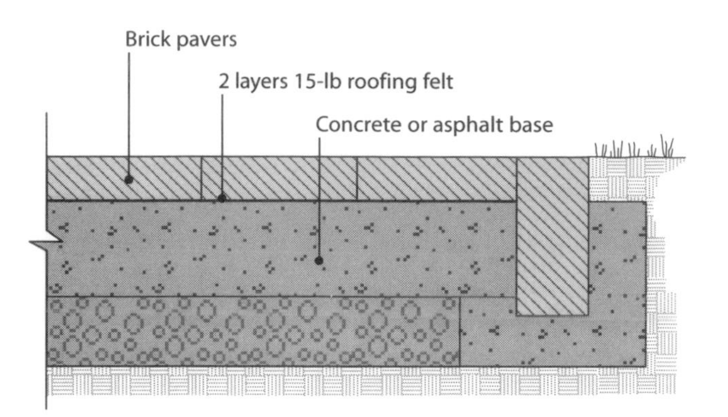

Figure 17-13
Concrete or asphalt base with flexible brick pavers on an existing rigid or semi-rigid base in a residential driveway

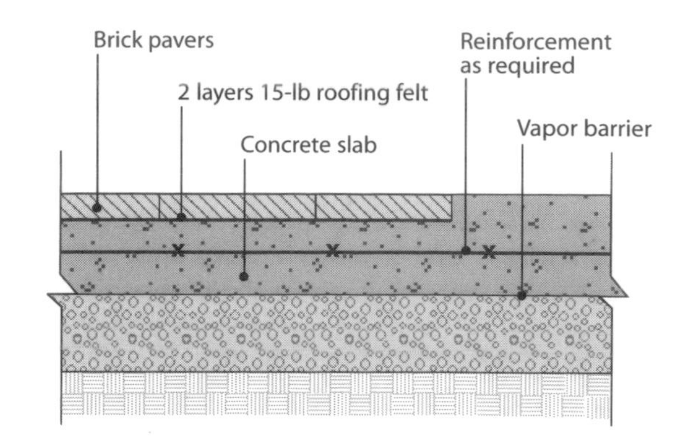

Figure 17-14
Concrete base for interior flooring

Compaction

To compact pavers and bases, use a metal plate tamper. You can rent one if you don't use it often enough to buy one. (When you rent any paving equipment, the insurance available from the rental company is well worth the price.) This compaction is necessary to set the pavers in the sand for mortarless applications. Go over the pavers at least two times in alternate directions, beginning in the center of the paved area and working out toward the edges. Don't work close to the edging if it's not solid. The pressure can move plastic material. If a unit drops lower than the surrounding units, use a screwdriver to pry it up into the correct position. Add a little sand under it and compact it. Don't apply sand in the joints before you compact the pavers.

Appendix: Construction Safety

Construction is one of the most hazardous of all occupations. Construction workers work with heavy machinery and often must climb high scaffolds, descend into excavations, and handle acids and other caustic materials.

Occupational Safety and Health Act

In 1970 Congress passed the Occupational Safety and Health Act (OSHA) ". . . to assure so far as possible every working man and woman in the nation safe and healthful working conditions and to preserve our human resources."

In general, coverage of the Act extends to all employers and their employees in the 50 states, the District of Columbia, Puerto Rico, the Canal Zone and all other territories under Federal Government jurisdiction.

As defined by the Act, an employer is "any person engaged in a business affecting commerce who has employees." The Act covers employers and employees in such varied fields as construction, longshoring, and agriculture.

The following employer/employee relationships are not covered under the act:

1) Self-employed persons.
2) Farms at which only immediate members of the farm employer's family are employed.
3) Workplaces already protected by other federal agencies under federal statutes.

Before the Act became effective, no centralized and systematic method existed for monitoring occupational safety and health problems. Statistics on job injuries and illnesses were collected by some states and by some private organizations; national figures were based on projections that weren't altogether reliable. With OSHA standards came the first basis for consistent nationwide procedures — a vital requirement for gauging problems and solving them.

Employers of eleven or more employees must maintain records of occupational injuries and illnesses as they occur. Employers with ten or fewer employees are exempt from keeping such records unless they're selected by the Bureau of Labor Statistics (BLS) to participate in periodic statistical surveys. The purpose of keeping records is to be able to comply with BLS survey needs, to help define high-hazard industries, and to inform employees of the status of their employer's record.

There's one exception to the exemption to maintain records: All employers, *regardless of the size of the company*, have to report any on-the-job accident which results in the death of an employee or in the hospitalization of five or more employees. The employer must make a report, in detail, and give it to the nearest OSHA office. In states with approved plans, employers report such accidents to the state agency responsible for safety and health programs.

You should maintain records on a calendar year basis. You don't have to send them to OSHA or any other agency, but you do need to keep them for five years and make them available upon request for inspection by representatives of OSHA, HEW, BLS, or any entitled state or local agency.

There are two forms that you need to keep on hand to comply with OSHA requirements:

1) OSHA Form 200, Log and summary of occupational injuries
2) OSHA Form 101, Supplementary record of occupational injuries and illness

If you want to set up a record-keeping system different from the one prescribed by OSHA regulations, you can apply for a record-keeping variance.

Occupational Injury and Illness

What is an occupational injury or illness? It's any injury such as a cut, fracture, sprain, or amputation that results from a work-related accident or from exposure involving a single incident in the work environment. An occupational illness is any abnormal condition or disorder, other than one resulting from an occupational injury, caused by exposure to environmental factors associated with employment. Included are acute and chronic illnesses which may be caused by inhalation, absorption, ingestion or direct contact with toxic substances or harmful agents.

You must report all occupational illnesses, regardless of their severity. And you must report all occupational injuries if they result in:

(1) Death (must be recorded regardless of length of time between the injury and death)

(2) One or more lost workdays

(3) Restriction of work or motion

(4) Loss of consciousness

(5) Transfer to another job

(6) Medical treatment (other than first aid)

Employer Responsibilities and Rights

Employers have certain responsibilities and rights under the Occupational Safety and Health Act of 1970. The checklists that follow review many of them.

Responsibilities

As an employer you must:

(1) Meet your general responsibility to provide a workplace free from recognized hazards that are causing or are likely to cause death or serious physical harm to employees, and to comply with standards, rules, and regulations issued under the Act.

(2) Be familiar with mandatory OSHA standards and make copies available to employees for review on request.

(3) Inform all employees about OSHA.

(4) Examine workplace conditions to make sure they conform to applicable standards.

(5) Minimize or reduce hazards.

(6) Make sure employees have and use safe tools and equipment (including personal protective equipment), and that such equipment is properly maintained.

(7) Use color codes, posters, labels, or signs to warn employees of potential hazards.

(8) Establish or update operating procedures and communicate them so all employees follow safety and health requirements.

(9) Provide medical examinations when required by OSHA standards.

(10) Report to the nearest OSHA office any fatal accident or one which results in the hospitalization of five or more employees.

(11) Keep OSHA-required records of work-related injuries and illnesses, and post a copy of the totals from the last page of OSHA No. 200 during the entire month of February of each year. (This applies to employers of eleven or more employees.)

(12) Pay employees for any time spent in taking part in an OSHA inspection.

(13) Provide employees, former employees and their representatives access to the Log and Summary of Occupational Injuries and Illnesses (OSHA No. 200).

(14) Cooperate with an OSHA compliance officer by furnishing names of authorized employee representatives who may be asked to accompany the compliance officer during an inspection. (If none, the compliance officer will consult with a reasonable number of employees concerning safety and health in the workplace.)

(15) Not discriminate against employees who properly exercise their rights under this Act.

(16) Post OSHA citations at or near the worksite involved. Each citation or copy thereof must remain posted until the violation has been abated or for three working days, whichever is longer.

(17) Abate cited violations within the prescribed period.

Rights

As an employer, you have the right to:

(1) Seek advice and off-site consultation as needed by writing, calling or visiting the nearest OSHA office. (OSHA will not inspect merely because an employer requests assistance.)

(2) Be active in the involvement of your industry association in job safety and health.

(3) Request and receive proper identification of the OSHA compliance officer prior to inspection.

(4) Be advised by the compliance office of the reason for an inspection.

(5) Have an opening and closing conference with the compliance officer.

(6) File a Notice of Contest with the OSHA area director within 15 working days of receipt of a notice of citation and proposed penalty.

(7) Apply to OSHA for a temporary variance from a standard if unable to comply because of the unavailability of materials, equipment or personnel needed to make necessary changes within the required time.

(8) Apply to OSHA for a permanent variance from a standard if you can furnish proof that your facilities or method of operation provide employee protection at least as effective as that required by the standard.

(9) Take an active role in developing safety and health standards through participation in OSHA Standards Advisory Committees, through nationally recognized standard-setting organizations and through evidence and view presenting in writing or an hearings.

(10) Avail yourself, if you are a small business employer, of long-term loans through the Small Business Administration to help bring your establishment into compliance, either before or after an OSHA inspection.

(11) Be assured of the confidentiality of any trade secrets observed by an OSHA compliance officer during an inspection.

OSHA Standards

OSHA standards fall into four major categories, one of which is construction.

The Federal Register is one of the best sources of information on standards, since all OSHA standards are published in the Federal Register when they're adopted. You'll also fine all amendments, corrections, insertions or deletions there. The Federal Register is available in many public libraries. You can get an annual subscription by contacting the Superintendent of Documents, U.S. Government Printing Office, Washington, DC 20402.

In masonry construction, you're concerned with the book of standards called 29 CFR PART 1926 SAFETY AND HEALTH REGULATIONS FOR CONSTRUCTION — 29 CFR PART 1910 GENERAL INDUSTRY SAFETY AND HEALTH REGULATIONS IDENTIFIED AS APPLICABLE TO CONSTRUCTION.

I recommend that you get and keep copies of the standards that apply to your company. But for your reference, here are most of the parts that deal with masonry construction.

1926.20 General Safety and Health Provisions

(A) Contractor Requirements.

(1) Section 107 of the Act requires that it shall be a condition of each contract which is entered into under legislation subject to Reorganization Plan No. 14 of 1950 (64 Stat. 1267), as defined in 1926.12, and is for construction, alteration, and/or repair, including painting and decorating, that no contractor or subcontractor for any part of the contract work shall require any laborer or mechanic employed in the performance of the contract to work in surroundings or under working conditions which are unsanitary, hazardous, or dangerous to his health or safety.

(B) Accident-Prevention Responsibilities.

(1) It shall be the responsibility of the employer to initiate and maintain such programs as may be necessary to comply with this part.

(2) Such programs shall provide for frequent and regular inspections of the job site, materials, and equipment to be made by competent persons designated by the employers.

(3) The use of any machinery, tool, material, or equipment which is not in compliance with any applicable requirement of this part is prohibited. Such machine, tool, material, or equipment shall be either identified as unsafe by tagging or locking the controls to render them inoperable or shall be physically removed from the place of operation.

(4) The employer shall permit only those employees qualified by training or experience to operate equipment or machinery.

(C) The standards contained in this part shall apply with respect to employment performed in a workplace in a State, the District of Columbia, the Commonwealth of Puerto Rico, the Virgin Islands, American Samoa, Guam, Trust Territory of the Pacific Islands, Wake Island, Outer Continental Shelf lands defined in the Outer Continental Shelf Lands Act, Johnston Island, and the Canal Zone.

(D) (1) If a particular standard is specifically applicable to a condition, practice, means, method, operation, or process, it shall prevail over any different general standard which might otherwise be applicable to the same condition, practice, means, method, operation, or process.

(2) On the other hand, any standard shall apply according to its terms to any employment and place of employment in any industry, even though particular standards are also prescribed for the industry to the extent that none of such particular standards applies.

(E) In the event a standard protects on its face a class of persons larger than employees, the standard shall be applicable under this part only to employees and their employment and places of employment.

1926.21 Safety Training and Education

(A) General Requirements.

The Secretary shall, pursuant to section 107(f) of the Act, establish and supervise programs for the education and training of employers and employees in the recognition, avoidance and prevention of unsafe conditions in employment covered by the act.

(B) Employer Responsibility.

(1) The employer should avail himself of the safety and health training programs the Secretary provides.

(2) The employer shall instruct each employee in the recognition and avoidance of unsafe conditions and the regulations applicable to his work environment to control or eliminate any hazards or other exposure to illness or injury.

(3) Employees required to handle or use poisons, caustics, and other harmful substances shall be instructed regarding the safe handling and use, and be made aware of the potential hazards, personal hygiene, and personal protective measures required.

(4) In job site areas where harmful plants or animals are present, employees who may be exposed shall be instructed regarding the potential hazards, and how to avoid injury, and the first aid procedures to be used in the event of injury.

(5) Employees required to handle or use flammable liquids, gases, or toxic materials shall be instructed in the safe handling and use of these materials and made aware of the specific requirements contained in Subparts D, F, and other applicable subparts of this part.

(6) (i) All employees required to enter into confined or enclosed spaces shall be instructed as to the nature of the hazards involved, the necessary precautions to be taken, and in the use of protective and emergency equipment required. The employer shall comply with any specific regulations that apply to work in dangerous or potentially dangerous areas.

(ii) For purposes of paragraph (B)(6)(i) of this section, "confined or enclosed space" means any space having a limited means of egress, which is subject to the accumulation of toxic or flammable contaminants or has an oxygen deficient atmosphere. Confined or enclosed spaces include, but are not limited to, storage tanks, process vessels, bins, boilers, ventilation or exhaust ducts, sewers, underground utility vaults, tunnels, pipelines, and open top spaces more than 4 feet in depth such as pits, tubs, vaults, and vessels.

1926.24 Fire Protection and Prevention

The employer shall be responsible for the development and maintenance of an effective fire protection and prevention program at the job site throughout all phases of the construction, repair, alteration, or demolition work. The employer shall ensure the availability of the fire protection and suppression equipment required by Subpart F of this part.

1926.25 Housekeeping

(A) During the course of construction, alteration, or repairs, form and scrap lumber with protruding nails and all other debris shall be kept clear from work areas, passageways, and stairs, in and around buildings or other structures.

(B) Combustible scrap and debris shall be removed at regular intervals during the course of construction. Safe means shall be provided to facilitate such removal.

(C) Containers shall be provided for the collection and separation of waste, trash, oily and used rags, and other refuse. Containers used for garbage and other flammable, or hazardous wastes, such as caustics, acids, harmful dusts, etc. shall be equipped with covers. Garbage and other waste shall be disposed of at frequent and regular intervals.

1926.26 Illumination

Construction areas, aisles, stairs, ramps, runways, corridors, offices, shops, and storage areas where work is in progress shall be lighted with either natural or artificial illumination. The minimum illumination requirements for work areas are contained in Subpart D of this part.

1926.28 Personal Protective Equipment

(A) The employer is responsible for requiring the wearing of appropriate personal protective equipment in all operations where there is an exposure to hazardous conditions or where this part indicates the need for using such equipment to reduce the hazards to the employees.

(B) Regulations governing the use, selection, and maintenance of personal protective and lifesaving equipment are described under Subpart E of this part.

1926.32 Definitions

The following definitions shall apply in the application of the regulations in this part:

(A) *Act* means section 107 of the Contract Work Hours and Safety Standards Act, commonly known as the Construction Safety Act (86 Stat. 96; 40 U.S.C. 333).

(B) *ANSI* means American National Standards Institute.

(C) *Approved* means sanctioned, endorsed, accredited, certified, or accepted as satisfactory by a duly constituted and nationally recognized authority or agency.

(D) *Authorized* person means a person approved or assigned by the employer to perform a specific type of duty or duties or to be at a specific location or locations at the jobsite.

(E) *Administration* means the Occupational Safety and Health Administration.

(F) *Competent person* means one who is capable of identifying existing and predictable hazards in the surroundings or working conditions which are unsanitary, hazardous, or dangerous to employees, and who has authorization to take prompt corrective measures to eliminate them.

(G) *Construction work* means, for purposes of this section, work for construction, alteration, and/or repair, including painting and decorating.

(H) *Defect* means any characteristic or condition which tends to weaken or reduce the strength of the tool, object, or structure of which it is a part.

(I) *Designated person* means "authorized person" as defined in paragraph (D) of this section.

(J) *Employee* means every laborer or mechanic under the Act regardless of the contractual relationship which may be alleged to exist between the laborer and mechanic and the contractor or subcontractor who engaged him. "Laborer" and "mechanic" are not defined in the Act, but the identical terms are used in the Davis-Bacon Act (40 U.S.C. 276a), which provides for minimum wage protection on Federal and federally-assisted construction contracts. The use of the same term in a statute which often applies concurrently with section 107 of the Act has considerable presidential value in ascertaining the meaning of "laborer" and "mechanic" as used in the Act. "Laborer" generally means one who performs manual labor or who labors at an occupation requiring physical strength; "mechanic" generally means a worker skilled with tools. See 18 Comp. Gen. 341.

(K) *Employer* means contractor or subcontractor within the meaning of the Act and of this part.

(L) *Hazardous substance* means a substance which, by reason of being explosive, flammable, poisonous, corrosive, oxidizing, irritating, or otherwise harmful, is likely to cause death or injury.

(M) *Qualified* means one who, by possession of a recognized degree, certificate, or professional standing, or who by extensive knowledge, training, and experience, has successfully demonstrated his ability to solve or resolve problems relating to the subject matter, the work, or the project.

(N) *Safety factor* means the ratio of the ultimate breaking strength of a member or piece of material or equipment to the actual working stress or safe load when in use.

(O) *Secretary* means the Secretary of Labor.

(P) *SAE* means Society of Automotive Engineers

(Q) *Shall* means mandatory.

(R) *Should* means recommended.

(S) *Suitable* means that which fits, and has the qualities or qualifications to meet a given purpose, occasion, condition, function, or circumstance.

1926.33 Access to Employee Exposure and Medical Records

(A) Purpose.

The purpose of this section is to provide employees and their designated representatives a right of access to relevant exposure and medical records; and to provide representatives of the Assistant Secretary a right of access to these records in order to fulfill responsibilities under the Occupational Safety and Health Act. Access by employees, their representatives, and the Assistant Secretary is necessary to yield both direct and indirect improvements in the detection, treatment, and prevention of occupational disease. Each employer is responsible for assuring compliance with this section, but the activities involved in complying with the access to medical records provisions can be carried out, on behalf of the employer, by the physician or other health care personnel in charge of employee medical records. Except as expressly provided, nothing in this section is intended to affect existing legal and ethical obligations concerning the maintenance and confidentiality of employee medical information, the duty to disclose information to a patient/employee or any other aspect of the medical-care relationship, or affect existing legal obligations concerning the protection of trade secret information.

(B) Scope and Application.

(1) This section applies to each general industry, maritime, and construction employer who makes, maintains, contracts for, or has access to employee exposure or medical records, or analyses thereof, pertaining to employees exposed to toxic substances or harmful physical agents.

(2) This section applies to all employee exposure and medical records, and analyses thereof, of such employees, whether or not the records are mandated by specific occupational safety and health standards.

(3) This section applies to all employee exposure and medical records, and analyses thereof, made or maintained in any manner, including on an in-house of contractual (e.g., fee-for-service) basis. Each employer shall assure that the preservation and access requirements of this section are complied with regardless of the manner in which the records are made or maintained.

(C) Definitions.

(1) *Access* means the right and opportunity to examine and copy.

(2) *Analysis using exposure or medical records* means any compilation of data or any statistical study based at least in part on information collected from individual employee exposure or medical records or information collected from health insurance claims records, provided that either the analysis has been reported to the employer or no further work is currently being done by the person responsible for preparing the analysis.

(3) *Designated representative* means any individual or organization to whom an employee gives written authorization to exercise a right of access. For the purposes of access to employee exposure records and analyses using exposure or medical records, a recognized or certified collective bargaining agent shall be treated automatically as a designated representative without regard to written employee authorization.

(4) *Employee* means a current employee, a former employee, or an employee being assigned or transferred to work where there will be exposure to toxic substances or harmful physical agents. In the case of a deceased or legally incapacitated employee, the employee's legal representative may directly exercise all the employee's rights under this section.

(5) *Employee exposure record* means a record containing any of the following kinds of information:

(i) Environmental (workplace) monitoring or measuring of a toxic substance or harmful physical agent, including personal, area, grab, wipe, or other form of sampling, as well as related collection and analytical methodologies, calculations, and other background data relevant to interpretation of the results obtained;

(ii) Biological monitoring results which directly assess the absorption of a toxic substance or harmful physical agent by body systems (e.g., the level of a chemical in the blood, urine, breath, hair, fingernails, etc.) but not including results which assess the biological effect of a substance or agent or which assess an employee's use of alcohol or drugs;

(iii) Material safety data sheets indicating that the material may pose a hazard to human health; or

(iv) In the absence of the above, a chemical inventory or any other record which reveals where and when used and the identity (e.g., chemical, common, or trade name) of a toxic substance or harmful physical agent.

(6) (i) *Employee medical record* means a record concerning the health status of an employee which is made or maintained by a physician, nurse, or other health care personnel or technician, including:

(a) Medical and employment questionnaires or histories (including job description and occupational exposures),

(b) The results of medical examinations (pre-employment, pre-assignment, periodic, or episodic) and laboratory tests (including chest and other X-ray examinations taken for the purposes of establishing a base-line or detecting occupational illness, and all biological monitoring not defined as an "employee exposure record"),

(c) Medical opinions, diagnoses, progress notes, and recommendations,

(d) First aid records,

(e) Descriptions of treatments and prescriptions, and

(f) Employee medical complaints.

(ii) "Employee medical record" *does not include* medical information in the form of:

(a) Physical specimens (e.g., blood or urine samples) which are routinely discarded as a part of normal medical practice; or

(b) Records concerning health insurance claims if maintained separately from the employer's medical program and its records, and not accessible to the employer by employee name or other direct personal identifier (e.g., social security number, payroll number, etc.); or

(c) Records created solely in preparation for litigation which are privileged from discovery under the applicable rules of procedure or evidence; or

(d) Records concerning voluntary employee assistance programs (alcohol, drug abuse, or personal counseling programs) if maintained separately from the employer's medical program and its records.

(7) *Employer* means a current employer, a former employer, or a successor employer.

(8) *Exposure or exposed* means that an employee is subjected to a toxic substance or harmful physical agent in the course of employment through any route of entry (inhalation, ingestion, skin contact or absorption, etc.), and includes post exposure and potential (e.g., accidental or possible) exposure, but does not include situations where the employer can demonstrate that the toxic substance or harmful physical agent is not used, handled, stored, generated, or present in the workplace in any manner different from typical non-occupational situations.

(9) *Health professional* means a physician, occupational health nurse, industrial hygienist, toxicologist, or epidemiologist, providing medical or other occupational health services to exposed employees.

(10) *Record* means any item, collection, or grouping of information regardless of the form or process by which it is maintained (e.g., paper document, microfiche, microfilm, X-ray film, or automated data processing).

(11) *Specific chemical identity* means the chemical name, Chemical Abstracts Service (CAS) Registry Number, or any other information that reveals the precise chemical designation of the substance.

(12) (i) *Specific written consent* means a written authorization concerning the following:

(a) The name and signature of the employee authorizing the release of medical information,

(b) The date of the written authorization,

(c) The name of the individual or organization that is authorized to release the medical information,

(d) The name of the designated representative (individual or organization) that is authorized to receive the released information,

(e) A general description of the medical information that is authorized to be released,

(f) A general description of the purpose for the release of the medical information, and

(g) A date or condition upon which the written authorization will expire (if less than one year).

(ii) A written authorization does not operate to authorize the release of medical information not in existence on the date of written authorization, unless the release of future information is expressly authorized, and does not operate for more than one year from the date of written authorization.

(iii) A written authorization may be revoked in writing prospectively at any time.

(13) *Toxic substance or harmful physical agent* means any chemical substance, biological agent (bacteria, virus, fungus, etc.), or physical stress (noise, heat, cold, vibration, repetitive motion, ionizing and non-ionizing radiation, hypo- or hyperbaric pressure, etc.) which:

(i) Is listed in the last printed edition of the National Institute for Occupational Safety and Health (NIOSH) Registry of Toxic Effects of Chemical Substances (RTECS); or

(ii) Has yielded positive evidence of an acute or chronic health hazard in testing conducted by, or known to, the employer; or

(iii) Is the subject of a material safety data sheet kept by or known to the employer indicating that the material may pose a hazard to human health.

(14) *Trade secret* means any confidential formula, pattern, process, device, or information or compilation of information that is used in an employer's business and that gives the employer an opportunity to obtain an advantage over competitors who do not know or use it.

(D) Preservation of Records.

(1) Unless a specific occupational safety and health standard provides a different period of time, each employer shall assure the preservation and retention of records as follows:

(i) *Employee medical records.* The medical record for each employee shall be preserved and maintained for at least the duration of employment plus thirty (30) years, except that the following types of records need not be retained for any specified period:

(a) Health insurance claims records maintained separately from the employer's medical program and its records.

(b) First aid records (not including medical histories) of one-time treatment and subsequent observation of minor scratches, cuts, burns, splinters, and the like which do not involve medical treatment, loss of consciousness, restriction of work or motion, or transfer to another job, if made on-site by a non-physician and if maintained separately from the employer's medical program and its records, and

(c) The medical records of employees who have worked for less than (1) year for the employer need not be retained beyond the term of employment if they are provided to the employee upon the termination of employment.

(ii) *Employee exposure records.* Each employee exposure record shall be preserved and maintained for at least thirty (30) years, except that:

(a) Background data to environmental (workplace) monitoring or measuring, such as laboratory reports and worksheets, need only be retained for one (1) year as long as the sampling results, the collection methodology (sampling plan), a description of the analytical and mathematical methods used, and a summary of other background data relevant to interpretation of the results obtained, are retained for at least thirty (30) years; and

(b) Material safety data sheets and paragraph (C)(5)(iv) records concerning the identity of a substance or agent need not be retained for any specified period as long as some record of the identity (chemical name if known) of the substance or agent, where it was used, and when it was used is retained for at least thirty (30) years;[1]

(c) Biological monitoring results designated as exposure records by specific occupational safety and health standards shall be preserved and maintained as required by the specific standard.

(iii) *Analyses using exposure or medical records.* Each analysis using exposure or medical records shall be preserved and maintained for at least thirty (30) years.

(2) Nothing in this section is intended to mandate the form, manner, or process by which an employer preserves a record as long as the information contained in the record is preserved and retrievable, except that chest X-ray films shall be preserved in their original state.

(E) Access to Records.

(1) General.

(i) Whenever an employee or designated representative requests access to a record, the employer shall assure that access is provided in

[1]Material safety data sheets must be kept for those chemicals currently in use that are affected by the Hazard Communica-tion Standard in accordance with 29 CFR 1926.59(g).

a reasonable time, place, and manner. If the employer cannot reasonably provide access to the record within fifteen (15) working days, the employer shall within the fifteen (15) working days appraise the employee or designated representative requesting the record of the reason for the delay and the earliest date when the record can be made available.

(ii) The employer may require of the requester only such information as should be readily known to the requester and which may be necessary to locate or identify the records being requested (e.g. dates and locations where the employee worked during the time period in question).

(iii) Whenever an employee or designated representative requests a copy of a record, the employer shall assure that either:

(a) A copy of the record is provided without cost to the employer or representative,

(b) The necessary mechanical copying facilities (e.g., photocopying) are made available without cost to the employee or representative for copying the record, or

(c) The record is loaned to the employee or representative for a reasonable time to enable a copy to be made.

(iv) In the case of an original X-ray, the employer may restrict access to on-site examination or make other suitable arrangements for the temporary loan of the X-ray.

(v) Whenever a record has been previously provided without cost to an employee or designated representative, the employer may charge reasonable, nondiscriminatory administrative costs (i.e., search and copying expenses but not including overhead expenses) for a request by the employee or designated representative for additional copies of the record, except that

(a) An employer shall not charge for an initial request for a copy of new information that has been added to a record which was previously provided; and

(b) An employer shall not charge for an initial request by a recognized or certified collective bargaining agent for a copy of an employee exposure record or an analysis using exposure or medical records.

(vi) Nothing in this section is intended to preclude employees and collective bargaining agents from collectively bargaining to obtain access to information in addition to that available under this section.

(2) Employee and designated representative access.

(i) *Employee exposure records.*

(a) Except as limited by paragraph (F) of this section, each employer shall, upon request, assure the access to each employee and designated representative to employee exposure records relevant to the employee. For the purpose of this section, an exposure record relevant to the employee consists of:

(1) A record which measures or monitors the amount of a toxic substance or harmful physical agent to which the employee is or has been exposed;

(2) In the absence of such directly relevant records, such records of other employees with past or present job duties or working conditions related to or similar to those of the employee to the extent necessary to reasonably indicate the amount and nature of the toxic substances or harmful physical agents to which the employee is or has been subjected, and

(3) Exposure records to the extent necessary to reasonably indicate the amount and nature of the toxic substances or harmful physical agents at workplaces or under working conditions to which the employee is being assigned or transferred.

(b) Requests by designated representatives for unconsented access to employee exposure records shall be in writing and shall specify with reasonable particularity:

(1) The records requested to be disclosed; and

(2) The occupational health need for gaining access to these records.

(ii) *Employee medical records.*

(a) Each employer shall, upon request, assure the access of each employee to employee medical records of which the employee is the subject, except as provided in paragraph (E)(2)(ii)(d) of this section.

(b) Each employer shall, upon request, assure the access of each designated representative to the employee medical records of any employee who has given the designated representative specific written consent. Appendix A to this section contains a sample form which may be used to establish specific written consent for access to employee medical records.

(c) Whenever access to employee medical records is requested, a physician representing the employer may recommend that the employee or designated representative:

(1) Consult with the physician for the purposes of reviewing and discussing the records requested.

(2) Accept a summary of material facts and opinions in lieu of the records requested, or

(3) Accept release of the requested records only to a physician or other designated representative.

(d) Whenever an employee requests access to his or her employee medical records, and a physician representing the employer believes that direct employee access to information contained in the records regarding a specific diagnosis of a terminal illness or a psychiatric condition could be detrimental to the employee's health, the employer may inform the employee that access will only be provided to a designated representative of the employee having specific written consent, and deny the employee's request for direct access to this information only. Where a designated representative with specific written consent requests access to information so withheld, the employer shall assure the access of the designated representative to this information, even when it is known that the designated representative will give the information to the employee.

(e) A physician, nurse, or other responsible health care personnel maintaining medical records may delete from requested medical records the identity of a family member, personal friend, or fellow employee who has provided confidential information concerning an employee's health status.

(iii) *Analyses using exposure or medical records.*

(a) Each employee shall, upon request, assure the access of each employee and designated representative to each analysis using exposure or medical records concerning the employee's working conditions or workplace.

(b) Whenever access is requested to an analysis which reports the contents of employee medical records by either direct identifier (name, address, social security number, payroll number, etc.) or by information which could reasonably be used under the circumstances indirectly to identify specific employees (exact age, height, weight, race, sex, date of initial employment, job title, etc.), the employer shall assure that personal identifiers are removed before access is provided. If the employer can demonstrate that removal of personal identifiers from an analysis is not feasible, access to the personally identifiable portions of the analysis need not be provided.

(3) OSHA access.

(i) Each employer shall, upon request, and without derogation of any rights under the Constitution or the Occupational Safety and Health Act of 1970, 29 U.S.C. 651 et seq. that the employee chooses to exercise, assure the prompt access of representatives of the Assistant Secretary of Labor for Occupational Safety and Health to employee exposure and medical records and to analysis using exposure or medical records. Rules of agency practice and procedure governing OSHA access to employee medical records are contained in 29 CFR 1913.10.

(ii) Whenever OSHA seeks access to personally identifiable employee medical information by presenting to the employer a written access order pursuant to 29 CFR 1913.10(d), the employer shall prominently post a copy of the written access order and its accompanying cover letter for at least fifteen (15) working days.

(F) Trade Secrets.

(1) Except as provided in paragraph (F)(2) of this section, nothing in this section precludes an employer from deleting from records requested by a health professional, employee, or designated representative any trade secret data which discloses manufacturing processes, or discloses the percentage of a chemical substance in mixture, as long as the health professional, employee, or designated representative is notified that information has been deleted. Whenever deletion of trade secret information substantially impairs evaluation of the place where or the time when exposure to a toxic substance or harmful physical agent occurred, the employer shall provide alternative information which is sufficient to permit the requesting party to identify where and when exposure occurred.

(2) The employer may withhold specific chemical identity, including the chemical name and other specific identification of a toxic substance from a disclosable record provided that:

(i) The claim that the information withheld is a trade secret can be supported;

(ii) All other available information on the properties and effects of the toxic substance is disclosed;

(iii) The employer informs the requesting party that the specific chemical identity is being withheld as a trade secret; and

(iv) The specific chemical identity is made available to health professionals, employees and designated representatives in accordance with the specific applicable provisions of this paragraph.

(3) Where a treating physician or nurse determines that a medical emergency exists and the specific chemical identity of a toxic substance is necessary for emergency or first-aid treatment, the employer shall immediately disclose the specific chemical identity of a trade secret chemical to the treating physician or nurse, regardless of the existence of a written statement of need or a confidentiality agreement. The employer may require a written statement of need and confidentiality agreement, in accordance with the provisions of paragraphs (F)(4) and (F)(5), as soon as circumstances permit.

(4) In non-emergency situations, an employer shall, upon request, disclose a specific chemical identity, otherwise permitted to be withheld under paragraph (F)(2) of this section, to a health professional, employee, or designated representative if:

(i) The request is in writing;

(ii) The request describes with reasonable detail one or more of the following occupational health needs for the information:

(a) To assess the hazards of the chemicals to which employees will be exposed;

(b) To conduct or assess sampling of the workplace atmosphere to determine employee exposure levels;

(c) To conduct pre-assignment or periodic medical surveillance of exposed employees;

(d) To provide medical treatment to exposed employees;

(e) To select or assess appropriate personal protective equipment for exposed employees;

(f) To design or assess engineering controls or other protective measures for exposed employees; and

(g) To conduct studies to determine the health effects of exposure.

(iii) The request explains in detail why the disclosure of the specific chemical identity is essential and that, in lieu thereof, the disclosure of the following information would not enable the health professional, employee or designated representative to provide the occupational health services described in paragraph (F)(4)(ii) of this section:

(a) The properties and effects of the chemical;

(b) Measures for controlling workers' exposure to the chemical;

(c) Methods of monitoring and analyzing worker exposure to the chemical; and

(d) Methods of diagnosing and treating harmful exposures to the chemical.

(iv) The request includes a description of the procedures to be used to maintain the confidentiality of the disclosed information.

(v) The health professional, employee, or designated representative and the employer or contractor of the services of the health professional or designated representative agree in a written confidentiality agreement that the health professional, employee or designated representative will not use the trade secret information for any purpose other than the health need(s) asserted and agree not to release the information under any circumstances other than to OSHA, as provided in paragraph (F)(9) of this section, except as authorized by the terms of the agreement or by the employer.

(5) The confidentiality agreement authorized by paragraph (F)(4)(iv) of this section:

(i) May restrict the use of the information to the health purposes indicated in the written statement of need;

(ii) May provide for appropriate legal remedies in the event of a breach of the agreement, including stipulation of a reasonable pre-estimate of likely damages; and

(iii) May not include requirements for the posting of a penalty bond.

(6) Nothing in this section is meant to preclude the parties from pursuing non-contractual remedies to the extent permitted by law.

(7) If the health professional, employee or designated representative receiving the trade secret information decides that there is a need to disclose it to OSHA, the employer who provided the information shall be informed by the health professional prior to, or at the same time as, such disclosure.

(8) If the employer denies a written request for disclosure of a specific chemical identity, the denial must:

(i) Be provided to the health professional, employee or designated representative within thirty days of the request;

(ii) Be in writing;

(iii) Include evidence to support the claim that the specific chemical identity is a trade secret;

(iv) State the specific reasons why the request is being denied; and

(v) Explain in detail how alternative information may satisfy the specific medical or occupational health need without reveling the specific chemical identity.

(9) The health professional, employee, or designated representative whose request for information is denied under paragraph (F)(4) of this section may refer the request and the written denial of the request to OSHA for consideration.

(10) When a health professional employee, or designated representative refers a denial to OSHA under paragraph (F)(9) of this section, OSHA shall consider the evidence to determine if:

(i) The employer has supported the claim that the specific chemical identity is a trade secret;

(ii) The health professional employee, or designated representative has supported the claim that there is a medical or occupational health need for the information; and

(iii) The health professional, employee or designated representative has demonstrated adequate means to protect the confidentiality.

(11) (i) If OSHA determines that the specific chemical identity requested under paragraph (F)(4) of this section is not a bona fide trade secret, or that it is a trade secret but the requesting health professional, employee or designated representatives has a legitimate medical or occupational health need for the information, has executed a written confidentiality agreement, and has shown adequate means for complying with the terms of such agreement, the employer will be subject to citation by OSHA.

(ii) If an employer demonstrates to OSHA that the execution of a confidentiality agreement would not provide sufficient protection against the potential harm from the unauthorized disclosure of a trade secret specific chemical identity, the Assistant Secretary may issue such orders or impose such additional limitations or conditions upon the disclosure of the requested chemical information as may be appropriate to assure that the occupational health needs are met without an undue risk of harm to the employer.

(12) Notwithstanding the existence of a trade secret claim, an employer shall, upon request, disclose to the Assistant Secretary any information which this section requires the employer to make available. Where there is a trade secret claim, such claim shall be made no later than at the time the information is provided to the Assistant Secretary so that suitable determinations of trade secret status can be made and the necessary protections can be implemented.

(13) Nothing in this paragraph shall be construed as requiring the disclosure under any circumstances of process or percentage of mixture information which is trade secret.

(G) Employee Information.

(1) Upon an employee's first entering into employment, and at least annually thereafter, each employer shall inform current employees covered by this section of the following:

(i) The existence, location, and availability of any records covered by this section;

(ii) The person responsible for maintaining and providing access to records; and

(iii) Each employee's rights of access to these records.

(2) Each employer shall keep a copy of this section and its appendices, and make copies readily available, upon request, to employees. The employer shall also distribute to current employees any informational materials concerning this section which are made available to the employer by the Assistant Secretary of Labor for Occupational Safety and Health.

(H) Transfer of Records.

(1) Whenever an employer is ceasing to do business, the employer shall transfer all records subject to this section to the successor employer. The successor employer shall receive and maintain these records.

(2) Whenever an employer is ceasing to do business and there is no successor employer to receive and maintain the records subject to this standard, the employer shall notify affected current employees of their rights of access to records at least three (3) months prior to the cessation of the employer's business.

(3) Whenever an employer either is ceasing to do business and there is no successor employer to receive and maintain the records, or intends to dispose of any records required to be preserved for at least thirty (30) years, the employer shall:

(i) Transfer the records to the Director of the National Institute for Occupational Safety and Health (NIOSH) if so required by a specific occupational safety and health standard: or

(ii) Notify the Director of NIOSH in writing of the impending disposal of records at least three (3) months prior to the disposal of the records.

(4) Where an employer regularly disposes of records required to be preserved for at least thirty (30) years, the employer may, with at least (3) months notice, notify the Director of NIOSH on an annual basis of the records intended to be disposed of in the coming year.

(I) Appendices.

The information contained in appendices A and B to this section is not intended, by itself, to create any additional obligations not otherwise imposed by this section nor detract from any existing obligation.

APPENDIX A TO ARTICLE 1926.33

SAMPLE AUTHORIZATION LETTER FOR THE RELEASE OF EMPLOYEE MEDICAL RECORD INFORMATION TO A DESIGNATED REPRESENTATIVE

(NON-MANDATORY)

I, ______________________________ *(full name of worker/patient)*, hereby authorize ______________________________ *(individual or organization holding the medical records)* to release to______________________________ *(individual or organization authorized to receive the medical information)*, the following medical information from my personal medical records:

____

____

(Describe generally the information desired to be released)

I give my permission for this medical information to be used for the following purpose:

____

____

(Describe generally the purpose)

but I do not give permission for any other use or re-disclosure of this information.

(Note: Several extra lines are provided below so that you can place additional restrictions on this authorization letter if you want to. You may, however, leave these lines blank. On the other hand, you may want to (1) specify a particular expiration date for this letter — if less than one year; (2) describe medical information to be created in the future that you intend to be covered by this authorization letter; or (3) describe portions of the medical information in your records which you do not intend to be released as a result of this letter.)

____

____

____

____

Full name of Employee or Legal Representative

____

Signature of Employee or Legal Representative

____

____

Date of Signature

APPENDIX B TO ARTICLE 1926.33

AVAILABILITY OF NIOSH REGISTRY OF TOXIC EFFECTS OF CHEMICAL SUBSTANCES (RTECS)

(NON-MANDATORY)

Section 1926.33 applies to all employee exposure and medical records, and analyses thereof, of employees exposed to toxic substances or harmful physical agents (paragraph (b)(2)). The term toxic substance or harmful physical agent is defined by paragraph (c)(13) to encompass chemical substances, biological agents, and physical stresses for which there is evidence of harmful health effects. The regulation uses the latest printed edition of the National Institute for Occupational Safety and Health (NIOSH) Registry of Toxic Effects of Chemical Substances (RTECS) as one of the chief sources of information as to whether evidence of harmful health effects exists. If a substance is listed in the latest printed RTECS, the regulation applies to exposure and medical records (and analyses of these records) relevant to employees exposed to the substance.

It is appropriate to note that the final regulation does not require that employers purchase a copy of RTECS, and many employers need not consult RTECS to ascertain whether their employee exposure or medical records are subject to the rule. Employers who do not currently have the latest printed edition of the NIOSH RTECS, however, may desire to obtain a copy. The RTECS is issued in an annual printed edition as mandated by section 20(a)(6) of the Occupational Safety and Health Act (29 U.S.C. 669(a)(6)).

The Introduction to the 1980 printed edition describes the RTECS as follows:

"The 1980 edition of the Registry of Toxic Effects of Chemical Substances, formerly known as the Toxic Substances list, is the ninth revision prepared in compliance with the requirements of Section 20(a)(6) of the Occupational Safety and Health Act of 1970 (Public Law 91-596). The original list was completed on June 28, 1971, and has been updated annually in book format. Beginning in October 1977, quarterly revisions have been provided in microfiche. This edition of the Registry contains 168,096 listings of chemical substances: 45,156 are names of different chemicals with their associated toxicity data and 122,940 are synonyms. This edition includes approximately 5,900 new chemical compounds that did not appear in the 1979 Registry. (p. xi)

"The Registry's purposes are many, and it serves a variety of users. It is a single source document for basic toxicity information and for other data, such as chemical identifiers and information necessary for the preparation of safety directives and hazard evaluations for chemical substances. The various types of toxic effects linked to literature citations provide researchers and occupational health scientists with an introduction to the toxicological literature, making their own review of the toxic hazards of a given substance easier. By presenting data on the lowest reported doses that produce effects by several routes of entry in various species, the Registry furnishes valuable information to those responsible for preparing safety data sheets for chemical substances in the workplace. Chemical and production engineers can use the Registry to identify the hazards which may be associated with chemical intermediates in the development of final products, and thus can more readily select substitutes or alternative processes which may be less hazardous. Some organizations, including health agencies and chemical companies, have included the NIOSH Registry accession numbers with the listing of chemicals in their files to reference toxicity information associated with those chemicals. By including foreign language chemical names, a start has been made toward providing rapid identification of substances produced in other countries.(p. xi)

"In this edition of the Registry, the editors intend to identify all known toxic substances which may exist in the environment and to provide pertinent data on the toxic effects from known doses entering an organism by any route described. (p. xi)

"It must be reemphasized that, the entry of a substance in the Registry does not automatically mean that it must be avoided. A listing does mean, however, that the substance has the documented potential of being harmful if misused, and care must be exercised to prevent tragic consequences. Thus, the Registry lists many substances that are common in everyday life and are in nearly every household in the United States. One can name a variety of such dangerous substances: prescription and non-prescription drugs; food additives; pesticide concentrates, sprays, and dusts; fungicides; herbicides; paints; glazes, dyes; bleaches and other household cleaning agents; alkalies; and various solvents and diluents. The list is extensive because chemicals have become an integral part of our existence."

The RTECS printed edition may be purchased from the Superintendent of Documents, U.S. Government Printing Office (GPO), Washington, DC 20402 (202-783-3238).

Some employers may desire to subscribe to the quarterly update to the RTECS which is published in a microfiche edition. An annual subscription to the quarterly microfiche may be purchased from the GPO (Order the "Microfiche Edition, Registry of Toxic Effects of Chemical Substances"). Both the printed edition and the microfiche edition of RTECS are available for review at many university and public libraries throughout the country. The latest RTECS editions may also be examined at the OSHA Technical Data Center, Room N2439 — Rear, U.S. Department of Labor, 200 Constitution Avenue, NW., Washington, DC 20210 (202-219-7500), or at any OSHA Regional or Area Office (See major city telephone directories under U.S. Government — Labor Department).

1926.34 Means of Egress

(A) General.

In every building or structure, exits shall be so arranged and maintained as to provide free and unobstructed egress from all parts of the building or structure at all times when it is occupied. No lock or fastening to prevent free escape from the inside of any building shall be installed except in mental, penal, or corrective institutions where supervisory personnel are continually on duty and effective provisions are made to remove occupants in case of fire or other emergency.

(B) Exit Marking.

Exits shall be marked by a readily visible sign. Access to exits shall be marked by readily visible signs in all cases where the exit or way to reach it is not immediately visible to the occupants.

(C) Maintenance and Workmanship.

Means of egress shall be continually maintained free of all obstructions or impediments to full instant use in the case of fire or other emergency.

1926.35 Employee Emergency Action Plans

(A) Scope and Application.

This section applies to all emergency action plans required by a particular OSHA standard. The emergency action plan shall be in writing (except as provided in the last sentence of paragraph (E)(3) of this section) and shall cover those designated actions employers and employees must take to ensure employee safety from fire and other emergencies.

(B) Elements.

The following elements, at a minimum, shall be included in the plan:

(1) Emergency escape procedures and emergency escape route assignments;

(2) Procedures to be followed by employees who remain to operate critical plant operations before they evacuate;

(3) Procedures to account for all employees after emergency evacuation has been completed;

(4) Rescue and medical duties for those employees who are to perform them;

(5) The preferred means of reporting fires and other emergencies; and

(6) Names or regular job titles of persons or departments who can be contacted for further information or explanation of duties under the plan.

(C) Alarm System.

(1) The employer shall establish an employee alarm system which complies with Article 1926.159.

(2) If the employee alarm system is used for alerting fire brigade members, or for other purposes, a distinctive signal for each purpose shall be used.

(D) Evacuation.

The employer shall establish in the emergency action plan the types of evacuation to be used in emergency circumstances.

(E) Training.

(1) Before implementing the emergency action plan, the employer shall designate and train a sufficient number of persons to assist in the safe and orderly emergency evacuation of employees.

(2) The employer shall review the plan with each employee covered by the plan at the following times;

(i) Initially when the plan is developed;

(ii) Whenever the employee's responsibilities or designated actions under the plan change; and

(iii) Whenever the plan is changed.

(3) The employer shall review with each employee upon initial assignment those parts of the plan which the employee must know to protect the employee in the event of an emergency. The written plan shall be kept at the workplace and made available for employee review. For those employers with 10 or fewer employees, the plan may be communicated orally to employees and the employer need not maintain a written plan.

1926.50 Medical Services and First Aid

(A) The employer shall ensure the availability of medical personnel for advice and consultation on matters of occupational health.

(B) Provisions shall be made prior to commencement of the project for prompt medical attention in case of serious injury.

(C) In the absence of an infirmary, clinic, hospital, or physician that is reasonably accessible in terms of time and distance to the worksite, a person who has valid certification in first-aid training from the Bureau of Mines, the American Red Cross, or equivalent training that can be verified by documentary evidence, shall be available at the worksite to render first aid.

(D) First-Aid Supplies.

(1) First-aid supplies approved by the consulting physician shall be easily accessible when required.

(2) The first-aid kit shall consist of materials approved by the consulting physician and placed in a weatherproof container with individual sealed packages for each item. The contents of the first aid kit shall be checked by the employer before being sent out on each job and at least weekly on each job to ensure that the expended items are replaced.

(E) Proper equipment for prompt transportation of the injured person to a physician or hospital, or a communication system for contacting necessary ambulance service, shall be provided.

(F) The telephone numbers of the physicians, hospitals, or ambulances, shall be conspicuously posted.

(G) Where the eyes or body of any person may be exposed to injurious corrosive materials, suitable facilities for quick drenching or flushing of the eyes and body shall be provided within the work area for immediate emergency use.

1926.51 Sanitation

(A) Potable Water.

(1) An adequate supply of potable water shall be provided in all places of employment.

(2) Portable containers used to dispense drinking water shall be capable of being tightly closed, and equipped with a tap. Water shall not be dipped from containers.

(3) Any container used to distribute drinking water shall be marked as to the nature of its contents and not used for any other purpose.

(4) The common drinking cup is prohibited.

(5) Where single-service cups (to be used but once) are supplied, both a sanitary container for the unused cups and a receptacle for disposing of the used cups shall be provided.

(6) *Potable water* means water which meets the quality standards prescribed in the U.S. Public Health Service Drinking Water Standards, published in 42 CRF part 72, or water which is approved for drinking purposes by the State or local authority having jurisdiction.

(B) Nonpotable Water.

(1) Outlets for nonpotable water, such as water for industrial or firefighting purposes only, shall be identified by signs meeting the requirements of Subpart G of this part, to indicate clearly that

the water is unsafe and is not to be used for drinking, washing, or cooking purposes

(2) There shall be no cross-connection, open or potential, between a system furnishing potable water and a system furnishing nonpotable water.

(C) Toilets at Construction Jobsites.

(1) Toilets shall be provided for employees according to the following table:

Table D-1

Number of employees	Minimum number of facilities
20 or less	1
20 or more	1 toilet seat and 1 urinal per 40 workers
200 or more	1 toilet seat and 1 urinal per 50 workers

(2) Under temporary field conditions, provisions shall be made to assure not less than one toilet facility is available.

(3) Job sites not provided with a sanitary sewer shall be provided with one of the following toilet facilities unless prohibited by local codes.

(i) Privies (where their use will not contaminate ground or surface water);

(ii) Chemical toilets;

(iii) Recirculating toilets;

(iv) Combustion toilets.

(4) The requirements of this paragraph (C) for sanitation facilities shall not apply to mobile crews having transportation readily available to nearby toilet facilities.

(D) Food Handling.

(1) All employee food service facilities and operations shall meet the applicable laws, ordinances, and regulations of the jurisdictions in which they are located.

(2) All employee food service facilities and operations shall be carried out in accordance with sound hygienic principles. In all places of employment where all or part of the food service is provided, the food dispensed shall be wholesome, free from spoilage, and shall be processed, prepared, handled, and stored in such a manner as to be protected against contamination.

(E) Temporary Sleeping Quarters.

When temporary sleeping quarters are provided, they shall be heated, ventilated, and lighted.

(F) Washing Facilities.

(1) The employer shall provide adequate washing facilities for employees engaged in the application of paints, coating, herbicides, or insecticides, or in other operations where contaminants may be harmful to the employees. Such facilities shall be in near proximity to the worksite and shall be so equipped as to enable employees to remove such substances.

(2) General. Washing facilities shall be maintained in a sanitary condition.

(3) Lavatories.

(i) Lavatories shall be made available in all places of employment. The requirements of this subdivision do not apply to mobile crews or to normally unattended work locations if employees working at these locations have transportation readily available to nearby washing facilities which meet the other requirements of this paragraph.

(ii) Each lavatory shall be provided with hot and cold running water, or tepid running water.

(iii) Hand soap or similar cleansing agents shall be provided.

(iv) Individual hand towels or sections thereof, of cloth or paper, warm air blowers or clean individual sections of continuous cloth toweling, convenient to the lavatories, shall be provided.

(4) Showers.

(i) Whenever showers are required by a particular standard, the showers shall be provided in accordance with paragraphs (F)(4)(ii) through (v) of this section,

(ii) One shower shall be provided for each 10 employees of each sex, or numerical fraction thereof, who are required to shower during the same shift.

(iii) Body soap or other appropriate cleansing agents convenient to the showers shall be provided as specified in paragraph (F)(3)(iii) of this section.

(iv) Showers shall be provided with hot and cold water feeding a common discharge line.

(v) Employees who use showers shall be provided with individual clean towels.

(G) Eating and Drinking Areas.

No employee shall be allowed to consume food or beverages in a toilet room nor in any area exposed to a toxic material.

(H) Vermin Control.

Every enclosed workplace shall be so constructed, equipped, and maintained, so far as reasonably practicable, as to prevent the entrance or harborage of rodents, insects, and other vermin. A continuing and effective extermination program shall be instituted where their presence is detected.

(I) Change Rooms.

Whenever employees are required by a particular standard to wear protective clothing because of the possibility of contamination with toxic materials, change rooms equipped with storage facilities for street clothes and separate storage facilities for the protective clothing shall be provided.

1926.52 Occupational Noise Exposure

(A) Protection against the effects of noise exposure shall be provided when the sound levels exceed those shown in Table D-2 of this section when measured on the A-scale of a standard sound level meter at slow response.

(B) When employees are subjected to sound levels exceeding those listed in Table D-2 of this section, feasible administrative or engineering controls shall be utilized. If such controls fail to reduce sound levels within the levels of the table, personal protective equipment as required in Subpart E, shall be provided and used to reduce sound levels within the levels of the table.

(C) If the variations in noise level involve maxima at intervals of 1 second or less, it is to be considered continuous.

(D (1) In all cases where the sound levels exceed the values shown herein, a continuing, effective hearing conservation program shall be administered.

Table D-2 – Permissible Noise Exposures

Duration per day, hours	Sound level dBA slow response
8	90
6	92
4	95
3	97
2	100
1½	102
1	105
½	110
¼ or less	115

1926.100 Head Protection

(A) Employees working in areas where there is a possible danger of head injury from impact, or from falling or flying objects, or from electrical shock and burns, shall be protected by protective helmets.

(B) Helmets for the protection of employees against impact and penetration of falling and flying objects shall meet the specifications contained in American National Standards Institute, Z89.1 1969, Safety Requirements for Industrial head Protection.

(C) Helmets for the head protection of employees exposed to high-voltage electrical shock and burns shall meet the specification contained in American National Standards Institute.

1926.101 Hearing Protection

(A) Whenever it is not feasible to reduce the noise levels or duration of exposures to those specified . . . ear protective devices shall be provided and used.

(B) Ear protective devices inserted in the ear shall be fitted or determined individually by competent persons.

(C) Plain cotton is not an acceptable protective device.

1926.102 Eye and Face Protection

(A) Employees shall be provided with eye and face protection equipment when machines or operations present potential eye or face injury from physical, chemical, or radiation agents.

(1) Eye and face protection equipment required by this part shall meet the requirements specified in American National Standards Institute Z87.1 1968, Practice for Occupational and Educational Eye and Face Protection.

(2) Employees whose vision requires the use of corrective lenses in spectacles, when required by this regulation to wear eye protection, shall be protected by goggles or spectacles of the following types:

(a) Spectacles whose protective lenses provide optical correction.

(b) Goggles that can be worn over corrective spectacles without disturbing the adjustment of the spectacles.

(c) Goggles that incorporate corrective lenses mounted behind the protective lenses.

(3) Face and eye protection equipment shall be kept clean and in good repair. The use of this type equipment with structural or optical defects shall be prohibited.

Subpart H. Material Handling, Storage, Use, And Disposal

1926.252

(A) All materials stored in tiers shall be stacked, racked, blocked, interlocked, or otherwise secured to prevent sliding, falling or collapsing.

(B) Bagged materials shall be stacked by stepping the layers and cross-keying the bags at least every 10 bags high.

(C) Materials shall not be stored on scaffolds or runways except for supplies needed for immediate operations.

(D) Brick stacks shall not be more than 7 feet in height. When a loose brick stack reaches a height of 4 feet, it shall be tapered back 2 inches in every foot of height above the 4-foot level.

(E) When masonry blocks are stacked higher than 6 feet, the stack shall be tapered back one-half block per tier above the 6-foot level.

Subpart I. Tools — Hand and Power

1926.300 General Requirements

(A) All hand and power tools and similar equipment, whether furnished by the employer or the employee, shall be maintained in a safe condition.

(B) When power-operated tools are designed to accommodate guards, they shall be equipped with such guards when in use.

Subpart K. Electrical

1926.401 Grounding and Bonding

(A) All temporary wiring shall be effectively grounded in accordance with the National Electrical Code. NFPA 70-1971; ANSI CI-1971 (Rev. of CI-1968) Articles 305 and 310.

(B) Construction-site precautions shall be taken to make any necessary open wiring inaccessible to unauthorized personnel.

(C) Temporary lights shall be equipped with heavy-duty electrical cords with connections and insulation maintained in safe condition. Temporary lights shall not be suspended by their electric cords unless cords and lights are designed for this means of suspension.

(D) Splices shall have insulation equal to that of the cable.

Subpart L. Ladders and Scaffolding

1926.450 Ladders

(A) Except where either permanent or temporary stairways or suitable ramps or runways are provided, ladders described in this subpart shall be used to give safe access to all elevations.

(B) The use of ladders with broken or missing rungs or steps, broken or split side rails, or other defective construction is prohibited. When ladders with such defects are discovered, they shall be immediately withdrawn from service. Inspection of metal ladders shall include checking for corrosion of interiors of open-end hollow rungs.

(C) Manufactured Ladders.

(1) Manufactured portable wooden ladders provided by the employer shall be in accordance with the provisions of the employer shall be in accordance with the provisions of the American National Standards Institute. A14.1-1968 Safety Code for Ladders.

(2) Portable metal ladders shall be of strength equivalent to that of wood ladders. Manufactured portable metal ladders provided by the employer shall be in accordance with the provisions of the American National Standards Institute. A 14.2-1956, Safety Code for Portable Metal Ladders.

(3) Fixed ladders shall be in accordance with the provisions of the American National Standards Institute, A 14.3 1956 Safety Code for Fixed Ladders.

(4) Portable ladder feet shall be placed on a substantial base, and the area around the bottom of the ladder shall be kept clear.

(5) Portable ladder shall be used at such a pitch that the horizontal distance from the top support to the foot of the ladder is about one-quarter of the working length of the ladder (the length of the ladder between the foot and the top support). Ladders shall not be used in a horizontal position as platforms, runways, or scaffolds.

(6) Ladders shall not be placed in passageways, doorways, driveways, or any location where they might be displaced by activities being conducted on any other work, unless protected by barricades or guards.

(7) The side rails shall extend not less than 36 inches above the landing. When this is not practical, grab rails which provide a secure grip for an employee moving to or from the point of access shall be installed.

(8) Portable ladders in use shall be tied, blocked, or otherwise secured to prevent their being displaced.

(9) Portable metal ladders shall not be used for electrical work or where they may contact electrical conductors.

(D) Job-Made Ladders.

(1) Job-made ladders shall be constructed for intended use. If a ladder is to provide the only means of access for exit from a working area for 25 or more employees, or if simultaneous two-way traffic is expected, a double-cleat ladder shall be installed.

(2) Double-cleat ladders shall not exceed 24 feet in length.

(3) Single-cleat ladders shall not exceed 30 feet in length between supports (base and top landing). If ladders are to connect different landings or if the length exceeds the maximum length, two or more separate ladders shall be used, offset with a platform between each ladder. Guardrails and toe boards shall be erected on the exposed side of the platforms. (See 1926.4521 [E].)

(4) The width of single-cleat ladders shall be at least 15 inches, but not more than 20 inches between rails at the top.

(5) Side rails shall be parallel or flared top to bottom by not more than one-quarter of an inch for each 2 feet of length.

(6) Wood side rails of ladders having cleats shall be not less than 1½ inches thick and 3½ inches deep (2 inches by 4 inches nominal).

(7) It is preferable that side rails be continuous. If a splice is necessary to attain the required length, the splice must develop the full strength of a continuous side rail of the same length.

(8) 2-inch by 4-inch lumber shall be used for side rails of single-cleat ladders up to 16 feet long; 3-inch by 6-inch lumber shall be used for single-cleat ladders from 16 feet to 30 feet in length.

(9) 2-inch by 4-inch lumber shall be used for side and middle rails of double-cleat ladders up to 12 feet in length; 2-inch by 6-inch lumber for double-cleat ladders from 12 to 24 feet in length.

(10) Wood cleats shall have the following dimensions:

Length of cleat (in.)	Thickness (in.)	Width (in.)
Up to and including 20	¾	3
Over 20 and up to and including 30	¾	3¾

(11) Cleats may be made of any wood type similar to the rails.

(12) Cleats shall be inset into the edges of the side rails one-half inch, or filler blocks shall be used on the rails between the cleats. The cleats shall be secured to each rail with three 10d common wire nails or other fasteners of equivalent strength. Cleats shall be uniformly spaced 12 inches top to top.

1926.451 Scaffolding

(A) General requirements.

(1) Scaffolds shall be erected in accordance with requirements of this section.

(2) The footing or anchorage for scaffolds shall be sound, rigid and capable of carrying the maximum load without settling or displacement. Unstable objects such as barrels, boxes, loose brick, or concrete blocks, shall not be used to support scaffolds or planks.

(3) No scaffold shall be erected, moved, dismantled or altered except under the supervision of a competent person.

(4) Guardrails and toeboards shall be installed on all open sides and ends of platforms more than 10 feet above the ground or floor, except needle-beam scaffolds and floats (see paragraphs (p) and (w) of this section). Scaffolds 4 feet to 10 feet in height, having a minimum horizontal dimension in either direction of less than 45 inches, shall have standard guardrails installed on all open sides and ends of the platform.

(5) Guardrails shall be 2 × 4 inches, or the equivalent, approximately 42 inches high, with a midrail, when required. Supports shall be at intervals not to exceed 8 feet. Toeboards shall be minimum of 4 inches in height.

(6) Where persons are required to work or pass under the scaffold, scaffolds shall be provided with a screen between the toeboard and the guardrail, extending along the entire opening, consisting of No. 18 gauge U.S. Standard wire ½-inch mesh or the equivalent.

(7) Scaffolds and their components shall be capable of supporting without failure at least 4 times the maximum intended load.

(8) Any scaffold including accessories such as braces, brackets, trusses, screw legs, ladders, etc. damaged or weakened from any cause shall be immediately repaired or replaced.

(9) All load-carrying timber members of scaffold framing shall be a minimum of 1,500 fiber (Stress Grade) construction-grade lumber.

All dimensions are nominal sizes as provided in the American Lumber Standards, except that where rough sizes are noted, only rough or undressed lumber of the size specified will satisfy minimum requirements.

(10) All planking shall be scaffold grades, or equivalent, as recommended by approved grading rules for the species of wood used. The maximum permissible spans for 2- × 10-inch or wider planks shall be as shown in Table L-3 on the following page.

(11) The maximum permissible span for 1¼- × 9-inch or wider plank of full thickness shall be 4 feet with a medium-duty loading of 50 p.s.f.

Table L-3 – Material

	Full-thickness undressed lumber			Nominal-thickness lumber[1]	
Working load (p.s.f.)	25	50	75	25	50
Permissible span (ft.)	10	8	6	8	6

[1]Nominal-thickness lumber not recommended for heavy-duty use.

(12) All planking of platforms shall be overlapped (minimum 12 inches), or secured from movement.

(13) An access ladder or equivalent safe access shall be provided.

(14) Scaffold planks shall extend over their end supports not less than 6 inches nor more than 12 inches.

(15) The poles, legs, or uprights of scaffolds shall be plumb and securely and rigidly braced to prevent swaying and displacement.

(16) Overhead protection shall be provided for men on a scaffold exposed to overhead hazards.

(17) Slippery conditions on scaffolds shall be eliminated as soon as possible after they occur.

(18) No welding, burning, riveting, or open flame work shall be performed on any staging suspended by means of fiber or synthetic rope. Only treated or protected fiber or synthetic rope. Only treated or protected fiber or synthetic shall be used for or near any work involving the use of corrosive substances or chemicals. Special requirements for boatswain's chairs and float or ship scaffolds are contained in paragraphs (l) and (w) of this section.

(19) Wire, synthetic, or fiber rope used for scaffold suspension shall be capable of supporting at least 6 times the rated load.

(20) The use of shore or lean-to scaffolds is prohibited.

(21) Lumber sizes, when used in this subpart, refer to nominal sizes except where otherwise stated.

(22) Materials being hoisted onto a scaffold shall have a tag line.

(23) Employees shall not work on scaffolds during storms or high winds.

(24) Tools, materials, and debris shall not be allowed to accumulate in quantities to cause a hazard.

(B) Wood Pole Scaffolds.

(1) Scaffold poles shall bear on a foundation of sufficient size and strength to spread the load from the pole over a sufficient area to prevent settlement. All poles shall be set plumb.

(2) Where wood poles are spliced, the ends shall be squared and the upper section shall rest squarely on the lower section. Wood splice plates shall be provided on at least two adjacent sides and shall be not less than 4 feet in length, overlapping the abutted ends equally, and have the same width and not less than the cross-sectional area of the pole. Splice plates or other materials of equivalent strength may be used.

(3) Independent pole scaffolds shall be set as near to the wall of the building as practicable.

(4) All pole scaffolds shall be securely guyed or tied to the building or structure. Where the height or length exceeds 25 feet, the scaffold shall be secured at intervals not greater than 25 feet vertically and horizontally.

(5) Putlogs or bearers shall be set with their greater dimension vertical, long enough to project over the ledgers of the inner and outer rows of poles at least 3 inches for proper support.

(6) Every wooden putlog on single pole scaffolds shall be reinforced with a 3/16- × 2-inch steel strip, or equivalent, secured to its lower edge throughout its entire length.

(7) Ledgers shall be long enough to extend over two pole spaces. Ledgers shall not be spliced between the poles. Ledgers shall be reinforced by bearing blocks securely nailed to the side of the pole to form a support for the ledger.

(8) Diagonal bracing shall be provided to prevent the poles from moving in a direction parallel with the wall of the building, or from buckling.

(9) Cross bracing shall be provided between the inner and outer sets of poles in independent pole scaffolds. The free ends of pole scaffolds shall be cross braced.

(10) Full diagonal face bracing shall be erected across the entire face of pole scaffolds in both directions. The braces shall be spliced at the poles. The inner row of poles on medium and heavy-duty scaffolds shall be braced in a similar manner.

(11) Platform planks shall be laid with their edges close together so the platform will be tight with no spaces through which tools or fragments of material can fall.

(12) Where planking is lapped, each plank shall lap its end supports at least 12 inches. Where the ends of planks abut each other to form a flush floor, the butt joint shall be at the centerline of a pole. The abutted ends shall rest on separate bearers. Intermediate beams shall be provided where necessary to prevent dislodgment of planks due to deflection, and the ends shall be secured to prevent their dislodgment.

(13) When a scaffold materially changes its direction, the platform planks shall be laid to prevent tipping. The planks meet the corner putlog at an angle shall be laid first, extending over the diagonally placed putlog far enough to have a good safe bearing, but not far enough to involve any danger from tipping. The planking running in the opposite direction at an angle shall be laid so as to extend over and rest on the first layer of planking.

(14) When moving platforms to the next level, the old platform shall be left undisturbed until the new putlogs or bearers have been set in place, ready to receive the platform planks.

(15) Guardrails, made of lumber not less than 2 × 4 inches (or other material providing equivalent protection), approximately 42 inches high, with a midrail of 1- × 6-inch lumber (or other material providing equivalent protection), and toeboards, shall be installed at all open sides and ends on all scaffolds more than 10 feet above the ground or floor. Toeboards shall be a minimum of 4 inches in height. Wire mesh shall be installed in accordance with paragraph (A)(6) of this section, when required.

(16) All wood pole scaffolds 60 feet or less in height shall be constructed and erected in accordance with Tables L-4 to 10. If they are over 60 feet in height, they shall be designed by a qualified engineer competent in this field, and it shall be constructed and erected in accordance with such design.

(C) Tube and Coupler Scaffolds.

(1) A light duty tube and coupler scaffold shall have all posts, bearers, runners, and bracing of nominal 2-inch O.D. steel tubing. The posts shall be spaced no more than 6 feet apart by 10 feet along the length of the scaffold. Other structural metals when used must be designed to carry an equivalent load. No dissimilar metals shall be used together.

(2) A medium duty tube and coupler scaffold shall have all posts, runners, and bracing of nominal 2-inch O.D. steel tubing. Posts spaced not more than 6 feet apart by 8 feet along the length of the scaffold shall have bearers of nominal 2½-inch O.D. steel tubing. Posts spaced not more than 5 feet apart by 8 feet along the length of the scaffold shall have bearers of nominal 2-inch O.D. steel tubing. Other structural metals, when used, must be designed to carry an equivalent load. No dissimilar metals shall be used together.

(3) A heavy duty tube and coupler scaffold shall have all posts, runners, and bracing of nominal 2-inch O.D. steel tubing, with the posts spaced not more than 6 feet by 6 feet-6 inches. Other structural metals, when used, must be designed to carry an equivalent load. No dissimilar metals shall be used together.

(4) Tube and coupler scaffolds shall be limited in heights and working levels to those permitted in Tables L-10, 11, and 12. Drawings and specifications of all tube and coupler scaffolds above the limitations in Tables L-10, 11, and 12 shall be designed by a qualified engineer competent in this field.

(5) All tube and coupler scaffolds shall be constructed and erected to support four times the maximum intended loads, as set forth in Tables L-10, 11, and 12, or as set forth in the specifications by a licensed professional engineer competent in this field.

Table L-4 – Minimum Nominal Size and Maximum Spacing of Members of Single Pole Scaffolds – Light Duty		
	Maximum height of scaffold	
	20 ft.	**60 ft.**
Uniformly distributed load	Not to exceed 25 p.s.f.	
Poles or uprights	2 × 4 in.	4 × 4 in.
Pole spacing (longitudinal)	6 ft. 0 in.	10 ft. 0 in.
Maximum width of scaffold	5 ft. 0 in.	5 ft. 0 in.
Bearers or putlogs 3 ft. 0 in. width	2 × 4 in.	2 × 4 in.
Bearers or putlogs 5 ft. 0 in. width	2 x 6 in. or 3 x 4 in.	2 × 6 in. or 3 × 4 in. (rough)
Ledgers	1 × 4 in.	1¼ × 9 in.
Planking	1¼ × 9 in. (rough)	2 × 10 in.
Vertical spacing of horizontal members	7 ft. 0 in.	9 ft. 0 in.
Bracing, horizontal and diagonal	1 × 4 in.	1 × 4 in.
Tie-ins	1 × 4 in.	1 × 4 in.
Toeboards	4 in. high (minimum)	4 in. high (minimum)
Guardrail	2 × 4 in.	2 × 4 in.
All members except planking are used on edge.		

Table L-5 – Minimum Nominal Size and Max. Spacing of Members of Single Pole Scaffolds – Medium Duty	
Uniformly distributed load	Not to exceed 50 p.s.f.
Maximum height of scaffold	60 ft.
Poles or uprights	4 × 4 in.
Pole spacing (longitudinal)	8 ft 0 in.
Maximum width of scaffold	5 ft. 0 in.
Bearers or putlogs	2 × 10 in. or 3 × 4 in.
Spacing of bearers or putlogs	8 ft. 0 in.
Ledgers	2 × 10 in.
Vertical spacing of horiz. members	7 ft. 0 in.
Bracing, horizontal	1 × 6 in. or 1¼ × 4 in.
Bracing, diagonal	1 × 4 in.
Tie-ins	1 × 4 in.
Planking	2 × 10 in.
Toeboards	4-in. high (minimum)
Guardrail	2 × 4 in.
All members except planking are used on edge.	

Table L-6 – Minimum Nominal Size and Max. Spacing of Members of Single Pole Scaffolds – Heavy Duty	
Uniformly distributed load	Not to exceed 75 p.s.f.
Maximum height of scaffold	60 ft.
Poles or uprights	4 × 6 in.
Pole spacing (longitudinal)	6 ft 0 in.
Maximum width of scaffold	5 ft. 0 in.
Bearers or putlogs	2 × 10 in. or 3 × 5 in.
Spacing of bearers or putlogs	6 ft. 0 in.
Ledgers	2 × 10 in.
Vertical spacing of horiz. members	6 ft. 6 in.
Bracing, horizontal and diagonal	2 × 4 in.
Tie-ins	1 × 4 in.
Planking	2 × 10 in.
Toeboards	4-in. high (minimum)
Guardrail	2 × 4 in.
All members except planking are used on edge.	

Table L-7 – Minimum Nominal Size and Maximum Spacing of Members of Independent Pole Scaffold – Light Duty		
	Maximum height of scaffold	
	20 ft.	**60 ft.**
Uniformly distributed load	Not to exceed 25 p.s.f.	
Poles or uprights	2 × 4 in.	4 × 4 in.
Pole spacing (longitudinal)	6 ft. 0 in.	10 ft. 0 in.
Pole spacing (transverse)	6 ft. 0 in.	10 ft. 0 in.
Ledgers	1¼ × 4 in.	1¼ × 9 in.
Bearers to 3 ft. 0 in. span	2 × 4 in.	2 × 4 in.
Bearers to 10 ft. 0 in. span	2 × 6 in. or 3 × 4 in.	2 × 10 in. (rough) or 3 x 8 in.
Planking	1¼ × 9 in.	2 × 10 in.
Vertical spacing of horizontal members	7 ft. 0 in.	7 ft. 0 in.
Bracing, horizontal and diagonal	1 × 4 in.	1 × 4 in.
Tie-ins	1 × 4 in.	1 × 4 in.
Toeboards	4 in. high	4 in. high (minimum)
Guardrail	2 × 4 in.	2 × 4 in.
All members except planking are used on edge.		

Table L-8 – Minimum Nominal Size and Max. Spacing of Members of Independent Pole Scaffold – Medium Duty	
Uniformly distributed load	Not to exceed 50 p.s.f.
Maximum height of scaffold	60 ft.
Poles or uprights	4 × 4 in.
Pole spacing (longitudinal)	8 ft. 0 in.
Pole spacing (transverse)	8 ft. 0 in.
Ledgers	2 × 10 in.
Vertical spacing of horiz. members	6 ft. 0 in.
Spacing of bearers	8 ft. 0 in.
Bearers	2 × 10 in.
Bracing, horizontal	1 × 6 in. or 1¼ × 4 in.
Bracing, diagonal	1 × 4 in.
Tie-ins	1 × 4 in.
Planking	2 × 10 in.
Toeboards	4-in. high (minimum)
Guardrail	2 × 4 in.
All members except planking are used on edge.	

Table L-9 – Minimum Nominal Size and Max. Spacing of Members of Independent Pole Scaffold – Heavy Duty	
Uniformly distributed load	Not to exceed 75 p.s.f.
Maximum height of scaffold	60 ft.
Poles or uprights	4 × 4 in.
Pole spacing (longitudinal)	6 ft. 0 in.
Pole spacing (transverse)	8 ft. 0 in.
Ledgers	2 × 10 in.
Vertical spacing of horiz. members	6 ft. 0 in.
Bearers	2 × 10 in. (rough)
Bracing, horizontal and diagonal	2 × 4 in.
Tie-ins	1 × 4 in.
Planking	2 × 10 in.
Toeboards	4-in. high (minimum)
Guardrail	2 × 4 in.
All members except planking are used on edge.	

Table L-10 – Tube and Coupler Scaffolds – Light Duty		
Uniformly distributed load		Not to exceed 25 p.s.f
Post spacing (longitudinal)		10 ft. 0 in.
Post spacing (transverse)		6 ft. 0 in.
Working levels	**Additional planked levels**	**Maximum height**
1	8	125 ft.
2	4	125 ft.
3	0	91 ft. 0 in.

Table L-11 – Tube and Coupler Scaffolds – Medium Duty		
Uniformly distributed load		Not to exceed 50 p.s.f
Post spacing (longitudinal)		8 ft. 0 in.
Post spacing (transverse)		6 ft. 0 in.
Working levels	**Additional planked levels**	**Maximum height**
1	6	125 ft.
2	0	78 ft. 0 in.

Table L-12 – Tube and Coupler Scaffolds – Heavy Duty		
Uniformly distributed load		Not to exceed 75 p.s.f
Post spacing (longitudinal)		6 ft. 6 in.
Post spacing (transverse)		6 ft. 0 in.
Working levels	**Additional planked levels**	**Maximum height**
1	6	125 ft.

(6) Posts shall be accurately spaced, erected on suitable bases, and maintained plumb.

(7) Runners shall be erected along the length of the scaffold, located on both the inside and the outside posts at even height. Runners shall be interlocked to the inside and the outside posts at even heights. Runners shall be interlocked to form continuous lengths and coupled to each post. The bottom runners shall be located as close to the base as possible. Runners shall be placed not more than 6 feet-6 inches on centers.

(8) Bearers shall be installed transversely between posts and shall be securely coupled to the posts bearing on the runner coupler. When coupled directly to the runners, the coupler must be kept as close to the posts as possible.

(9) Bearers shall be at least 4 inches but not more than 12 inches longer than the post spacing or runner spacing.

(10) Cross bracing shall be installed across the width of the scaffold at least every third set of posts horizontally and every fourth runner vertically. Such bracing shall extend diagonally from the inner and outer runners upward to the next outer and inner runners.

(11) Longitudinal diagonal bracing on the inner and outer rows of poles shall be installed at approximately a 45 degree angle from near the base of the first outer post upward to the extreme top of the scaffold. Where the longitudinal length of the scaffold permits, such bracing shall be duplicated beginning at every fifth post. In a similar manner, longitudinal diagonal bracing shall also be installed from the last post extending back and upward toward the first post. Where conditions preclude the attachment of this bracing to the posts, it may be attached to the runners.

(12) The entire scaffold shall be tied to and securely braced against the building at intervals not to exceed 30 feet horizontally and 26 feet vertically.

(13) Guardrails, made of lumber not less than 2 × 4 inches (or other material providing equivalent protection), approximately 42 inches high, with a midrail of 1 × 6 inch lumber (or other material providing equivalent protection), and toeboard shall be installed at all open sides and

ends on all scaffolds more than 10 feet above the ground or floor. Toeboards shall be a minimum of 4 inches in height. Wire mesh shall be installed in accordance with paragraph (A)(6) of this section.

(D) Tubular Welded Frame Scaffolds.

(1) Metal tubular frame scaffolds, including accessories such as braces, brackets, trusses, screw legs, ladders, etc., shall be designed, constructed, and erected to safely support four times the maximum rated load.

(2) Spacing of panels or frames shall be consistent with the loads imposed.

(3) Scaffolds shall be properly braced by cross bracing or diagonal braces, or both, for securing vertical members together laterally, and the cross braces shall be of such length as will automatically square and align vertical members so that the erected scaffold is always plumb, square, and rigid. All brace connections shall be made secure.

(4) Scaffold legs shall be set on adjustable bases or plain bases placed on mud sills or other foundations adequate to support the maximum rated load.

(5) The frames shall be placed one on top of the other with coupling or stacking pins to provide proper vertical alignment of the legs.

(6) Where uplift may occur, panels shall be locked together vertically by pins or other equivalent suitable means.

(7) To prevent movement, the scaffold shall be secured to the building or structure at intervals not to exceed 30 feet horizontally and 26 feet vertically.

(8) Maximum permissible spans or planking shall be in conformity with paragraph (A)(10) of this section.

(9) Drawings and specifications for all frame scaffolds over 125 feet in height above the base plates shall be designed by a registered professional engineer.

(10) Guardrails made of lumber, not less than 2 × 4 inches (or other material providing equivalent protection), and approximately 42 inches high, with a midrail of 1- × 6-inch lumber (or other material providing equivalent protection), and toeboards, shall be installed at all open sides and ends on all scaffolds more than 10 feet above the ground or floor. Toeboards shall be a minimum of 4 inches in height. Wire mesh shall be installed in accordance with paragraph (A)(6) of this section.

(E) Manually Propelled Mobile Scaffolds.

(1) When free-standing mobile scaffold towers are used, the height shall not exceed four times the minimum base dimension.

(2) Casters shall be properly designed for strength and dimensions to support four times the maximum intended load. All casters shall be provided with a positive locking device to hold the scaffold in position.

(3) Scaffolds shall be properly braced by cross bracing and horizontal bracing conforming with paragraph (D)(3) of this section.

(4) Platforms shall be tightly planked for the full width of the scaffold except for necessary entrance opening. Platforms shall be secured in place.

(5) A ladder or stairway shall be provided for proper access and exit and shall be affixed or built into the scaffold and so located that when in use it will not have a tendency to tip the scaffold. A landing platform must be provided at intervals not to exceed 35 feet.

(6) The force necessary to move the mobile scaffold shall be applied near or as close to the base as practicable and provision shall be made to stabilize the tower during movement from one location to another. Scaffolds shall only be moved on level floors, free of obstructions and openings.

(7) The employer shall not allow employees to ride on manually propelled scaffolds unless the following conditions exist:

(i) The floor or surface is within 3 degrees of level, and free from pits, holes, or obstructions;

(ii) The minimum dimension of the scaffold base when ready for rolling, is at least one-half of the height. Outriggers, if used, shall be installed on both sides of staging;

(iii) The wheels are equipped with rubber or similar resilient tires;

(iv) All tools and materials are secured or removed from the platform before the mobile scaffold is moved.

(8) Scaffolds in use by any persons shall rest upon a suitable footing and shall stand plumb. The casters or wheels shall be locked to prevent any movement.

(9) Mobile scaffolds constructed of metal members shall also conform to applicable provisions of paragraphs (B), (C), or (D) of this section, depending on the material of which they are constructed.

(10) Guardrails made of lumber, not less than 2 × 4 inches (or other material providing equivalent protection), approximately 42 inches high, with a midrail, of 1- × 6-inch lumber (or other material providing equivalent protection), and toeboards, shall be installed at all open sides and ends on all scaffolds more than 10 feet above the ground or floor. Toeboards shall be a minimum of 4 inches in height. Wire mesh shall be installed in accordance with paragraph (A)(6) of this section.

(F) Elevating and Rotating Work Platforms.

Applicable requirements of American National Standards Institute A92.2-1969, Vehicle Mounted Elevating and Rotating Work Platforms, shall be complied with for such equipment, as required by the provisions of Article 1926.556.

(G) Outrigger Scaffolds.

(1) Outrigger beams shall extend not more than 6 feet beyond the face of the building. The inboard end of outrigger beams, measured from the fulcrum point to anchorage point, shall be not less than 1½ times the outboard end in length. The beams shall rest on edge, the sides shall be plumb, and the edges shall be horizontal. The fulcrum point of the beam shall rest on a secure bearing at least 6 inches in each horizontal dimension. The beam shall be secured in place against movement and shall be securely braced at the fulcrum point against tipping.

Table L-13 – Minimum Nominal Size and Maximum Spacing of Members of Outrigger Scaffolds

Maximum scaffold load	Light duty 25 p.s.f.	Medium duty 50 p.s.f.
Outrigger size	2 × 10 in.	3 × 10 in.
Maximum outrigger spacing	10 ft. 0 in.	6 ft. 0 in.
Planking	2 × 10 in.	2 × 10 in.
Guardrail	2 × 4 in.	2 × 4 in.
Guardrail uprights	2 × 4 in.	2 × 4 in.
Toeboards	4 in. (minimum)	4 in. (minimum)

(2) The inboard ends of outrigger beams shall be securely anchored either by means of struts bearing against sills in contact with the overhead beams or ceiling, or by means of tension members secured to the floor joists underfoot, or by both if necessary. The inboard ends of outrigger beams shall be secured against tipping and the entire supporting structure shall be securely braced in both directions to prevent any horizontal movement.

(3) Unless outrigger scaffolds are designed by a registered professional engineer competent in this field, they shall be constructed and erected in accordance with Table L-13. Outrigger scaffolds, designed by a registered professional engineer, shall be constructed and erected in accordance with such design.

(4) Planking shall be laid tight and shall extend to within 3 inches of the building wall. Planking shall be secured to the beams.

(5) Guardrails made of lumber, not less than 2 × 4 inches (or other material providing equivalent protection), approximately 42 inches high, with a midrail of 1- × 6-inch lumber (or other material providing equivalent protection), and toeboards, shall be installed at all open sides and ends on all scaffolds more than 10 feet above the ground or floor. Toeboards shall be a minimum of 4 inches in height. Wire mesh shall be installed in accordance with paragraph (A)(6) of this section.

(H) Masons' Adjustable Multiplepoint Suspension Scaffolds.

(1) The scaffold shall be capable of sustaining a working load of 50 pounds per square foot and shall not be loaded in excess of that figure.

(2) The scaffold shall be provided with hoisting machines that meet the requirements of Underwriters' Laboratories or Factory Mutual Engineering Corporation.

(3) The platform shall be supported by wire ropes, capable of supporting at least six times the intended load, suspended from overhead outrigger beams.

(4) The scaffold outrigger beams shall consist of structural metal securely fastened or anchored to the frame or floor system of the building or structure.

(5) Each outrigger beam shall be equivalent in strength to at least a standard 7-inch, 15.3-pound steel I-beam, at least 15 feet long, and shall not project more than 6 feet 6 inches beyond the bearing point.

(6) Where the overhang exceeds 6 feet 6 inches, outrigger beams shall be composed of stronger beams or multiple beams and be installed under the supervision of a competent person,

(7) All outrigger beams shall be set and maintained with their webs in a vertical position.

(8) A stop bolt shall be placed at each end of every outrigger beam.

(9) The outrigger beam shall rest on suitable wood bearing blocks.

(10) The free end of the suspension wire ropes shall be equipped with proper size thimbles and secured by splicing or other equivalent means. The running ends shall be securely attached to the hoisting drum and at least four turns of wire rope shall at all times remain on the drum. The use of fiber rope is prohibited.

(11) Where a single outrigger beam is used, the steel shackles or clevises with which the wire ropes are attached to the outrigger beams shall be placed directly over the hoisting drums.

(12) The scaffold platform shall be equivalent in strength to at least 2-inch planking. (For maximum planking spans, see paragraph (A)(11) of this section.)

(13) When employees are at work on the scaffold and an overhead hazard exists, overhead protection shall be provided on the scaffold, not more than 9 feet above the platform, consisting of 2-inch planking, or material of equivalent strength, laid tight, and extending not less than the width of the scaffold.

(14) Each scaffold shall be installed or relocated under the supervision of a competent person.

(15) Guardrails made of lumber, not less than 2 × 4 inches (or other material providing equivalent protection), approximately 42 inches high, with a midrail, and toeboards, shall be installed at all open sides and ends on all scaffolds more than 10 feet above the ground or floor. Toeboards shall be a minimum of 4 inches in height. Wire mesh shall be installed in accordance with paragraph (A)(6) of this section.

(I) (Swinging Scaffolds) Two-Point Suspension Scaffolds.

(1) Two-point suspension scaffold platforms shall be not less than 20 inches nor more than 36 inches wide overall. The platform shall be securely fastened to the hangers by U-bolts or by other equivalent means.

(2) The hangers of two-point suspension scaffolds shall be made of mild steel, or other equivalent materials, having a cross-sectional area capable of sustaining four times the maximum rated load, and shall be designed with a support for guardrail, intermediate rail, and toeboard.

(3) When hoisting machines are used on two-point suspension scaffolds, such machines shall be of a design tested and approved by Underwriters' Laboratories or Factory Mutual Engineering Corporation.

(4) The roof irons or hooks shall be of mild steel, or other equivalent material, of proper size and design, securely installed and anchored. Tiebacks of ¾-inch manila rope, or the equivalent, shall serve as a secondary means of anchorage, installed at right angles to the face

Table L-14 – Schedule for Ladder-Type Platforms

	Length of platform (feet)				
	12	14 and 16	18 and 20	22 and 24	28 and 30
Side stringers, minimum cross section (finished sizes):					
At ends (inches)	1¾ × 2¾	1¾ × 2¾	1¾ × 3	1¾ × 3	1¾ × 3½
At middle (inches)	1¾ × 3¾	1¾ × 3¾	1¾ × 4	1¾ × 4¼	1¾ × 5
Reinforcing strip (minimum)	1	1	1	1	1
Rungs	2	2	2	2	2
Tie Rods:					
Number (minimum)	3	4	4	5	6
Diameter (minimum)	¼ in.	¼ in.	¼ in.	¼ in.	¼ in.
Flooring, minimum finished size (inches)	½ × 2¾	½ × 2¾	½ × 2¾	½ × 2¾	½ × 2¾

[1]A ⅛ x ⅞-inch steel reinforcing strip or its equivalent shall be attached to the side or underside, full length.
[2]Rungs shall be 1⅛-inches minimum diameter with at least ⅞-inch diameter tenons, and the maximum spacing shall be 12 inches center to center.

of the building, whenever possible, and secured to a structurally sound portion of the building.

(5) Two-point suspension scaffolds shall be suspended by wire, synthetic, or fiber ropes capable of supporting at least six times the rated load. All other components shall be capable of supporting at least four times the rated load.

(6) The sheaves of all blocks, consisting of at least one double and one single block, shall fit the size and type of rope used.

(7) All wire ropes, fiber and synthetic ropes, slings, hangers, platforms, and other supporting parts shall be inspected before every installation. Periodic inspections shall be made while the scaffold is in use.

(8) On suspension scaffolds designed for a working load of 500 pounds, no more than two men shall be permitted to work at one time. On suspension scaffolds with a working load of 750 pounds, no more than three men shall be permitted to work at one time. Each employee shall be protected by an approved safety life belt attached to a lifeline. The lifeline shall be securely attached to substantial members of the structure (not scaffold), or to securely rigged lines, which will safely suspend the employee in case of a fall. In order to keep the lifeline continuously attached, with a minimum of slack, to a fixed structure, the attachment point of the lifeline shall be appropriately changed as the work progresses.

(9) Two-point suspension scaffolds shall be securely lashed to the building or structure to prevent them from swaying. Window cleaners' anchors shall not be used for this purpose.

(10) The platform of every two-point suspension scaffold shall be one of the following types:

(i) *Ladder-type platforms.* The side stringer shall be of clear straight-grained spruce or materials of equivalent strength and durability. The rungs shall be of straight-grained oak, ash, or hickory, at least 1⅛ inch in diameter, with ⅞-inch tenons mortised into the side stringers at least ⅞ of an inch. The stringers shall be tied together with tie rods not less than ¼ inch in diameter, passing through the stringers and riveted up tight against washers on both ends. The flooring strips shall be spaced not more than ⅝ inch apart except at the side rails where the space may be 1 inch. Ladder type platforms shall be constructed in accordance with Table L-14.

(ii) *Plank-type platforms.* Plank-type platforms shall be composed of not less than nominal 2- × 10-inch unspliced planks, properly cleated together on the underside, starting 6 inches from each end; intervals in between shall not exceed 4 feet. The plank-type platform shall not extend beyond the hangers more than 12 inches. A bar or other effective means shall be securely fastened to the platform at each end to prevent its slipping off the hanger. The span between hangers for plank-type platforms shall not exceed 8 feet.

(iii) *Beam-type platforms*. Beam platforms shall have side stringers of lumber not less than 2 × 6 inches set on edge. The span between hangers shall not exceed 12 feet when beam platforms are used. The flooring shall be supported on 2- × 6-inch cross beams, laid flat and set into the upper edge of the stringers with a snug fit, at intervals of not more than 4 feet, securely nailed in place. The flooring shall be of 1- × 6-inch material properly nailed. Floor boards shall not be spaced more than ½ inch apart.

(iv) *Light metal-type platforms*. Light metal-type platforms, when used, shall be tested and listed according to Underwriters' Laboratories or Factory Mutual Engineering Corporation.

(11) Guardrails made of lumber, not less than 2 × 4 inches (or other material providing equivalent protection), approximately 42 inches high, with a midrail, and toeboards, shall be installed at all open sides and ends on all scaffolds more than 10 feet above the ground or floor. Toeboards shall be a minimum of 4 inches in height. Wire mesh shall be installed in accordance with paragraph (A)(6) of this section.

(J) Stone Setters' Adjustable Multiplepoint Suspension Scaffolds.

(1) The scaffold shall be capable of sustaining a working load of 25 pounds per square foot and shall not be overloaded. Scaffolds shall not be used for storage of stone or other heavy materials.

(2) When used, the hoisting machine and its supports shall be of a type tested and listed by Underwriters' Laboratories or Factory Mutual Engineering Corporation.

(3) The platform shall be securely fastened to the hangers by U-bolts or other equivalent means. (For materials and spans, see subdivision (ii) of paragraph (I)(10), *Plank-type platforms*, and Table L-14 of this section.)

(4) The scaffold unit shall be suspended from metal outriggers, iron brackets, wire rope slings, or iron hooks.

(5) Outriggers, when used, shall be set with their webs in a vertical position, securely anchored to the building or structure and provided with stop bolts at each end.

(6) The scaffold shall be supported by wire rope capable of supporting at least six times the rated load. All other components shall be capable of supporting at least four times the rated load.

(7) The free ends of the suspension wire ropes shall be equipped with proper size thimbles, secured by splicing or other equivalent means. The running ends shall be securely attached to the hoisting drum and at least four turns of wire rope shall remain at the drum at all times.

(8) When two or more scaffolds are used on a building or structure, they shall not be bridged one to the other, but shall be maintained at even height with platforms abutting closely.

(9) Guardrails made of lumber, not less than 2 × 4 inches (or other material providing equivalent protection), approximately 42 inches high, with a midrail, and toeboards, shall be installed at all open sides and ends on all scaffolds more than 10 feet above the ground or floor. Toeboards shall be a minimum of 4 inches in height. Wire mesh shall be installed in accordance with paragraph (A)(6) of this section.

(K) Single-Point Adjustable Suspension Scaffolds.

(1) The scaffolding, including power units or manually operated winches, shall be of a type tested and listed by Underwriters' Laboratories or Factory Mutual Engineering Corporation.

(2) The power units may be either electrically or air motor driven.

(3) All power-operated gears and brakes shall be enclosed.

(4) In addition to the normal operating brake, all power-driven units shall have an emergency brake which engages automatically when the normal speed of descent is exceeded.

(5) The hoisting machines, cables, and equipment shall be regularly serviced and inspected.

(6) The units may be combined to form a two-point suspension scaffold. Such scaffold shall then comply with paragraph (I) of this section.

(7) The supporting cable shall be vertical for its entire length, and the basket shall not be swayed nor the cable fixed to any intermediate points to change the original path of travel.

(8) Suspension methods shall conform to applicable provisions of paragraphs (H) and (I) of this section.

(9) Guards, midrails, and toeboards shall completely enclose the cage or basket. Guardrails shall be no less than 2- × 4-inch lumber, or the equivalent, approximately 42 inches above the platform. Midrails shall be 1- × 6-inch lumber, or the equivalent, installed equidistant between the guardrail and the platform. Toeboards shall be a minimum of 4 inches in height.

(10) For additional details not covered in this paragraph, applicable technical portions of American National Standards Institute, A120.1-1970, Power-Operated Devices for Exterior Building Maintenance Powered Platforms, shall be used.

(L) Boatswain's Chairs.

(1) The chair seat shall not be less than 12 × 24 inches, and 1-inch thickness. The seat shall be reinforced on the underside by cleats securely fastened to prevent the board from splitting.

(2) The two fiber rope seat slings shall be of 5/8-inch diameter, reeved through the four seat holes so as to cross each other on the underside of the seat.

(3) Seat slings shall be of at least 3/8-inch wire rope when an employee is conducting a heat-producing process, such as gas or arc welding.

(4) The employee shall be protected by a safety belt and lifeline in accordance with Article 1926.104. The attachment point of the lifeline to the structure shall be appropriately changed as the work progresses.

(5) The tackle shall consist of correct size ball bearing or bushed blocks and properly spliced 5/8-inch diameter first-grade manila rope, or equivalent.

(6) The roof irons, hooks, or the object to which the tackle is anchored, shall be securely installed. Tiebacks, when used, shall be installed at right angles to the face of the building and securely fastened.

(M) Carpenters' Bracket Scaffolds.

(1) The brackets shall consist of a triangular wood frame not less than 2 × 3 inches in cross section, or of metal of equivalent strength. Each member shall be properly fitted and securely joined.

(2) Each bracket shall be attached to the structure by means of one of the following:

(i) A bolt, no less than 5/8 inch in diameter, which shall extend through to the inside of the building wall;

(ii) A metal stud attachment device;

(iii) Welding to steel tanks;

(iv) Hooking over a well-secured and adequately strong supporting member.

(3) The brackets shall be spaced no more than 8 feet apart.

(4) No more than two employees shall occupy any given 8 feet of a bracket scaffold at any one time. Tools and materials shall not exceed 75 pounds in addition to the occupancy.

(5) The platform shall consist of not less than two 2- × 10-inch nominal size planks extending not more than 12 inches or less than 6 inches beyond each end support.

(6) Guardrails made of lumber, not less than 2 × 4 inches (or other material providing equivalent protection), approximately 42 inches high, with a midrail, of 1- × 6-inch lumber (or other material providing equivalent protection), and toeboards, shall be installed at all open sides and

ends on all scaffolds more than 10 feet above the ground or floor. Toeboards shall be a minimum of 4 inches in height. Wire mesh shall be installed in accordance with paragraph (A)(6) of this section.

(N) Bricklayers' Square Scaffolds.

(1) The squares shall not exceed 5 feet in width and 5 feet in height.

(2) Members shall be not less than those specified in Table L-15.

(3) The squares shall be reinforced on both sides of each corner with 1-×6-inch gusset pieces. They shall also have diagonal braces 1 × 8 inches on both sides running from center to center of each member, or other means to secure equivalent strength and rigidity.

(4) The squares shall be set not more than 5 feet apart for medium duty scaffolds, and not more than 8 feet apart for light duty scaffolds. Bracing, 1 × 8 inches, extending from the bottom of each square to the top of the next square, shall be provided on both front and rear sides of the scaffold.

(5) Platform planks shall be at least 2- × 10-inch nominal size. The ends of the planks shall overlap the bearers of the squares and each plank shall be supported by not less than three squares.

(6) Bricklayers' square scaffolds shall not exceed three tiers in height and shall be so constructed and arranged that one square shall rest directly above the other. The upper tiers shall stand on a continuous row of planks laid across the next lower tier and be nailed down or otherwise secured to prevent displacement.

(7) Scaffolds shall be level and set upon a firm foundation.

(O) Horse Scaffolds.

(1) Horse scaffolds shall not be constructed or arranged more than two tiers or 10 feet in height.

(2) The members of the horses shall be not less than those specified in Table L-16.

Table L-15 – Minimum Dimensions for Bricklayer's Square Scaffold Members

Members	Dimensions
Bearers or horizontal members	2 × 6 in.
Legs	2 × 6 in.
Braces at corners	1 × 6 in.
Braces diagonally from center frame	1 × 8 in.

Table L-16 – Minimum Dimensions for Horse Scaffold Members

Members	Dimensions
Horizontal members or bearers	3 × 4 in.
Legs	1¼ × 4½ in.
Longitudinal brace between legs	1 × 6 in.
Gusset brace at top of legs	1 × 8 in.
Half diagonal braces	1¼ × 4½ in.

(3) Horses shall be spaced not more than 5 feet for medium duty and not more than 8 feet for light duty.

(4) When arranged in tiers, each horse shall be placed directly over the horse in the tier below.

(5) On all scaffolds arranged in tiers, the legs shall be nailed down or otherwise secured to the planks to prevent displacement or thrust and each tier shall be substantially cross braced.

(6) Horses or parts which have become weak or defective shall not be used.

(7) Guardrails made of lumber, not less than 2 × 4 inches (or other material providing equivalent protection), approximately 42 inches high, with a midrail, of 1- × 6-inch lumber (or other material providing equivalent protection), and toeboards, shall be installed at all open sides and ends on all scaffolds more than 10 feet above the ground or floor. Toeboards shall be a minimum of 4 inches in height. Wire mesh shall be installed in accordance with paragraph (A)(6) of this section.

(P) Needle Beam Scaffold.

(1) Wood needle beams shall be not less than 4 × 6 inches in size, with the greater dimension placed in a vertical direction. Metal beams or the equivalent, conforming to paragraphs (A)(8) and (10) of this section, may be used and shall not be altered or moved horizontally while they are in use.

(2) Ropes or hangers shall be provided for supports. The span between supports on the needle beam shall not exceed 10 feet for 4- × 6-inch timbers. Rope supports shall be equivalent in strength to 1-inch diameter first-grade manila rope.

(3) The ropes shall be attached to the needle beams by a scaffold hitch or a properly made eye splice. The loose end of the rope shall be tied by a bowline knot or by a round turn and a half hitch.

(4) The scaffold hitch shall be arranged so as to prevent the needle beam from rolling or becoming otherwise displaced.

(5) The platform span between the needle beams shall not exceed 8 feet when using 2-inch scaffold plank. For spans greater than 8 feet, platforms shall be designed based on design requirements for the special span. The overhang of each end of the platform planks shall be not less than 6 inches and not more than 12 inches.

(6) When needle beam scaffolds are used, the planks shall be secured against slipping.

(7) All unattached tools, bolts, and nuts used on needle beam scaffolds shall be kept in suitable containers, properly secured.

(8) One end of a needle beam scaffold may be supported by a permanent structural member conforming to paragraphs (A)(8) and (10) of this section.

(9) Each employee working on a needle beam scaffold shall be protected by a safety belt and lifeline in accordance with Article 1926.104.

(Q) Plasterers', Decorators', and Large Area Scaffolds.

(1) Plasterers', lathers', and ceiling workers' inside scaffolds shall be constructed in accordance with the general requirements set forth for independent wood pole scaffolds. (See paragraph (B) and Tables L-7, 8, and 9 of this section.)

(2) All platform planks shall be laid with the edges close together.

(3) When independent pole scaffold platforms are erected in sections, such sections shall be provided with connecting runways equipped with substantial guardrails.

(4) Guardrails made of lumber, not less than 2 × 4 inches (or other material providing equivalent protection), approximately 42 inches high, with a midrail of 1- × 6-inch lumber (or other material providing equivalent protection), and toeboards, shall be installed on all open sides and ends of all scaffolds more than 10 feet above the ground or floor. Toeboards shall be a minimum of 4 inches in height. Wire mesh shall be installed in accordance with paragraph (A)(6) of this section.

(R) Interior Hung Scaffolds.

(1) An interior hung scaffold shall be hung or suspended from the roof structure or ceiling beams.

(2) The suspending wire or fiber rope shall be capable of supporting at least six times the rated load. The rope shall be wrapped at least twice around the supporting members and twice around the bearers of the scaffold, with each end of the wire rope secured by at least three standard wire rope clips properly installed.

(3) For hanging wood scaffolds, the following minimum nominal size material shall be used:

(i) Supporting bearers 2 × 10 inches on edge;

(ii) Planking 2 ×10 inches, with maximum span 7 feet for heavy duty and 10 feet for light duty or medium duty.

(4) Steel tube and coupler members may be used for hanging scaffolds with both types of scaffold designed to sustain a uniform distributed working load up to heavy duty scaffold loads with a safety factor of four.

(5) Guardrails made of lumber, not less than 2 × 4 inches (or other material providing equivalent protection), approximately 42 inches high, with a midrail of 1- × 6-inch lumber (or other mate-

rial providing equivalent protection), and toeboards, shall be installed at all open sides and ends on all scaffolds more than 10 feet above the ground or floor. Toeboards shall be a minimum of 4 inches in height. Wire mesh shall be installed in accordance with paragraph (A)(6) of this section.

(S) Ladder Jack Scaffolds.

(1) All ladder jack scaffolds shall be limited to light duty and shall not exceed a height of 20 feet above the floor or ground.

(2) All ladders used in connection with ladder jack scaffolds shall be heavy-duty ladders and shall be designed and constructed in accordance with American National Standards Institute A 14.1-1968, Safety Code for Portable Wood Ladders, and A 14.2-1968, Safety Code for Portable Metal Ladders. Cleated ladders shall not be used for this purpose.

(3) The ladder jack shall be so designed and constructed that it will bear on the side rails in addition to the ladder rungs, or if bearing on rungs only, the bearing area shall be at least 10 inches on each rung.

(4) Ladders used in conjunction with ladder jacks shall be so placed, fastened, held, or equipped with devices so as to prevent slipping.

(5) The wood platform planks shall be not less than 2 inches nominal in thickness. Both metal and wood platform planks shall overlap the bearing surface not less than 12 inches. The span between supports for wood shall not exceed 8 feet. Platform width shall be not less than 18 inches.

(6) Not more than two employees shall occupy any given 8 feet of any ladder jack scaffold at any one time.

(T) Window Jack Scaffolds.

(1) Window jack scaffolds shall be used only for the purpose of working at the window opening through which the jack is placed.

(2) Window jacks shall not be used to support planks placed between one window jack and another or for other elements of scaffolding.

(3) Window jack scaffolds shall be provided with guardrails unless safety belts with lifelines are attached and provided for employees.

(4) Not more than one employee shall occupy a window jack scaffold at any one time.

(U) Roofing Brackets.

(1) Roofing brackets shall be constructed to fit the pitch of the roof.

(2) Brackets shall be secured in place by nailing in addition to the pointed metal projections. When it is impractical to nail brackets, rope supports shall be used. When rope supports are used, they shall consist of first-grade manila of at least ¾-inch diameter, or equivalent.

(3) A catch platform shall be installed below the working area of roofs more than 16 feet from the ground to eaves with a slope greater than 4 inches in 12 inches without a parapet. In width, the platform shall extend 2 feet beyond the protection of the eaves and shall be provided with a guardrail, midrail, and toeboard. This provision shall not apply where employees engaged in work upon such roofs are protected by a safety belt attached to a lifeline.

(V) Crawling Boards or Chicken Ladders.

(1) Crawling boards shall be not less than 10 inches wide and 1 inch thick, having cleats 1 × 1½ inches. The cleats shall be equal in length to the width of the board and spaced at equal intervals not to exceed 24 inches. Nails shall be driven through and clinched on the underside. The crawling board shall extend from the ridge pole to the eaves when used in connection with roof construction, repair, or maintenance.

(2) A firmly fastened lifeline of at least ¾-inch diameter rope, or equivalent, shall be strung beside each crawling board for a handhold.

(3) Crawling boards shall be secured to the roof by means of adequate ridge hooks or other effective means.

(W) Float or Ship Scaffolds.

(1) Float or ship scaffolds shall not be used to support more than three men and a few light tools, such as those needed for riveting, bolting, and

welding. They shall be constructed as designed in paragraphs (W) (2) through (6) of this section unless substitute designs and materials provide equivalent strength, stability, and safety.

(2) The platform shall be not less than 3 feet wide and 6 feet long, made of ¾-inch plywood, equivalent to American Plywood Association Grade B-B, Group 1, Exterior, or other similar material.

(3) Under the platform, there shall be two supporting bearers made from 2- × 4-inch, or 1- × I0-inch, rough, "selected lumber," or better. They shall be free of knots or other flaws and project 6 inches beyond the platform on both sides. The ends of the platform shall extend 6 inches beyond the outer edges of the bearers. Each bearer shall be securely fastened to the platform.

(4) An edging of wood not less than ¾ × 1½ inches or equivalent shall be placed around all sides of the platform to prevent tools from rolling off.

(5) Supporting ropes shall be 1-inch diameter manila rope or equivalent, free from deterioration, chemical damage, flaws, or other imperfections. Rope connections shall be such that the platform cannot shift or slip. If two ropes are used with each float, they shall be arranged so as to provide four ends which are to be securely fastened to an overhead support. Each of the two supporting ropes shall be hitched around one end of bearer and pass under the platforms to the other end of the bearer where it is hitched again, leaving sufficient rope at each end for the supporting ties.

(6) Each employee shall be protected by an approved safety lifebelt and lifeline, in accordance with Article 1926.104.

(X) Form Scaffolds.

(1) Form scaffolds shall be constructed of wood or other suitable materials, such as steel or aluminum members of known strength characteristics. All scaffolds shall be designed and erected with a minimum safety factor of 4, computed on the basis of the maximum rated load.

(2) All scaffold planking shall be a minimum of 2- × 10-inch nominal Scaffold Grade, as recognized by approved grading rules for the species of lumber used, or equivalent material. Maximum permissible spans shall not exceed 8 feet on centers for 2- × 10-inch nominal planking. Scaffold planks shall be either nailed or bolted to the ledgers or of such length that they overlap the ledgers at least 6 inches. Unsupported projecting ends of scaffolding planks shall be limited to a maximum overhang of 12 inches.

(3) Scaffolds shall not be loaded in excess of the working load for which they were designed.

(4) Figure-four form scaffolds:

(i) Figure-four scaffolds are intended for light duty and shall not be used to support loads exceeding 25 pounds per square foot unless specifically designed for heavier loading. For minimum design criteria, see Table L-17.

(ii) Figure-four form scaffold frames shall be spaced not more than 8 feet on centers and constructed from sound lumber, as follows: The outrigger ledger shall consist of two pieces of 1- × 6-inch or heavier material nailed on opposite sides of the vertical form support. Ledgers shall project not more than 3 feet 6 inches from the outside of the form support and shall be substantially braced and secured to prevent tipping or turning. The knee or angle brace shall intersect the ledger at least 3 feet from the form at an angle of approximately 45 degrees, and the lower end shall be nailed to a vertical support. The platform shall consist of two or more 2- × 10-inch planks, which shall be of such length that they extend at least 6 inches beyond ledgers at each end unless secured to the ledgers. When planks are secured to the ledgers (nailed or bolted), a wood filler strip shall be used between the ledgers. Unsupported projecting ends of planks shall be limited to an overhang of 12 inches.

(5) Metal bracket form scaffolds:

(i) Metal brackets or scaffold jacks which are an integral part of the form shall be securely bolted or welded to the form. Folding type brackets shall be either bolted or secured with a locking-type pin when extended for use.

Table L-17 – Minimum Design Criteria for Figure-Four Form Scaffolds

Members	Dimensions
Uprights	2 × 4 in. or 2 × 6 in.
Outriggers ledgers (two)	1 × 6 in.
Braces	1 × 6 in.
Guardrails	2 × 4 in.
Guardrail height	Approximately 42 in.
Intermediate guardrails	1 × 6 in.
Toeboards	4 in. (minimum)
Maximum length of ledgers	3 ft. 6 in. (unsupported)
Planking	2 × 10 in.
Upright spacing	8 ft. 0 in. (on centers)

Table L-18 – Minimum Design Criteria for Metal Bracket Form Scaffolds

Members	Dimensions
Uprights	2 × 4 in.
Guardrails	2 × 4 in.
Guardrail height	Approximately 42 in.
Intermediate guardrails	1 × 6 in.
Toeboards	4 in. (minimum)
Planking	2 × 9 in.

Table L-19 – Minimum Design Criteria for Wooden Bracket Form Scaffolds

Members	Dimensions
Uprights	2 × 4 in. or 2 × 6 in.
Support ledgers	2 × 6 in.
Maximum scaffold width	3 ft. 6 in.
Braces	1 × 6 in.
Guardrails	2 × 4 in.
Guardrail height	Approximately 42 in.
Intermediate guardrails	1 × 6 in.
Toeboards	4 in. (minimum)
Upright spacing	8 ft. 0 in. (on centers)

(ii) "Clip-on" or "hook-over" brackets may be used, provided the form walers are bolted to the form or secured by snap ties or shea-bolt extending through the form and securely anchored.

(iii) Metal brackets shall be spaced not more than 8 feet on centers.

(iv) Scaffold planks shall be either bolted to the metal brackets or of such length that they overlap the brackets at each end by at least 6 inches. Unsupported projecting ends of scaffold planks shall be limited to a maximum overhang of 12 inches.

(v) Metal bracket form scaffolds shall be equipped with wood guardrails, intermediate rails, toeboards, and scaffold planks meeting the minimum dimensions shown in Table L-18. (Metal may be substituted for wood, providing it affords equivalent or greater design strength.)

(6) Wooden bracket form scaffolds:

(i) Wooden bracket form scaffolds shall be an integral part of the form panel. The minimum design criteria set forth herein and in Table L-19 cover scaffolding intended for light duty and shall not be used to support loads exceeding 25 pounds per square foot, unless specifically designed for heavier loading.

(ii) Scaffold planks shall be either nailed or bolted to the ledgers or of such length that they overlap the ledgers at each end by at least 6 inches. Unsupported projecting ends of scaffold planks shall be limited to a maximum overhang of 12 inches.

(iii) Guardrails and toeboards shall be installed on all open sides and ends of platforms and scaffolding over 10 feet above floor or ground. Guardrails shall be made of lumber, 2- × 4-inch nominal dimension (or other material providing equivalent protection), approximately 42 inches high, supported at intervals not to exceed 8 feet. Guardrails shall be equipped with midrails constructed of 1- × 6-inch nominal lumber (or other material providing equivalent protection). Toeboards shall extend not less than 4 inches above the scaffold plank.

(Y) Pump Jack Scaffolds.

(1) Pump jack scaffolds shall:

(i) Not carry a working load exceeding 500 pounds; and

(ii) Be capable of supporting without failure at least four times the maximum intended load.

(iii) The manufactured components shall not be loaded in excess of the manufacturer's recommended limits.

(2) Pump jack brackets, braces, and accessories shall be fabricated from metal plates and angles. Each pump jack bracket shall have two positive gripping mechanisms to prevent any failure or slippage.

(3) The platform bracket shall be fully decked and the planking secured. Planking, or equivalent, shall conform with paragraph (A) of this section.

(4) When wood scaffold planks are used as platforms:

(i) Poles used for pump jacks shall not be spaced more than 10 feet center to center. When fabricated platforms are used that fully comply with all other provisions of this paragraph (Y), pole spacing may exceed 10 feet center to center.

(ii) Poles shall not exceed 30 feet in height.

(iii) Poles shall be secured to the work wall by rigid triangular bracing, or equivalent, at the bottom, top, and other points as necessary, to provide a maximum vertical spacing of not more than 10 feet between braces. Each brace shall be capable of supporting a minimum of 225 pounds tension or compression.

(iv) For the pump jack bracket to pass bracing already installed, an extra brace shall be used approximately 4 feet above the one to be passed until the original brace is reinstalled.

(5) All poles shall bear on mud sills or other adequate firm foundations.

(6) Pole lumber shall be two 2 × 4s, of Douglas fir or equivalent, straight-grained, clear, free of cross-grain, shakes, large loose or dead knots, and other defects which might impair strength.

(7) When poles are constructed of two continuous lengths, they shall be 2 × 4s, spiked together with the seam parallel to the bracket, and with 10d common nails, no more than 12 inches center to center, staggered uniformly from opposite outside edges.

(8) If 2 × 4s are spliced to make up the pole, the splices shall be so constructed as to develop the full strength of the member.

(9) A ladder, in accordance with Article 1926.1053, shall be provided for access to the platform during use.

(10) Not more than two persons shall be permitted at one time upon a pump jack scaffold between any two supports.

(11) Pump jacks scaffolds shall be provided with standard guardrails as defined in Article 1926.451(A)(15), but no guardrail is required when safety belts with lifelines are provided for employees.

(12) When a work bench is used at an approximate height of 42 inches, the top guardrail may be eliminated if the work bench is fully decked, the planking secured, and it is capable of withstanding 200 pounds pressure in any direction.

(13) Employees shall not be permitted to use a work bench as a scaffold platform.

1926.452 Definitions Applicable to this Subpart.

(A) (Reserved)

(B) Scaffolding.

(1) *Bearer* — A horizontal member of a scaffold upon which the platform rests and which may be supported by ledgers.

(2) *Boatswain's chair* — A seat supported by slings attached to a suspended rope, designed to accommodate one workman in a sitting position.

(3) *Brace* — A tie that holds one scaffold member in a fixed position with respect to another member.

(4) *Bricklayers' square scaffold* — A scaffold composed of framed wood squares that support a platform, limited to light and medium duty.

(5) *Carpenters' bracket scaffold* — A scaffold consisting of wood or metal brackets supporting a platform.

(6) *Coupler* — A device for locking together the component parts of a tubular metal scaffold. (The material used for the couplers shall be of a structural type, such as a drop-forged steel, malleable iron, or structural grade aluminum.)

(7) *Crawling board or chicken ladder* — A plank with cleats spaced and secured at equal intervals, for use by a worker on roofs, not designed to carry any material.

(8) *Double pole or independent pole scaffold* — A scaffold supported from the base by a double row of uprights, independent of support from the walls and constructed of uprights, ledgers, horizontal platform bearers, and diagonal bracing.

(9) *Float or ship scaffold* — A scaffold hung from overhead supports by means of ropes and consisting of a substantial platform having diagonal bracing underneath, resting upon and securely fastened to two parallel plank bearers at right angles to the span.

(10) *Guardrail* — A rail secured to uprights and erected along the exposed sides and ends of platforms.

(11) *Heavy duty scaffold* — A scaffold designed and constructed to carry a working load not to exceed 75 pounds per square foot.

(12) *Horse scaffold* — A scaffold for light or medium duty, composed of horses supporting a work platform.

(13) *Interior hung scaffold* — A scaffold suspended from the ceiling or roof structure.

(14) *Ladder jack scaffold* — A light duty scaffold supported by brackets attached to ladders.

(15) *Ledgers (stringers)* — A horizontal scaffold member which extends from post to post and which supports the putlogs or bearers forming a tie between the posts.

(16) *Light duty scaffold* — A scaffold designed and constructed to carry a working load not to exceed 25 pounds per square foot.

(17) *Manually propelled mobile scaffold* — A portable rolling scaffold supported by casters.

(18) *Masons' adjustable multiple-point suspension scaffold* — A scaffold having a continuous platform supported by bearers suspended by wire rope from overhead supports, so arranged and operated as to permit the raising or lowering of the platform to desired working positions.

(19) *Maximum rated load* — The total of all loads including the working load, the weight of the scaffold, and such other loads as may be reasonably anticipated.

(20) *Medium duty scaffold* — A scaffold designed and constructed to carry a working load not to exceed 50 pounds per square foot.

(21) *Midrail* — A rail approximately midway between the guardrail and platform, secured to the uprights, erected along the exposed sides and ends of platforms.

(22) *Needle beam scaffold* — A light duty scaffold consisting of needle beams supporting a platform.

(23) *Outrigger scaffold* — A scaffold supported by outriggers or thrustouts projecting beyond the wall or face of the building or structure, the inboard ends of which are secured inside of such building or structure.

(24) *Putlog* — A scaffold member upon which the platform rests.

(25) *Roofing or bearer bracket* — A bracket used in slope roof construction having provisions for fastening to the roof or supported by ropes fastened over the ridge and secured to some suitable object.

(26) *Runner* — The lengthwise horizontal bracing or bearing members, or both.

(27) *Scaffold* — Any temporary elevated platform and its supporting structure used for supporting workmen or materials, or both.

(28) *Single-point adjustable suspension scaffold* — A manually or power-operated unit designed for light duty use, supported by a single wire rope from an overhead support so arranged and operated as to permit the raising or lowering of platform to desired working positions.

(29) *Single-pole scaffold* — Platforms resting on putlogs or cross beams, the outside ends of which are supported on ledgers secured to a single row of posts or uprights, and the inner ends of which are supported on or in a wall.

(30) *Stone setters' adjustable multiplepoint suspension scaffold* — A swinging type scaffold having a platform supported by hangers suspended at four points so as to permit the raising or lowering of the platform to the desired working position by the use of hoisting machines.

(31) *Toeboard* — A barrier secured along the sides and ends of a platform to guard against the falling of material.

(32) *Tube and coupler scaffold* — An assembly consisting of tubing which serves as posts, bearers, braces, ties, and runners, a base supporting the posts, and special couplers which serve to connect the uprights and to join the various members.

(33) *Tubular welded frame scaffold* — A sectional panel or frame metal scaffold substantially built up of prefabricated welded sections that consist of posts and horizontal bearer with intermediate members.

(34) *Two-point suspension scaffold (swinging scaffold)* — A scaffold, the platform of which is supported by hangers (stirrups) at two points, suspended from overhead supports so as to permit the raising or lowering of the platform to the desired working position by tackle or hoisting machines.

(35) *Window jack scaffold* — A scaffold, the platform of which is supported by a bracket or jack which projects through a window opening.

(36) *Working load* — The load imposed by men, materials, and equipment.

1926.453 Manually Propelled Mobile Ladder Stands and Scaffolds (Towers)

(A) General Requirements.

(1) Application. This section is intended to prescribe rules and requirements for the design, construction, and use of mobile work platforms (including ladder stands but not including aerial ladders) and rolling (mobile) scaffolds (towers). This standard is promulgated to aid in providing for the safety of life, limb, and property, by establishing minimum standards for structural design requirements and for the use of mobile work platforms and towers.

(2) Working loads.

(i) Work platforms and scaffolds shall be capable of carrying the design load under varying circumstances depending upon the conditions of use. Therefore, all parts and appurtenances necessary for their safe and efficient utilization must be integral parts of the design.

(ii) Specific design and construction requirements are not a part of this section because of the wide variety of materials and design possibilities. However, the design shall be such as to produce a mobile ladder stand or scaffold that will safely sustain the specified loads. The material selected shall be of sufficient strength to meet the test requirements and shall be protected against corrosion or deterioration.

(a) The design working load of ladder stands shall be calculated on the basis of one or more 200-pound (90.6 kg) persons together with 50 pounds (22.65 kg) of equipment each.

(b) The design load of all scaffolds shall be calculated on the basis of:

Light — Designed and constructed to carry a working load of 25 pounds per square foot (1.05 kg M2).

Medium — Designed and constructed to carry a working load of 50 pounds per square foot (2.1 kg M2).

Heavy — Designed and constructed to carry a working load of 75 pounds per square foot (3.15 kg M2).

All ladder stands and scaffolds shall be capable of supporting at least four times the design working load.

(iii) The materials used in mobile ladder stands and scaffolds shall be of standard manufacture and conform to standard specifications of strength, dimensions, and weights, and shall be selected to safely support the design working load.

(iv) Nails, bolts, or other fasteners used in the construction of ladders, scaffolds, and towers shall be of adequate size and in sufficient numbers at each connection to develop the designed strength of the unit. Nails shall be driven full length. (All nails should be immediately withdrawn from dismantled lumber.)

(v) All exposed surfaces shall be free from sharp edges, burrs or other safety hazards.

(3) Work levels.

(i) The maximum work level height shall not exceed four (4) times the minimum or least base dimensions of any mobile ladder stand or scaffold. Where the basic mobile unit does not meet this requirement, suitable outrigger frames shall be employed to achieve this least base dimension, or provisions shall be made to guy or brace the unit against tipping.

(ii) The minimum platform width for any work level shall not be less than 20 inches (50.8 cm) for mobile scaffolds (towers). Ladder stands shall have a minimum step width of 16 inches (40.64 cm).

(iii) The supporting structure for the work level shall be rigidly braced, using adequate cross bracing or diagonal bracing with rigid platforms at each work level.

(iv) The steps of ladder stands shall be fabricated from slip resistant treads.

(v) The work level platform of scaffolds (towers) shall be of wood, aluminum, or plywood planking, steel or expanded metal, for the full width of the scaffold, except for necessary openings. Work platforms shall be secured in place. All planking shall be 2-inch (5.08 cm) (nominal) scaffold grade minimum 1,500 f. (stress grade) construction grade lumber or equivalent.

(vi) All scaffold work levels 10 feet (3.04 m) or higher above the ground or floor shall have a standard 4-inch (10.16 cm) (nominal) toeboard.

(vii) All work levels 10 feet (3.04 m) or higher above the ground or floor shall have a guardrail of 2- × 4-inch nominal lumber or the equivalent installed no less than 36 inches (0.912 m) or more than 42 inches (106.68 cm) high, with a midrail, when required, of 1- × 4-inch nominal lumber or equivalent.

(viii) A climbing ladder or stairway shall be provided for proper access and egress, and shall be affixed or built into the scaffold and so located that its use will not have a tendency to tip the scaffold. A landing platform shall be provided at intervals not to exceed 30 feet (9.12 m).

(4) Wheels or casters.

(i) Wheels or casters shall be properly designed for strength and dimensions to support four (4) times the design working load.

(ii) All scaffold casters shall be provided with a positive wheel and/or swivel lock to prevent movement. Ladder stands shall have at least two (2) of the four (4) casters and shall be of the swivel type.

(iii) Where leveling of the elevated work platform is required, screw jacks or other suitable means for adjusting the height shall be provided in the base section of each mobile unit.

(B) Mobile Tubular Welded Sectional Folding Scaffolds.

(1) *General.* Units including sectional stairway and sectional ladder scaffolds shall be designed to comply with the requirements of paragraph (A) of this section.

(2) *Stairway.* An integral stairway and work platform shall be incorporated into the structure of each sectional folding stairway scaffold.

(3) *Bracing.* An integral set of pivoting and hinged folding diagonal and horizontal braces and a detachable work platform shall be incorporated into the structure of each sectional folding ladder scaffold.

(4) *Sectional folding stairway scaffolds.* Sectional folding stairway scaffolds shall be designed as medium duty scaffolds except for high clearance. These special base sections shall be designed as light duty scaffolds. When upper sectional folding stairway scaffolds are used

with a special high clearance base, the load capacity of the entire scaffold shall be reduced accordingly. The width of a sectional folding stairway scaffold shall not exceed 4½ feet (1.368 m). The maximum length of a sectional folding stairway scaffold shall not exceed 6 feet (1.824 m).

(5) *Sectional folding ladder scaffolds.* Sectional folding ladder scaffolds shall be designed as light duty scaffolds including special base (open end) sections which are designed for high clearance. For certain special applications the 6-foot (1.824 m) folding ladder scaffolds, except for special high clearance base sections, shall be designed for use as medium duty scaffolds. The width of a sectional folding ladder scaffold shall not exceed 4½ feet (1.368 m). The maximum length of a sectional folding ladder scaffold shall not exceed 6 feet 6 inches (1.976 m) for a 6-foot-long (1.824 m) unit, 8 feet 6 inches (2.584 m) for an 8-foot-long (2.432 m) unit, or 10 feet 6 inches (3.192 m) for a 10-foot-long (3.04 m) unit.

(6) *End frames.* The end frames of sectional ladder and stairway scaffolds shall be designed so that the horizontal bearers provide supports for multiple planking levels.

(7) *Erection.* Only the manufacturer of the scaffold or his qualified designated agent shall be permitted to erect or supervise the erection of scaffolds exceeding 50 feet (15.2 m) in height above the base, unless such structure is approved in writing by a licensed professional engineer, or erected in accordance with instructions furnished by the manufacturer.

Subpart Q - Concrete and Masonry Construction

1926.700 Scope, Application, and Definitions Applicable to this Subpart

(A) Scope and Application.

This subpart sets forth requirements to protect all construction employees from the hazards associated with concrete and masonry construction operations performed in workplaces covered under 29 CFR Part 1926. In addition to the requirements in Subpart Q, other relevant provisions in Parts 1910 and 1926 apply to concrete and masonry construction operations.

(B) Definitions Applicable to this Subpart.

In addition to the definitions set forth in Article 1926.32, the following definitions apply to this subpart.

(1) *Bull float* means a tool used to spread out and smooth concrete.

(2) *Formwork* means the total system of support for freshly placed or partially cured concrete, including the mold or sheeting (form) that is in contact with the concrete as well as all supporting members including shores, reshores, hardware, braces, and related hardware.

(3) *Lift slab* means a method of concrete construction in which floor and roof slabs are cast on or at ground level and, using jacks, lifted into position.

(4) *Limited access zone* means an area alongside a masonry wall, which is under construction, and which is clearly demarcated to limit access by employees.

(5) *Precast concrete* means concrete members (such as walls, panels, slabs, columns, and beams) which have been formed, cast, and cured prior to final placement in a structure.

(6) *Reshoring* means the construction operation in which shoring equipment (also called reshores or reshoring equipment) is placed, as the original forms and shores are removed, in order to support partially cured concrete and construction loads.

(7) *Shore* means a supporting member that resists a compressive force imposed by a load.

(8) *Vertical slip* forms means forms which are jacked vertically during the placement of concrete.

(9) *Jacking operation* means the task of lifting a slab (or group of slabs vertically from one location to another (e.g., from the casting location to a temporary (parked) location, or from a temporary location to another temporary location, or to its final location in the structure), during the construction of a building/structure where the lift-slab process is being used.

1926.701 General requirements

(A) Construction Loads.

No construction loads shall be placed on a concrete structure or portion of a concrete structure unless the employer determines, based on information received from a person who is qualified in structural design, that the structure or portion of the structure is capable of supporting the loads.

(B) Reinforcing Steel.

All protruding reinforcing steel, onto and into which employees could fall, shall be guarded to eliminate the hazard of impalement.

(C) Post-Tensioning Operations.

(1) No employee (except those essential to the post-tensioning operations) shall be permitted to be behind the jack during tensioning operations.

(2) Signs and barriers shall be erected to limit employee access to the post-tensioning area during tensioning operations.

(D) Riding Concrete Buckets.

No employee shall be permitted to ride concrete buckets.

(E) Working Under Loads.

(1) No employee shall be permitted to work under concrete buckets while buckets are being elevated or lowered into position.

(2) To the extent practical, elevated concrete buckets shall be routed so that no employee, or the fewest number of employees, are exposed to the hazards associated with failing concrete buckets.

(F) Personal Protective Equipment.

No employee shall be permitted to apply a cement, sand, and water mixture through a pneumatic hose unless the employee is wearing protective head and face equipment.

1926.702 Requirements for Equipment and Tools

(A) Bulk Cement Storage.

(1) Bulk storage rag bins, containers, and silos shall be equipped with the following:

(i) Conical or tapered bottoms; and

(ii) Mechanical or pneumatic means of starting the flow of material.

(2) No employee shall be permitted to enter storage facilities unless the ejection system has been shut down, locked out, and tagged to indicate that the ejection system is not to be operated.

(B) Concrete Mixers.

Concrete mixers with one cubic yard (.8 M^3) or larger loading skips shall be equipped with the following:

(1) A mechanical device to clear the skip of materials; and

(2) Guardrails installed on each side of the skip.

(C) Power Concrete Trowels.

Powered and rotating type concrete troweling machines that are manually guided shall be equipped with a control switch that will automatically shut off the power whenever the hands of the operator are removed from the equipment handles.

(D) Concrete Buggies.

Concrete buggy handles shall not extend beyond the wheels on either side of the buggy.

(E) Concrete Pumping Systems.

(1) Concrete pumping systems using discharge pipes shall be provided with pipe supports designed for 100 percent overload.

(2) Compressed air hoses used on concrete pumping system shall be provided with positive failsafe joint connectors to prevent separation of sections when pressurized.

(F) Concrete Buckets.

(1) Concrete buckets equipped with hydraulic or pneumatic gates shall have positive safety latches or similar safety devices installed to prevent premature or accidental dumping.

(2) Concrete buckets shall be designed to prevent concrete from hanging up on top and the sides.

(G) Tremies.

Sections of tremies and similar concrete conveyances shall be secured with wire rope (or equivalent materials) in addition to the regular couplings or connections.

(H) Bull Floats.

Bull float handles used where they might contact energized electrical conductors, shall be constructed of nonconductive material or insulated with a nonconductive sheath whose electrical and mechanical characteristics provide the equivalent protection of a handle constructed of nonconductive material.

(I) Masonry Saws.

(1) Masonry saws shall be guarded with a semicircular enclosure over the blade.

(2) A method for retaining blade fragments shall be incorporated in the design of the semicircular enclosure.

(J) Lockout/Tagout Procedures.

(1) No employee shall be permitted to perform maintenance or repair activity on equipment (such as compressors, mixers, screens or pumps used for concrete and masonry construction activities) where the inadvertent operation of the equipment could occur and cause injury, unless all potentially hazardous energy sources have been locked out and tagged.

(2) Tags shall read Do Not Start or similar language to indicate that the equipment is not to be operated.

1926.703 Requirements for Cast-In-Place Concrete

(A) General Requirements for Formwork.

(1) Formwork shall be designed, fabricated, erected, supported, braced and maintained so that it will be capable of supporting without failure all vertical and lateral loads that may reasonably be anticipated to be applied to the formwork. Formwork which is designed, fabricated, erected, supported, braced and maintained in conformance with the Appendix to this section will be deemed to meet the requirements of this paragraph.

(2) Drawings or plans, including all revisions, for the jack layout, formwork (including shoring equipment), working decks, and scaffolds, shall be available at the jobsite.

(B) Shoring and Reshoring.

(1) All Shoring equipment (including equipment used in reshoring operations) shall be inspected prior to erection to determine that the equipment meets the requirements specified in the formwork drawings.

(2) Shoring equipment found to be damaged such that its strength is reduced to less than that required by 1926.703(A)(1) shall not be used for shoring.

(3) Erected shoring equipment shall be inspected immediately prior to, during, and immediately after concrete placement.

(4) Shoring equipment that is found to be damaged or weakened after erection, such that its strength is reduced to less than that required by 1926.703(A)(1), shall be immediately reinforced.

(5) The sills for shoring shall be sound, rigid, and capable of carrying the maximum intended load.

(6) All base plates, shore heads, extension devices, and adjustment screws shall be in firm contact, and secured when necessary, with the foundation and the form.

(7) Eccentric loads on shore heads and similar members shall be prohibited unless these members have been designed for such loading.

(8) Whenever single post shores are used one on top of another (tiered), the employer shall comply with the following specific requirements in addition to the general requirements for formwork:

(i) The design of the shoring shall be prepared by a qualified designer and the erected shoring shall be inspected by an engineer qualified in structural design.

(ii) The single post shores shall be vertically aligned.

(iii) The single post shores shall be spliced to prevent misalignment.

(iv) The single post shores shall be adequately braced in two mutually perpendicular directions at the splice level. Each tier shall also be diagonally braced in the same two directions.

(9) Adjustment of single post shores to raise formwork shall not be made after the placement of concrete.

(10) Reshoring shall be erected, as the original forms and shores are removed, whenever the concrete is required to support loads in excess of its capacity.

(C) Vertical Slip Forms.

(1) The steel rods or pipes on which jacks climb or by which the forms are lifted shall be:

(i) Specifically designed for that purpose; and

(ii) Adequately braced where not encased in concrete.

(2) Forms shall be designed to prevent excessive distortion of the structure during the jacking operation.

(3) All vertical slip forms shall be provided with scaffolds or work platforms where employees are required to work or pass.

(4) Jacks and vertical supports shall be positioned in such a manner that the loads do not exceed the rated capacity of the jacks.

(5) The jacks or other lifting devices shall be provided with mechanical dogs or other automatic holding devices to support the slip forms whenever failure of the power supply or lifting mechanism occurs.

(6) The form structure shall be maintained within all design tolerances specified for plumbness during the jacking operation.

(7) The predetermined safe rate of lift shall not be exceeded.

(D) Reinforcing Steel.

(1) Reinforcing steel for walls, piers, columns, and similar vertical structures shall be adequately supported to prevent overturning and to prevent collapse.

(2) Employers shall take measures to prevent unrolled wire mesh from recoiling. Such measures may include, but are not limited to, securing each end of the roll or turning over the roll.

(E) Removal of Formwork.

(1) Forms and shores (except those used for slabs on grade and slip forms) shall not be removed until the employer determines that the concrete has gained sufficient strength to support its weight and superimposed loads. Such determination shall be based on compliance with one of the following:

(i) The plans and specifications stipulate conditions for removal of forms and shores, and such conditions have been followed, or

(ii) The concrete has been properly tested with an appropriate ASTM standard test method designed to indicate the concrete compressive strength, and the test results indicate that the concrete has gained sufficient strength to support its weight and superimposed loads.

(2) Reshoring shall not be removed until the concrete being supported has attained adequate strength to support its weight and all loads in place upon it.

APPENDIX TO 1926.703(A)(1)

GENERAL REQUIREMENTS FOR FORMWORK

(This Appendix is non-mandatory.)

This appendix serves as a non-mandatory guideline to assist employers in complying with the formwork requirements in 1926.703(c)(1). Formwork which has been designed, fabricated, erected, braced, supported and maintained in accordance with Sections 6 and 7 of the American National Standard for Construction and Demolition Operations, Concrete and Masonry Work, ANSI A 10.9-1983, shall be deemed to be in compliance with the provision of 1926.703(a)(1).

1926.704 Requirements for Precast Concrete

(A) Precast concrete wall units, structural framing, and tilt-up wall panels shall be adequately supported to prevent overturning and to prevent collapse until permanent connections are completed.

(B) Lifting inserts which are embedded or otherwise attached to tilt-up precast concrete members shall be capable of supporting at least two times the maximum intended load applied or transmitted to them.

(C) Lifting inserts which are embedded or otherwise attached to precast concrete members, other than the tilt-up members, shall be capable of supporting at least four times the maximum intended load applied or transmitted to them.

(D) Lifting hardware shall be capable of supporting at least five times the maximum intended load applied transmitted to the lifting hardware.

(E) No employee shall be permitted under precast concrete members being lifted or tilted into position except those employees required for the erection of those members.

1926.705 Requirements for Lift-Slab Construction Operations

(A) Lift-slab operations shall be designed and planned by a registered professional engineer who has experience in lift-slab construction. Such plans and designs shall be implemented by the employer and shall include detailed instructions and sketches indicating the prescribed method of erection. These plans and designs shall also include the provisions for ensuring lateral stability of the building/structure during construction.

(B) Jacks/lifting units shall be marked to indicate their rated capacity as established by the manufacturer.

(C) Jacks/lifting units shall not be loaded beyond their rated capacity as established by the manufacturer.

(D) Jacking equipment shall be capable of supporting at least two and one-half times the load being lifted during jacking operations and the equipment shall not be overloaded. For the purpose of this provision, jacking equipment includes any load-bearing component which is used to carry out the lifting operation(s). Such equipment includes, but is not limited, to the following: threaded rods, lifting attachments, lifting nuts, hook-up collars, T-caps, shearheads, columns, and footings.

(E) Jacks/lifting units shall be designed and installed so that they will neither lift nor continue to lift when they are loaded in excess of their rated capacity.

(F) Jacks/lifting units shall have a safety device installed which will cause the jacks/lifting units to support the load in any position in the event any jack/lifting unit malfunctions or loses its lifting ability.

(G) Jacking operations shall be synchronized in such a manner to ensure even and uniform lifting of the slab. During lifting, all points at which the slab is supported shall be kept within ½ inch of that needed to maintain the slab in a level position.

(H) If leveling is automatically controlled, a device shall be installed that will stop the operation when the ½-inch tolerance set forth in paragraph (G) of this section is exceeded or where there is a malfunction in the jacking (lifting) system.

(I) If leveling is maintained by manual controls, such controls shall be located in a central location and attended by a competent person while lifting is in progress. In addition to meeting the definition in 1926.32(f), the competent person must be experienced in lifting operations and with the lifting equipment being used.

(J) The maximum number of manually controlled jacks/lifting units on one slab shall be limited to a number that will permit the operator to maintain the slab level within specified tolerances of paragraph (G) of this section, but in no case shall that number exceed 14.

(K) (1) No employee, except those essential to the jacking operation, shall be permitted in the building/structure while any jacking operation is taking place unless the building/structure has been reinforced sufficiently to ensure its integrity during erection. The phrase "reinforced sufficiently to ensure its integrity" used in this paragraph means that a registered professional engineer, independent of the engineer who designed and planned the lifting operation, has determined from the plans that if there is a loss of support at any jack location, that loss will be confined to that location and the structure as a whole will remain stable.

(2) Under no circumstances, shall any employee who is not essential to the jacking operation be permitted immediately beneath a slab while it is being lifted.

(3) For the purpose of paragraph (K) of this section, a jacking operation begins when a slab or group of slabs is lifted and ends when such slabs are secured (with either temporary connections or permanent connections).

(4) Employers who comply with Appendix A to 1926.705 shall be considered to be in compliance with the provisions of paragraphs (K)(1) through (K)(3) of this section.

(L) When making temporary connections to support slabs, wedges shall be secured by tack welding, or an equivalent method of securing the wedges to prevent them from falling out of position. Lifting rods may not be released until the wedges at that column have been secured.

(M) All welding on temporary and permanent connections shall be performed by a certified welder familiar with the welding requirements specified in the plans and specifications for the lift-slab operation.

(N) Load transfer from jacks/lifting units to building columns shall not be executed until the welds on the column shear plates (weld blocks) are cooled to air temperature.

(O) Jacks/lifting units shall be positively secured to building columns so that they do not become dislodged or dislocated.

(P) Equipment shall be designed and installed so that the lifting rods cannot slip out of position or the employer shall institute other measures, such as the use of locking or blocking devices, which will provide positive connection between the lifting rods and attachments and will prevent components from disengaging during lifting operations.

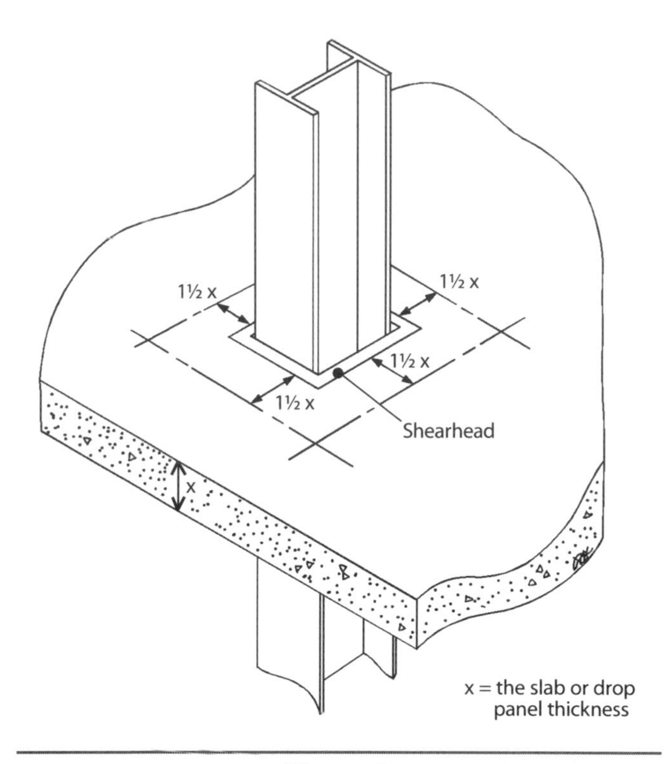

Figure 1
Column head area

APPENDIX TO 1926.705

LIFT SLAB OPERATIONS

(This Appendix is Non-Mandatory.)

In paragraph 1926.705(K), OSHA requires employees to be removed from the building/structure during jacking operations unless an independent registered professional engineer, other than the engineer who designed and planned the lifting operation, has determined that the building/structure has been sufficiently reinforced to insure the integrity of the building/structure. One method to comply with this provision is for the employer to ensure that continuous bottom steel is provided in every slab and in both directions through every wall or column head area. (Column head area means the distance between lines that are 1½ times the thickness of the slab or drop panel. These lines are located outside opposite faces of the outer edges of the shearhead sections. See Figure 1.) The amount of bottom steel shall be established by assuming loss of support at a given lifting jack and then determining the steel necessary to carry, by catenary action over the span between surrounding supports, the slab service dead load plus any service dead and live loads likely to be acting on the slab during jacking. In addition, the surrounding supports must be capable of resisting any additional load transferred to them as a result of the loss of support at the lifting jack considered.

1926.706 Requirements for Masonry Construction

(A) A limited access zone shall be established whenever a masonry wall is being constructed. The limited access zone shall conform to the following.

(1) The limited access zone shall be established prior to the start of construction of the wall.

(2) The limited access zone shall be equal to the height of the wall to be constructed plus four feet, and shall run the entire length of the wall.

(3) The limited access zone shall be established on the side of the wall which will be unscaffolded.

(4) The limited access zone shall be restricted to entry by employees actively engaged in constructing the wall. No other employees shall be permitted to enter the zone.

(5) The limited access zone shall remain in place until the wall is adequately supported to prevent overturning and to prevent collapse unless the height of wall is over 8 feet, in which case, the limited access zone shall remain in place until the requirements of paragraph (B) of this section have been met.

(B) All masonry walls over 8 feet in height shall be adequately braced to prevent overturning and to prevent collapse unless the wall is adequately supported so that it will not overturn or collapse. The bracing shall remain in place until permanent supporting elements of the structure are in place.

Glossary

Absorption: The weight of water a brick unit absorbs, when immersed in either cold or boiling water for a stated length of time, expressed as percentage of the weight of the dry unit. *See ASTM Specification C67.*

Abutment: The masonry mass that supports the thrust of an arch or bridge.

Accelerator: Any chemical or other substance added to cement during the mixing process which increases the rate of hydration, shortens set time, and/or increases the rate of hardening or strength development.

Addition: A change in the design of a building to increase the overall dimensions; also the original design of a building constructed with connecting parts joined together to make one whole structure.

Admixtures: Materials added to mortar to impart special properties to the mortar.

Aggregate: Various hard materials such as sand, gravel, or crushed stone, added to cement to make concrete.

Air brick: A ceramic or metal unit about the size of a standard brick, open on the ends to permit the entrance of air into the building.

Air-entrained concrete: Portland cement that has had an ingredient added to cause millions of tiny air bubbles to be trapped in the concrete. The air increases the volume of the paste and improves the workability of the concrete. Air-entrained concrete can be finished sooner than ordinary concrete.

Anchor: A piece or assemblage, usually metal, used to attach building parts (plates, joists, trusses, etc.) to masonry or masonry materials.

Anchor bolts: Bolts that are used to fasten sills or plates to masonry.

Angle iron: A piece of iron that forms a right angle and is used to span openings and support masonry at these openings. In brick veneer they are used to secure the veneer to the foundation or to the structure being veneered.

ANSI: American National Standards Institute. This group publishes the American National Standards, which are the approved standards and specifications in all the areas of building construction.

Apex: The architectural term designating the topmost part of a structure.

Apprentice: A person who has entered into an agreement with a trade committee and employers to work for a period of time to learn the trade.

Arch: A curved compressive structural member, spanning openings or recesses; also built flat.

back arch: A concealed arch carrying the backing of a wall where the exterior facing is carried by a lintel.

jack arch: One having horizontal or nearly-horizontal upper and lower surfaces, also called a *flat* or *straight arch.*

major arch: Arch with spans greater than 6 feet and equivalent to uniform loads greater than 1,000 pounds per foot. Typically known as a Tudor arch, semicircular arch, Gothic arch or parabolic arch. Has a rise-to-span ratio greater than 0.15.

minor arch: Arch with a maximum span of 6 feet and loads not exceeding 1,000 pounds per foot. Typically known as a jack arch, segmented arch or multi-centered arch. It has a rise-to-span ratio less than or equal to 0.15.

receiving arch: One built over a lintel, flat arch, or smaller arch to divert loads, thus relieving the lower member from excessive loading; also known as a *discharging* or *safety* arch.

trimmer arch: Usually a low-rise arch of brick used for supporting a fireplace hearth.

Asbestos cement: A cement made by combining portland cement with asbestos. It is fire resistant and waterproof.

Ash dump: A metal frame placed in the floor of a fireplace for the disposal of ashes.

Ashlar masonry: Masonry composed of rectangular units of burned clay, shale, or stone, generally larger in size than brick, and properly bonded, having sawed, dressed, or squared beds, and joints laid in mortar.

ASHRAE: The American Society of Heating, Refrigerating & Air-Conditioning Engineers, Inc.

ASTM: The American Society for Testing and Materials, a scientific and technical organization formed for "the development of standards and characteristics and performance of materials, products, systems, and services; and the promotion of related knowledge."

Autogenous healing: A natural process occurring over a long time in which hairline cracks in masonry are filled up by the recarbonization of lime in the remaining mortar. Rainwater and atmospheric carbon dioxide react with lime in the mortar; the hydrated lime dissolves, and then becomes recarbonated by the carbon dioxide.

Back hearth: The floor of a fireplace.

Backfilling: 1. Rough masonry built behind a facing or between two faces. 2. Filling over the extrados of an arch. 3. Brick work in spaces between structural timbers, sometimes called *nogging*. 4. Fill (dirt, stone, or similar material) used to build up the ground between the foundation and the unexcavated part of the surrounding area.

Backup: The part of a masonry wall behind the exterior facing.

Bar chairs: A small device used to hold up reinforcing wire or rods in concrete while the concrete is being poured.

Bat: A small piece of brick, usually less than a half.

Batter: Recessing or sloping masonry back in successive courses; the opposite of corbel.

Batterboard: A board set up outside the building line to hold the lines both prior to the excavation and after the excavation to relocate the building lines.

Beam and slab construction: A method of supporting a reinforced concrete floor by a system of reinforced concrete beams or girders.

Bearing capacity: The maximum pressure that a soil or other material can withstand without failure. Specifically, with regard to foundations, the maximum pressure which a soil can withstand without settlement of an amount that compromises the integrity or function of the structure.

Bearing plate: A metal plate placed under a beam, girder, or column to spread its weight over a larger area for support.

Bed joint: The horizontal layer of mortar on which a masonry unit is laid.

Bed stone: A large stone sometimes used to support a beam.

Belt course: A narrow horizontal course of masonry, sometimes slightly projected, such as window sills; sometimes called *string* or *sill course*.

Bench marks: Permanent marks such as a tree or a fire hydrant used as reference for elevation checks.

Blocking: A method of bonding two adjoining or intersecting walls not built at the same time, by means of offsets whose vertical dimensions are not less than 8 inches. The 8-inch dimension allows installation of a second wall in the same coursing.

Bluestone: A stone quarried in the southern New York area; commonly used for sills, treads, and lintels.

Bond: 1. A method of tying various parts of a masonry wall by lapping the units one over another or by connecting with metal ties. 2. Patterns formed by exposed parts of masonry units. 3. Adhesion between mortar or grout and masonry units or reinforcement.

Bond beam: Course or courses of masonry wall grouted and usually reinforced in the horizontal direction. Serves as a horizontal tie of the wall, a bearing course for structural members, or as a flexural member itself.

Bond course: The course consisting of units which overlap more than one wythe (width) of masonry.

Bond stone: In stone wall construction, the stones that run through the thickness of the wall to tie the wall together.

Bonder: A bonding unit. *See header.*

Breaking joints: Any arrangement of masonry units which prevents continuous vertical joints from occurring in adjacent courses.

Brick: A solid masonry unit of clay or shale formed into a rectangular prism while plastic and burned or fired in a kiln.

acid-resistant brick: Brick suitable for use in contact with chemicals.

adobe brick: Large roughly-molded, sun-dried clay brick of varying size.

arch brick: 1. Wedge-shaped brick for special use in an arch. 2. Extremely hard-burned brick from an arch of a stove kiln.

building brick: Brick for building purposes not especially treated for texture or color. Formerly called *common brick. See ASTM Specification C62.*

clinker brick: A very hard-burned brick whose shape is distorted or bloated owing to nearly complete vitrification. It's seldom used for exterior face work.

common brick: *See building brick.*

dry press brick: Brick formed in molds under high pressures from relatively dry clay (5 to 7 percent moisture content).

economy brick: Brick whose nominal dimensions are 4" × 4" × 8".

engineered brick: Brick whose nominal dimensions are 4" × 3.2" × 8".

facing brick: Brick made especially for facing purposes. Often treated for surface texture and color. *See ASTM Specification C216.*

fire brick: Brick made of refractory ceramic material which will resist high temperatures.

floor brick: Smooth, dense brick, highly resistant to abrasion, used as finished floor surfaces. *See ASTM Specification C410.*

gauged brick: 1. Brick which have been ground to accurate dimensions. 2. A tapered arch brick.

hollow brick: A masonry unit of clay or shale whose net cross area in any plane parallel to the bearing surface is not less than 60 percent of its gross cross-sectional area measured in the same plane. The holes decrease delivery weight but don't reduce strength. *See ASTM Specification C652.*

jumbo brick: A generic term indicating a brick larger in size than the standard.

Norman brick: A brick whose nominal dimensions are 4" × 2²/₃" × 12".

paving brick: Vitrified brick used where resistance to abrasion is important. *See ASTM Specification C7.*

roman brick: Brick whose nominal dimensions are 4" × 2" × 12".

salmon brick: Generic term for underburned brick which are more porous, slightly larger, and lighter in color than hard-burned brick.

"SCR brick": Brick whose nominal dimensions are 6" × 2²/₃" × 12".

sewer brick: Low absorption abrasion-resistant brick used in drainage structures.

soft mud brick: Brick produced by molding relatively wet clay (20 to 30 percent moisture).

stiff mud brick: Brick produced by extruding a stiff but plastic clay (12 to 15 percent moisture) through a die.

Brick and brick: A method of laying brick so that units touch each other with only enough mortar to fill irregularities.

Brick grade: Designation for durability of the unit expressed as SW for severe weathering, MW for moderate weathering, or NW for negligible weathering. *See ASTM Specifications C216 and C265.*

Brick type: Designation for facing brick which controls tolerance, chippage, and distortion. Expressed as FBS, FBX, and FBA for solid brick, and HBS, HBX and HBB for hollow brick. *See ASTM Specification C216 and C652.*

Brick veneer: A building of masonry in which the brick facing is attached to a surface of a frame with wall ties, and is not bonded to the veneered wall.

Building code: A set of regulations that are adopted by a city or town for the construction of buildings.

Building line: The outside line of the building.

Building permits: A permission form obtained from a state or local government to permit construction of a structure.

Bull float: A wide float with a long handle.

Bull header: A brick having one corner that is rounded; used for window sills and for coping.

Bull nose: A masonry unit that is rounded on one side of one end of the unit for use on corners.

Buttering: Troweling mortar onto a masonry unit.

Buttress: A piece of masonry built against a wall to give the wall more strength.

Caisson: A concrete pile constructed for use under water.

Calcium chloride: A granulated salt added to water in mixing concrete and mortar to accelerate the setting time.

Cant brick: A brick that is made with one side beveled.

Capacity insulation: The ability of masonry to store heat as a result of its mass, density and specific heat.

Capping brick: Brick that are made for capping the top of a wall.

Cavity wall: A hollow wall built of masonry units so arranged as to provide a continuous air space within the wall (with or without insulating material). Both the inner and the outer wythes of the wall are reinforced so they separately resist seismic forces in proportion to how rigid they are.

C/B ratio: The ratio of the weight of water absorbed by a masonry unit during immersion in cold water to weight absorbed during immersion in boiling water. Also called *saturation coefficient. See ASTM Specification C67.*

Cement key: An open U-shaped channel in the top of the concrete footing, which helps interlock the footing with the wall built on top of it.

Centering: Temporary formwork for the support of masonry arches or lintels during construction; also called *centers.*

Ceramic color glaze: An opaque colored glaze of stain or gloss finish. *See ASTM Specification C126.*

Ceramic mosaic: Small ceramic tiles in sheets, usually for use on floors.

Ceramic tile: A flat piece of fired clay in a variety of shapes and compositions.

Chase: A continuous recess built into a wall to receive pipes, ducts, etc.

Clay: A natural, mineral aggregate consisting essentially of hydrous aluminum silicate; it is plastic when sufficiently wetted, rigid when dried, and vitrified when fired to a sufficiently high temperature.

Clay mortar mix: Finely-ground clay used as a plasticizer for masonry mortars.

Clear ceramic glaze: Same as ceramic color glaze except that it is translucent or slightly tinted, with a gloss finish.

Clip: A portion of a brick cut to length.

Closer: The last masonry unit laid in a course. It may be whole or a portion of a unit.

Closure: Supplementary or short length units used at corners or jambs to maintain bond patterns.

Cold joint: A joint or a discontinuity formed when a concrete surface hardens before the next pour is placed to it.

Collar joint: The continuous vertical, longitudinal joint between two widths of masonry designed to accommodate movements resulting from temperature and moisture changes.

Colorimetric test: A test used to judge the amount of organic material present in fine aggregate. The test results are read by comparing the color of a solution with a standard scale of possible colors that have been assigned specific interpretations.

Column: A vertical member whose horizontal dimension measured at right angles to the thickness does not exceed three times its thickness.

Compass brick: A factory-made brick for use in curved work.

Compressive strength: The measured maximum resistance to axial loading expressed in pounds per square inch (psi).

Construction Specifications Institute (C.S.I.): An organization that established a format for construction specifications.

Control joint: A groove that is cut or tooled in the surface of concrete to predetermine the place where a crack will occur due to shrinkage of the concrete. Also called *expansion joint.*

Coping: The material or masonry units forming a cap or finish on top of a wall, pier, pilaster, chimney, etc.

Corbel: A shelf or ledge formed by projecting successive courses of masonry out from the face of the wall.

Counterflashing: The flashing that projects from the masonry wall over the base flashing to protect the upper end of the flashing.

Course: One of the continuous horizontal layers of units bonded with mortar in masonry.

Coursed ashlar: Method of arranging various units in the wall according to height to form courses in stone masonry.

Coursed rubble: Method of laying roughly-shaped stones in approximately level beds in stone masonry.

Cricket: A small watershed built behind a chimney to make water run away from the chimney.

CRSI: Concrete Reinforcing Steel Institute.

Culls: Masonry units which do not meet the standards or specifications and have been rejected.

Curing: The process in which mortar and concrete harden.

Curtain wall: An exterior, nonbearing wall built outside the building frame, generally with vertical support at ground level only, but may be (and generally is) laterally supported at each story level by anchoring to floors, roof or spandrel beams.

Damp course: A course or layer of impervious material which prevents capillary action from the ground or lower course.

Dampproofing: Prevention of moisture penetration by capillary action.

Darby: A flat trowel-like tool used to smooth out the surface of concrete soon after the surface has been bull floated.

Datum point: A point that has been established by a city or town to use as a reference point in measuring elevations and distances.

Dog's tooth: Brick laid with their corners projecting from the wall.

Drip: A projecting piece of material, shaped to throw off water to prevent it from running down the face of the wall.

Dry stone wall: A stone wall that is laid without the aid of mortar.

Dutch bond: Brick laid in alternate stretchers and headers.

Ears: The extended ends found on most concrete stretcher blocks.

Edging: The process of finishing the edges of a concrete slab with a tool that has a radius on its edge.

Efflorescence: A powder or stain sometimes found on the surface of masonry caused by water soluble salts leaching from within the wall.

Engineered brick masonry: Masonry in which design is based on a rational structural analysis.

English bond: A masonry bond where courses alternate between headers and stretchers.

English cross bond: Sometimes called *Old English bond*, it is similar to or the same as Dutch bond.

Expansion joint: *See control joint.*

Exposed aggregate: A concrete finish that has the top of the surface cement washed off to show the stone aggregate.

Exterior wall: Any outer wall serving as a vertical enclosure of the building.

Face: 1. The exposed surface of a wall or masonry unit. 2. The surface of a unit designed to be exposed in the finished masonry.

Face brick: Good quality brick used for the exposed surface of a brick wall.

Faced masonry: Masonry construction in which the structural-bonded facing and backing are of different materials.

Facing: Any material forming a part of the wall used as a finished surface.

False header: A header that does not tie two walls together. As when a half brick is used.

Field: The expanse of wall between openings, corners, etc., principally composed of stretchers.

Filled-cell masonry: Single-wythe masonry construction composed of hollow units in which all voids are filled with grout after the wall is laid.

Filler wall: A nonbearing wall in skeleton frame construction, built between steel or concrete columns and wholly supported at each story.

Filter block: A hollow, vitrified clay masonry unit, sometimes salt glazed, designed for trickling filter floors in sewage disposal plants. *See ASTM Specification C159.*

Finish grade: The surface elevation after the grading is completed.

Fire clay: A clay used to make brick which can resist high heat without deforming.

Fire resistant material: A noncombustible material.

Firebrick: A brick that can withstand the effects of heat extremes.

Fireproofing: Any material or combination of materials that increases fire resistance.

Flash set: A process by which concrete sets faster than normal because of too much heat.

Flashing: 1. A thin impervious material placed in mortar joints and through air spaces in masonry to prevent water penetration and/or to provide water drainage. 2. Manufacturing method to produce specific color tones in brick.

Flemish bond: A brick bond in which headers and stretchers alternate on every course.

Fly ash: The fine residue resulting from the burning of ground or powdered coal, used as an additive to improve workability of concrete or mortar.

Footing drain: Drain tile run installed around the building footing, inside or outside, to carry off ground water.

Form oil: A nonstaining oil usually painted on the forms to act as a nonstick surface between the concrete and the forms.

Frog: A depression on the top side of a brick.

Furring: A method of finishing the interior face of masonry walls to provide a space for insulation, prevent moisture transmittal, or to provide a level surface for finishing.

Gingerbread: Fancy brickwork on old buildings.

Glazed brick: A brick that has a glazing material fused to the face, usually providing a smooth, glassy surface.

Grade: The level of the ground around a building.

Grade beam: A reinforced concrete beam, supported on piers, used to hold up the walls of a building.

Green cement: Masonry that has not yet set up.

Grounds: Nailed strips placed in masonry walls as a means of attaching trim or furring.

Grout: A mixture of portland cement, aggregates and water which is proportioned to produce pouring or pumping consistency without segregation of the constituents. Used to fill voids and cells, or collar joints in masonry walls so as to encase steel and bond units together for composite action.

high lift grouting: The technique of grouting masonry lifts up to 12 feet.

low lift grouting: The technique of grouting as a wall is constructed.

Grouted masonry: Multi-wythe masonry construction in which the space between wythes is solidly filled with grout.

Hard burned: Nearly vitrified clay products which have been fired at high temperatures.

Head joint: The vertical joint between masonry units.

Header: A masonry unit which overlaps two or more adjacent wythes of masonry to tie them together.

blind header: A concealed brick header on the interior of a brick wall.

clipped header: A bat placed to look like a header for the purpose of establishing a pattern; also called a false header. The header extends into the backup course.

flare header: A header of darker color than the rest of the wall.

Heading course: A continuous bonding course of header brick; also called a *header course.*

Heavyweight concrete: Concrete that is constructed from heavyweight aggregates and weighing about 390 pounds per cubic foot: used in the construction of laboratories as radiation shields.

High chairs: A manufactured product that is used to hold up reinforcing wire in concrete as the concrete is poured.

High early cement: A portland cement sold as Type III; sets up to its full strength faster than other types.

High-lift grouting method: Indicates that grout will be pumped into all wall voids after the masonry units, reinforcing steel and embedded items are built to full story height. High-lift grout is placed in one continuous pour by lifts, which allows time for consolidation and loss of water, but placed at such a rate as not to form intermediate construction joints or blowouts.

Hog: 1. A wall built to different height (number of courses) on the two sides of an opening; such as on either side of a doorway or a window frame. 2. An uneven course in a masonry wall.

Hollow masonry: Single-wythe masonry construction composed of hollow units in which cells and voids containing reinforcing bars or embedded items are filled in with grout as the work progresses.

Honeycomb: Method by which concrete is poured and not puddled or vibrated, allowing the edges to have voids or holes after the forms are removed.

Hydrated lime: Material that is left after quicklime is added to water; also called *slaked lime*.

Hydraulic: The ability of cement to harden when mixed with or under water.

Initial rate of absorption: The weight of water that is absorbed expressed in grams per 30 square inches of contact surface when a brick is partially immersed for one minute. *See ASTM Specification C67.*

Isolation joint: A joint that completely separates one piece of concrete from another.

Joint reinforcement: An assemblage of steel reinforcing wires designed for use in masonry bed joints, serving to distribute stresses and to tie separate wythes together.

Journeyman: A mason who has learned his trade through an apprenticeship.

Keyway: A recess in one piece of concrete to improve the shear strength of the next pour by tying the pours together mechanically.

Kiln: A furnace oven or heated enclosure used for burning or firing brick.

Kiln run: Brick from one kiln which have not been sorted or graded for size or color variation.

King closer: A brick cut diagonally to have one 2-inch end and one full-width end.

Ladder bar: Prefabricated joint reinforcing wires to which parallel deformed side rods are connected by perpendicular cross wires, forming a ladder design.

Laitance: A soft or weak layer in concrete or mortar; a surface layer sometimes brought to the surface by bleeding caused by excess water in the mix.

Lateral support: These are members such as cross walls, columns, pilasters, buttresses, floors, roofs, or spandrel beams, which brace walls either vertically or horizontally, and which have sufficient strength and stability to resist the horizontal forces transmitted to them.

Layout: Process of measuring and marking building material to indicate placement of other materials, as in preparation for the initial course of masonry units for a wall.

Lead: The section of wall built up and racked back on successive courses.

Lime, hydrated: Quicklime with water added to it in a hydrator that converts the oxides to hydroxides. Hydrated lime is added to mortar to increase the mortar's bond strength, workability, water retention, tensile strength, flexibility and autogenous healing capabilities. These are essential qualities in an all-purpose mortar mix.

Lime putty: Hydrated lime in plastic form ready for addition to mortar.

Lintel: A beam placed over an opening in a wall.

Load-bearing wall: Any wall which, in addition to supporting its own weight, supports the structure above it without benefit of a complete load-carrying space frame in structural steel or reinforced concrete.

Low-lift grouting method: Indicates that grout will be poured in small increments as the masonry work progresses.

Masonry cement: A mill-mixed cementitious material to which sand and water must be added. *See ASTM Specification C91.*

Masonry unit: Natural or manufactured building units of burned clay, concrete, stone, glass, gypsum, etc.

hollow masonry unit: One whose net cross-sectional area in any plane parallel to the bearing surface is 74 percent or less of the gross.

solid masonry unit: One whose net cross-sectional area in every plane parallel to the bearing surface is 75 percent or more of the gross.

Modular masonry: Masonry in which the materials and dimensions fit the 4-inch modular grid.

Moisture barrier: Materials used to retard the flow of vapor or moisture and thus prevent condensation on the slab or foundation surfaces or within walls. There are two types of barriers, the membrane type that comes in rolls and is applied as a unit of the construction, and the paint type that is applied with a brush. Also called *vapor barrier*.

Mortar: A plastic mixture of cementitious materials, fine aggregate and water.

fat mortar: A very sticky mortar containing a high percentage of cementitious components.

high bond mortar: Mortar which develops higher bond strengths with masonry units than is normally developed with conventional mortar.

lean mortar: Mortar which is deficient in cementitious materials; sandy and difficult to spread.

Nonbearing partition: A wall that does not support the structure above it; usually a partition wall or a filler wall.

Noncombustible material: Any material which will neither ignite nor actively support combustion in air at a temperature of 1200 degrees F when exposed to fire.

OSHA: Occupational Safety and Health Act of 1970.

Overhand work: Method of laying brick or block while standing inside the finished wall and reaching over the wall.

Overturn force: Any of several kinds of force, or a combination of forces, most commonly but not only wind, that have a tendency to overcome the stable equilibrium of a structure.

Pargeting: The process of applying a coat of cement mortar to masonry. Often spelled and/or pronounced *parging*.

Partition: Any interior wall, one story or less in height.

Pick and dip: A method of laying brick whereby the bricklayer simultaneously picks up a brick with one hand and, with the other hand, gathers enough mortar on a trowel to lay the brick.

Pier: An isolated column of masonry.

Pilaster: A wall portion projecting from either or both faces and serving as a vertical column and/or beam.

Plasticity: The property of fresh cement paste, concrete or mortar which makes it adhere to the masonry units if mixed correctly.

Plug: A piece of unit masonry smaller than one half unit.

Plumb rule: A mason's hand level. It is used in a horizontal position as a level and in a vertical position to determine if a wall is plumb.

Pointing: Troweling mortar into a joint after masonry units are laid.

Ponding: Curing concrete by flooding the surface with water.

Prefabricated brick masonry: Masonry construction fabricated in a location other than its final in-service location in the structure. Also known as preassembled, panelized and sectionalized brick masonry.

Prestressed concrete: Concrete poured around a steel member that is under tension when the pour is made.

Queen closure: A cut brick having a nominal 2-inch horizontal dimension, usually used on Flemish bond courses.

Quoin: A projecting right-angle masonry corner, usually three bricks long and four to six bricks high. Bricks project 3/4 to 1 inch out of the wall with a normal coursing of brickwork between (two — four courses).

Racking: A method entailing stepping back successive courses of masonry. This is usually done at openings when the window or door units are not available at that stage of construction, but are added later.

RBM: Reinforced brick masonry.

Reglet: A groove or channel in masonry designed to accept flashing or some other type of masonry attachment. Reglets are cast in concrete and embedded in other forms of masonry.

Reinforced masonry: Masonry units, reinforcement, grout, and mortar combined in such a manner that the component materials act together in resisting seismic forces.

Reinforcement: Structural steel shapes, deformed reinforcing bars or joint reinforcement embedded or encased in unit masonry in such a manner that it works with the masonry in resisting stress.

Return: Any surface turned back from the face of the principal surface, such as window and door openings.

Reveal: That portion of the jamb or recess which is visible from the face of the wall.

Rowlock: a brick laid on its face edge so that the normal bedding area is visible in the face of the wall.

Scaling: The peeling away of the surface of concrete.

Segregation: The tendency, as concrete is made to flow laterally, for coarse aggregate and any dryer material to remain behind while mortar, cement or wetter material flows ahead. Also can occur vertically when wet concrete is overvibrated or dropped into forms — mortar and wetter material will rise to the top.

Shale: Clay which has been subjected to high pressure until it has hardened.

Shear wall: Any wall which resists a horizontal force applied in the plane of the wall (i.e. any wall not isolated along three edges.)

Shotcreting: Method of placing concrete on curved surfaces such as swimming pools under pneumatic pressure through a nozzle.

Shoved joints: Vertical joints filled by shoving a brick against the next brick when it is being laid in a bed of mortar.

Sieve number: A number used to designate the size of a sieve, usually the approximate number of sieve crosswires per linear inch.

Slenderness ratio: Ratio of the effective height of a member to its effective thickness.

Slip form: A form that permits constant movement during the placing of concrete; used on dams, towers, etc.

Slump: A measure of consistency of freshly mixed concrete measured to the nearest 1/4 inch immediately after removal of the slump cone mold.

Slushed joints: Vertical joints filled after units are laid by "throwing" mortar in with the edge of the trowel. This is a bad practice that can cause failures later.

Soap: A masonry unit of normal face dimensions, having a nominal 2-inch thickness. A thin brick.

Soft-burned: Clay products that have been fired at low temperature ranges, producing relatively high absorption and low compressive strengths.

Solar screen: A perforated wall used as a sunshade.

Soldier: A stretcher set on end with the face showing on the surface.

Sound: A characteristic of solid materials. The material is free of cracks, flaws, fissures or variations from an accepted standard. Specifically, with regard to aggregate, the ability to withstand the aggressive action to which concrete might be exposed, particularly weather exposure.

Spall: A small fragment removed from the face of a masonry unit by a blow or by action of the elements.

Spandrel: 1. The triangular areas on each side of a masonry or steel arch. 2. The portion of wall between two windows and between the supporting columns.

Stack: Any structure or part thereof which contains a flue or flues for the discharge of gases.

Story pole: A marked pole for measuring masonry coursing during construction.

Stretcher: A masonry unit laid with its greatest dimension horizontal and its face parallel to the wall face.

Stringing mortar: The procedure of spreading enough mortar to lay several brick at a time.

Struck joint: Any joint that has been finished with a trowel.

Structural member: Any part of a structure that, in addition to its own weight, carries the weight of forces of other parts of the building. Examples include: footings, foundations, piers, beams, lintels, exterior walls and some interior walls.

Structural wall: Any wall which supports vertical loads other than its own weight or which resists lateral movement from horizontal forces such as those caused by an earthquake.

Subgrade: The prepared and compacted soil that functions as a base for a concrete slab or foundation.

Tensile strength: The maximum stress that a material is able to resist under axial tensile loading, before failing.

Temper: To moisten and mix clay, plaster or mortar to its proper consistency.

Tie: Any unit of material which connects masonry to masonry or to other materials.

Tooling: Compressing or shaping the face of the mortar joint with a metal tool other than a trowel.

Toothing: Constructing a temporary end of a wall with the end stretcher of every other course projecting.

Traditional masonry: Masonry in which the design is based on empirical rules which control minimum thickness, lateral support requirements, and height, without a structural analysis.

Tuck pointing: The filling in with fresh mortar to cutout or defective mortar joints in masonry.

Uplift force: An upward force on a structure caused by water, frost heave, or wind force on the side of a structure.

Vapor barrier: *See moisture barrier.*

Veneer: A masonry facing which is attached to the backup but not so bonded as to intentionally act with it under load or movement. To limit potential damage, the use of veneer construction will be restricted to masonry veneers of less than $1^1/_2$ inches nominal thickness, such as ceramic tile. Otherwise, use faced construction.

Vitrification: The condition resulting when kiln temperatures are sufficient to fuse grains and close pores of a clay product, making the mass impervious.

Wall: A vertical member of a structure whose horizontal dimension measured at right angles to the thickness exceeds three times its thickness.

apron wall: That part of a panel wall between window sill and wall support.

area wall: 1. The masonry surrounding or partly surrounding an area. 2. The retaining wall around basement windows below grade.

bearing wall: One which supports a vertical load in addition to its own weight.

cavity wall: A wall built of masonry units so arranged as to provide a continuous air space within the wall (with or without insulating material), and in which the inner and outer wythes of the wall are tied together with metal ties.

composite wall: A multiple wythe wall in which at least one of the wythes is dissimilar to the other wythe or wythes with respect to type or grade of masonry unit or mortar.

curtain wall: An exterior nonbearing wall not wholly supported at each story.

dwarf wall: A wall or partition which does not extend to the ceiling.

enclosure wall: An exterior nonbearing wall in skeleton frame construction.

exterior wall: Any outside wall or vertical enclosure of a building other than a party wall.

faced wall: A composite wall in which the masonry facing and backing are so bonded as to exert a common reaction under load.

fire wall: Any wall which subdivides a building to resist the spread of fire and which extends continuously from the foundation through the roof.

foundation wall: That portion of a load-bearing wall below the level of the adjacent grade, or below the first floor beams or joists.

hollow wall: A wall built of masonry units arranged to provide an air space within the wall.

insulated cavity wall: A cavity wall that contains insulation of some kind.

load bearing wall: A wall which supports any vertical load in addition to its own weight.

nonbearing wall: A wall which supports no vertical load other than its own weight.

panel wall: An exterior, nonbearing wall wholly supported at each story.

parapet wall: That part of any wall entirely above the roof line.

party wall: A wall used for joint service by adjoining buildings.

perforated wall: One which contains a considerable number of relatively small openings; also called a *pierced wall* or *screen wall.*

shear wall: A wall which resists horizontal forces applied in the plane of the wall.

single wythe wall: A wall only one masonry unit in thickness.

solid masonry wall: A wall built of solid masonry units, laid continuously, with mortar joints completely filled with mortar or grout.

spandrel wall: That part of a curtain wall above the top of a window in one story and below the sill of the window in the story above.

veneered wall: A wall having a facing of masonry units or other weather-resistant noncombustible materials securely attached to the backing, but not so bonded as to intentionally exert common action under load.

Wall plate: A horizontal member anchored to a masonry wall to which other structural elements may be attached; also called a *head plate*.

Wall tie: A bonder or metal piece which connects wythes of masonry to each other or to other materials.

Wall tie, cavity: A rigid, corrosion-resistant metal tie which bonds two wythes of a cavity wall.

Wall tie, veneer: A strip of metal used to tie facing veneer to the backing.

Water reducing agent: An admixture material which either increases the workability of freshly-mixed mortar or concrete without increasing water content, or maintains workability with a reduced amount of water.

Water retentivity: That property of a mortar which prevents the rapid loss of water to masonry units with high absorption.

Water table: A projection of lower masonry on the outside of the wall slightly above the ground to prevent upward penetration of ground water.

Waterproofing: Prevention of moisture flow through masonry due to water pressure.

Weep holes: Openings placed in mortar joints of facing materials at the level of the flashing to permit the escape of water or moisture.

Workability: The property of fresh concrete or mortar which determines the ease with which it can be mixed, placed and finished.

Wow: A bow in a masonry wall.

Wythe: 1. Each continuous vertical section of a masonry wall, one unit in thickness. 2. The thickness of masonry units separating flues in a chimney.

Index

C

D

K

L

M

R

S

T

U

V

W

X

Y

Z

Markup & Profit: A Contractor's Guide

In order to succeed in a construction business, you have to be able to price your jobs to cover all labor, material and overhead expenses, and make a decent profit. The problem is knowing what markup to use. You don't want to lose jobs because you charge too much, and you don't want to work for free because you've charged too little. If you know how to calculate markup, you can apply it to your job costs to find the right sales price for your work. This book gives you tried and tested formulas, with step-by-step instructions and easy-to-follow examples, so you can easily figure the markup that's right for your business. Includes a CD-ROM with forms and checklists for your use. **320 pages, 8½ x 11, $32.50**

Greenbook Standard Specifications For Public Works Construction

Since 1967, eleven previous editions of the popular "Greenbook" have been used as the official specification, bidding and contract document for many cities, counties and public agencies throughout the West. New federal regulations mandate that all public construction use metric documentation. This complete reference, which meets this new requirement, provides uniform standards of quality and sound construction practice easily understood and used by engineers, public works officials, and contractors across the U.S. Includes hundreds of charts and tables.
480 pages, 8½ x 11, $59.95

Excavation & Grading Handbook Revised

Explains how to handle all excavation, grading, compaction, paving and pipeline work: setting cut and fill stakes (with bubble and laser levels), working in rock, unsuitable material or mud, passing compaction tests, trenching around utility lines, setting grade pins and string line, removing or laying asphaltic concrete, widening roads, cutting channels, installing water, sewer, and drainage pipe. This is the completely revised edition of the popular guide used by over 25,000 excavation contractors.
384 pages, 5½ x 8½, $22.75

Plumber's Handbook Revised

This new edition shows what will and won't pass inspection in drainage, vent, and waste piping, septic tanks, water supply, graywater recycling systems, pools and spas, fire protection, and gas piping systems. All tables, standards, and specifications are completely up-to-date with recent plumbing code changes. Covers common layouts for residential work, how to size piping, select and hang fixtures, practical recommendations, and trade tips. It's the approved reference for the plumbing contractor's exam in many states. Includes an extensive set of multiple choice questions after each chapter, and in the back of the book, the answers and explanations. Also in the back of the book, a full sample plumber's exam.
352 pages, 8½ x 11, $32.00

CD Estimator Heavy

CD Estimator Heavy has a complete 780-page heavy construction cost estimating volume for each of the 50 states. Select the cost database for the state where the work will be done. Includes thousands of cost estimates you won't find anywhere else, and in-depth coverage of demolition, hazardous materials remediation, tunneling, site utilities, precast concrete, structural framing, heavy timber construction, membrane waterproofing, industrial windows and doors, specialty finishes, built-in commercial and industrial equipment, and HVAC and electrical systems for commercial and industrial buildings. **CD Estimator Heavy is $69.00**

Contractor's Guide to the Building Code Revised

This new edition was written in collaboration with the International Conference of Building Officials, writers of the code. It explains in plain English exactly what the latest edition of the *Uniform Building Code* requires. Based on the 1997 code, it explains the changes and what they mean for the builder. Also covers the *Uniform Mechanical Code* and the *Uniform Plumbing Code*. Shows how to design and construct residential and light commercial buildings that'll pass inspection the first time. Suggests how to work with an inspector to minimize construction costs, what common building shortcuts are likely to be cited, and where exceptions may be granted. **320 pages, 8½ x 11, $39.00**

Craftsman Book Company
6058 Corte del Cedro
P.O. Box 6500
Carlsbad, CA 92018

☎ 24 hour order line
1-800-829-8123
Fax (760) 438-0398

Name

Company

Address

City/State/Zip

❍ This is a residence

Total enclosed______________________(In California add 7.25% tax)

We pay shipping when your check covers your order in full.

In A Hurry?

We accept phone orders charged to your
❍ Visa, ❍ MasterCard, ❍ Discover or ❍ American Express

Card#______________________

Exp. date______________Initials______________

Tax Deductible: Treasury regulations make these references tax deductible when used in your work. Save the canceled check or charge card statement as your receipt.

Order online http://www.craftsman-book.com
Free on the Internet! Download any of Craftsman's estimating costbooks for a 30-day free trial! http://costbook.com

10-Day Money Back Guarantee

- ❍ 39.50 Basic Concrete Engineering for Builders
- ❍ 40.00 Basic Construction Management: The Superintendent's Job, 4th Edition
- ❍ 34.00 Basic Engineering for Builders
- ❍ 19.00 Building Layout
- ❍ 68.50 CD Estimator
- ❍ 69.00 CD Estimator Heavy
- ❍ 25.00 Concrete Construction & Estimating
- ❍ 39.50 Construction Estimating Reference Data with FREE *National Estimator* on a CD-ROM.
- ❍ 41.75 Construction Forms & Contracts with a CD-ROM for *Windows*™ and Macintosh.
- ❍ 39.00 Contractor's Guide to the Building Code Revised
- ❍ 39.50 Estimating Excavation
- ❍ 22.75 Excavation & Grading Handbook Revised
- ❍ 59.95 Greenbook Standard Specifications for Public Works Construction
- ❍ 32.50 Markup & Profit: A Contractor's Guide
- ❍ 47.50 National Construction Estimator with FREE *National Estimator* on a CD-ROM.
- ❍ 32.00 Plumber's Handbook Revised
- ❍ 19.50 Stair Builder's Handbook
- ❍ 39.75 Steel-Frame House Construction
- ❍ 25.50 Wood-Frame House Construction
- ❍ 28.50 Masonry & Concrete Construction Revised
- ❍ FREE Full Color Catalog

Prices subject to change without notice

NO POSTAGE
NECESSARY
IF MAILED
IN THE
UNITED STATES

BUSINESS REPLY MAIL

FIRST CLASS MAIL PERMIT NO. 271 CARLSBAD, CA

POSTAGE WILL BE PAID BY ADDRESSEE

Craftsman Book Company
6058 Corte del Cedro
P.O. Box 6500
Carlsbad, CA 92018-9974

NO POSTAGE
NECESSARY
IF MAILED
IN THE
UNITED STATES

BUSINESS REPLY MAIL

FIRST CLASS MAIL PERMIT NO. 271 CARLSBAD, CA

POSTAGE WILL BE PAID BY ADDRESSEE

Craftsman Book Company
6058 Corte del Cedro
P.O. Box 6500
Carlsbad, CA 92018-9974

NO POSTAGE
NECESSARY
IF MAILED
IN THE
UNITED STATES

BUSINESS REPLY MAIL

FIRST CLASS MAIL PERMIT NO. 271 CARLSBAD, CA

POSTAGE WILL BE PAID BY ADDRESSEE

Craftsman Book Company
6058 Corte del Cedro
P.O. Box 6500
Carlsbad, CA 92018-9974